本书由云南省教育厅资助项目"几类特殊的丢番图方程
（2018JS608）、"丽江师范高等专科学校师资队伍建设经费、……、十月
年学术带头人）"资助出版。

从一到哥德巴赫猜想
From one to Goldbach conjecture

——整除性的典型问题与方法
——Typical problems and methods of divisibility

赵建红◎著

云南大学出版社
YUNNAN UNIVERSITY PRESS

图书在版编目（CIP）数据

从一到哥德巴赫猜想：整除性的典型问题与方法 /
赵建红著. —— 昆明：云南大学出版社，2018
ISBN 978-7-5482-3541-5

Ⅰ.①从… Ⅱ.①赵… Ⅲ.①数学 - 研究 Ⅳ.
①O1

中国版本图书馆CIP数据核字(2018)第215224号

策划编辑：万　斌
责任编辑：万　斌
封面设计：王婳一

从一到哥德巴赫猜想
From one to Goldbach conjecture
——整除性的典型问题与方法
——Typical problems and methods of divisibility

赵建红◎著

出版发行：云南大学出版社
印　　装：廊坊市海涛印刷有限公司
开　　本：787mm×1092mm　1/16
印　　张：21.75
字　　数：546千
版　　次：2018年12月第1版
印　　次：2018年12月第1次印刷
书　　号：ISBN 978-7-5482-3541-5
定　　价：78.00元

社　　址：昆明市一二一大街182号（云南大学东陆校区英华园内）
邮　　编：650091
发行电话：0871-65033244　65031071
网　　址：http://www.ynup.com
E - mail：market@ynup.com

若发现本书有印装质量问题，请与印厂联系调换，联系电话：0316-2507000。

序　一

　　数论是一门古老而又常新的学科，从数学史的角度看数论是经典的纯粹数学，从现代日益广泛的应用的角度看数论是新兴的应用数学。

　　初等数论以整除理论为基础研究整数和方程（组）的一些基本性质及相关整数解，展示由古至今数学中最典型、最基本的概念、思想、方法和技巧。

　　《从一到哥德巴赫猜想——整除性的典型问题与方法》的逻辑结构和行文是国内首创的，具有以下两个最基本的特点和优势：

　　第一，《从一到哥德巴赫猜想——整除性的典型问题与方法》从初等数论的基本概念到数论的经典运算——加减乘除入手，进而详细讨论整数的整除性，由整除性引出奇数偶数、素数、合数以及最大公因数和最小公倍数，并讨论了数的进位制，然后进一步过渡到算术基本定理，由此探讨了相关的几个典型问题——勾股数组、费马大定理和哥德巴赫猜想。

　　第二，全书行文口语化与数学化相结合，既重视初等数论这一数学分支的数学性，又注重读者的可读性。将生涩难懂的数学用一种平和的语言娓娓道来，通读全书有种让人既身处其中又不感其难的感觉。另一方面，从数学的角度来说，书中介绍了初等数论中整除性的很多典型问题，并从方法论的角度进行了相应的归纳，最后又介绍了作者对相关研究的最新成果，如 *One kind hybrid character sums and their upper bound estimates* 和 *Some symmetric identities involving Fubini polynomials and Euler numbers* 是发表在 SCI 期刊的两篇创新性较高的文章，还有其他一些发表在 EI 或国内数学类权威核心期刊的文章。这些都是作者长期从事数论，尤其是初等数论教学和研究工作的成果，是值得祝贺的，也是值得数学界同行参考学习的。

西北大学博士生导师、二级教授　张文鹏

2018 年 12 月

序　二

　　收到《从一到哥德巴赫猜想——整除性的典型问题与方法》的书稿，感到非常欣慰。2014 年，赵建红的处女作——《数学教学技能训练实证研究》也请我作了序，当时感觉他是一个"数学教育"的实践者。于是在后期的讨论中，我和他谈到了一个问题——数学与数学教育的关系，没想到才三年多时间，赵建红就能找到切入点并深入推进数论的研究。

　　通读《从一到哥德巴赫猜想——整除性的典型问题与方法》全书，发现一个很有意思的地方：用教学法将初等数论的核心内容进行了变革。具体来说，几乎所有的初等数论方面的教材或著作，都具有传统纯数学的基本特征——抽象性、严谨性。而这部著作却从阅读者的角度出发，用通俗易懂的语言来表述抽象的数学概念和讨论一些定理的证明，并以此来介绍数学中的方法论，从另一个角度来看初等数论。

　　那么是不是《从一到哥德巴赫猜想——整除性的典型问题与方法》就不够"数学"了？书中以初等数论中的一些典型问题来体现其数学本质，但数学类相关专业的大学师生亦可从中汲取知识和智慧。在最后一部分，以附录的形式将作者近期研究的典型成果展示给读者，其中有一些成果甚至是数学类专业的硕士、博士也要花费些时间和精力才能完全理解的。

　　简言之，这部著作的可读性非常强，不管你是中小学生或其他没有数学专业背景的读者，也不管你是数学专业背景下的大学生或更高层次的学者，都可以有选择地读一读、学一学这部著作中的另一角度的数学。

云南大学博士生导师、二级教授　李永昆

2018 年 12 月

目　　录

第1章 绪 论

数学本身，也有无穷的美妙. 认为数学枯燥无味，没有艺术性，这看法是不正确的. 就像站在花园外面，说花园里枯燥乏味一样. 只要你们踏进了大门，你们随时随地都会发现数学上也有许许多多趣味的东西.

——华罗庚

有史以来，人类就对自己赖以生存的环境进行探索，进而思考和探索大至宇宙空间，小到人类自我的奥秘，每个人如此自然地进行着，像一份天职：由于对未知世界的不确定，人类就会害怕，害怕就需要寄托，于是就寄托于加倍的探索．为了弄清楚这一过程，哲学产生了．基于哲学这一基础，人类开始了各种研究，最终走向艺术．

数学，就是其中一项具有特殊地位的研究．有学者认为在哲学之下，先有数学，随后出现的科学分为社会科学和自然科学．也有学者认为哲学之下，可分社会科学、数学、自然科学．如柏拉图就认为"数学是了解宇宙本身而不是它的表面现象的真正训练"．不管怎样划分，数学在人类认识世界的过程中的地位都是显而易见且不可撼动的．

这一章主要是对本书研究的范畴进行"剥洋葱"式的介绍：认识世界—哲学—数学—数论—初等数论—整数最基本的性质．

1.1　数论是什么

数学，可以粗略地分为纯粹数学和应用数学．我们无意对数学进行概念化阐述，但有一点是明确的——数学绝对不是课程中或教科书中所指的那种肤浅观察和寻常诠释，而应该具有更为丰富有趣的内涵．

图 1.1 是有人提出的一种有关纯粹数学和应用数学关系的观点．

数论（Number Theory），顾名思义，就是指有关数的理论．其中的"数"主要是"整数"，理论主要是整数的性质和不定方程（组）的整数解．从学科的角度说，数论是纯粹数学中研究整数性质的分支之一．

> 纯粹数学和应用数学就像是一座冰山——水面上的是应用数学，因为它有用，大家都看得见；水底下的是纯粹数学．

图 1.1

1.2　初等数论及其研究

按研究方法不同分类，数论可以分为初等数论和高等数论，如图 1.2 所示．简单地说，用整除、同余、连分数等初等方法研究的就是初等数论．用微积分、复分析等高等方法研究的就是高等数论，典型的高等数论主要有解析数论、代数数论、计算数论等．

数论 { 初等数论：用整除、同余、连分数等初等方法进行研究
　　　 高等数论：用微积分、复分析等高等方法进行研究

图 1.2

1.2.1　初等数论的研究对象

初等数论的研究对象诸如：

$$\cdots -5,\ -4,\ -3,\ -2,\ -1,\ 0,\ 1,\ 2,\ 3,\ 4,\ 5\cdots$$

这样的数，也就是整数．以 0 为界，整数又可以分为正整数、零和负整数，如图 1.3 所示．

$$整数\begin{cases}正整数\\0\\负整数\end{cases}$$

图 1.3

所有整数的总体称为整数集，一般记作 Z，其中的正整数部分可以记作 Z^+、N^*、N_+，其中 N 指的是自然数集[①]．

事实上，整数是我们最先认识的，也是最熟悉的数．还是小孩子的时候，父母在家里就教会我们一个人有 2 只手，一只手有 5 个手指头，以及用 1，2，3，4，5……数数，甚至还有单数、双数等．在人类文明的起始阶段，整数往往只是为人们提供数据，事实上，当时的人们没有也不需要研究和应用整数的一些其他性质：比如 8 的因数个数是多少，97 是不是素数等．这些都不能引起人们的重视．毕竟，他们在对食物、货币等进行计数的时候整数性质都是无关紧要的．

直到后来，人们逐渐意识到基础数学和应用数学的区别，才逐渐开始研究整数的性质．直至今天，整数的性质都一直引领着数学研究的潮流．于是，在学校里我们开始学习素数、合数等．初等数论还将讨论这样一些有趣的数，比如：

奇数：1，3，5，7，9，11，13，15，…

偶数：2，4，6，8，10，12，14，16，…

素数：2，3，5，7，11，13，17，19，…

合数：4，6，8，9，10，12，14，15，…

三角数（如图 1.4）：1，3，6，10，15，21，28，36，…

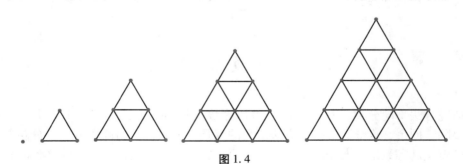

图 1.4

① 自然数集是否包含"0"的争议：有人认为按照"自然数是用以计量事物的件数或表示事物次序的数"及其出现顺序来说，自然数应为正整数，即"0"不属于自然数集．也有人认为自然数为非负整数，即"0"属于自然数集．到 21 世纪关于这个问题也尚无一致意见．在国外，有些国家的教科书是把"0"也算作自然数的．这本是一种人为的规定，我国为了推行国际标准化组织（ISO）制定的国际标准，早日和国际接轨，定义自然数集包含元素"0"．现行九年义务教育教科书和高级中学教科书等都把非负整数集叫作自然数集，明确指出"0"也是自然数集的一个元素．

平方数（也叫正方形数，如图 1.5）：1，4，9，16，25，36，49，64，…

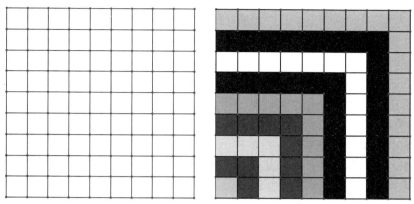

图 1.5

立方数（如表 1.1 所示）：1，8，27，64，125，216，343，512，…

表 1.1

x	1	2	3	4	5	6	7	8	9	10
x^3	1	8	27	64	125	216	343	512	729	1000

完美数：6，28，496，8128，33550336，…

之所以称之为完美数，是因为 6 的因数有 1，2，3，而 $1+2+3=6$；28 的因数有 1，2，3，7，14 而 $1+2+3+7+14=28$……

你能把 496，8128，33550336 的因数全找出来吗？尝试着把这些因数全部相加，看看是不是相同的结论？

下一个完美数是多少呢？

1.2.2　初等数论的研究内容

初等数论主要研究整数最基本的性质，如整除性.

例 1.1　显然，2 整除 4，4 整除 12，而 2 也整除 12；3 整除 9，9 整除 18，而 3 也整除 18. 其中是否有某种关系呢？比如是因为 "2 整除 4，4 整除 12"，所以 "2 整除 12"。这样的因果关系是否成立？如果成立，能否推而广之？比如对任意整数 a,b,c 都有：若 b 是 a 的倍数，c 是 b 的倍数，则 c 是 a 的倍数，也就是：

$$a\mid b, b\mid c \Rightarrow a\mid c$$

等等. 事实上，这就是整除的传递性.

数论的中心内容是以算术基本定理和最大公约数理论为基础的整除理论，核心是同余理论.

例 1.2　我们知道 4，6，8，9，10，12 是合数，而：

$$4 = 2 \times 2$$
$$6 = 2 \times 3$$
$$8 = 2 \times 2 \times 2$$
$$9 = 3 \times 3$$
$$10 = 2 \times 5$$
$$12 = 2 \times 2 \times 3$$

由此猜想是不是每一个合数都可以分解为几个素数的乘积，也就是：任何一个大于 1 的整数 n，如果 n 不是素数，那么 n 就可以分解成有限个素数的乘积．如果把那些素因数按从大到小或从小到大的顺序排列后再相乘，则表示方法是唯一的．

换句话说，因为乘法满足交换律，所以不计较素因数的先后顺序的话可以将任意合数进行唯一分解．

例 1.3　看到日历（如图 1.6）知道 2017 年 9 月 1 日是星期五，你能据此说出你出生那天是星期几吗？

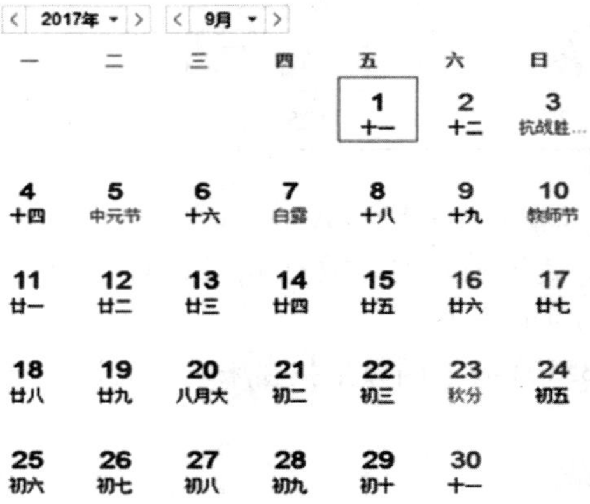

图 1.6

事实上，因为 9 月 1 日是星期五，那么 9 月 8 日、9 月 15 日、9 月 22 日、9 月 29 日都是星期五．不仅如此，任何月份的 1、8、15、22、29 日（若有）是星期几是相同的．同理 2、9、16、23、30 日（若有）是星期几是相同的，3、10、17、24、31 日（若有）是星期几是相同的，4、11、18、25 日是星期几是相同的，5、12、19、26 日是星期几是相同的，6、13、20、27 日是星期几是相同的，7、14、21、28 日是星期几是相同的，为什么呢？

因为这些数除以每周的天数 7 的余数是相同的，也即同余．依据同余理论，我们可以解决许多不定方程的问题．

1.2.3　初等数论的研究方法

初等数论主要用初等方法研究，从本质上说，就是利用整数的整除性质，如整除理论、同余理论等进行研究．

例 1.4 150 盏亮着的电灯，各有一个拉线开关控制，被顺序编号为 1，2，3，4，…，150. 将编号为 3 的倍数的灯的拉线各拉一下，再将编号为 5 的倍数的拉线各拉一下，拉完后亮着的灯数为几盏?

1.3 整数最基本的性质

由于初等数论的研究对象是"整数"，研究以"运算"为主，于是出现了整数的加减乘除运算. 根据整数环的概念，我们知道，整数的加减乘运算都是封闭运算，但是除法运算却不是封闭运算. 这里所谓的封闭与否讲的是在特定集合中定义的某一运算，经过这个运算之后，其结果仍然在这一集合中；而不封闭也就是它的反面，即运算后的结果不一定在这一集合中. 也就是说，一个整数除以一个整数的运算有特殊性，事实上，其结果往往不是整数，只有在特殊情况下才有可能刚好是一个整数，这个时候就是我们要研究的核心内容——整除.

整数的整除所具有的性质就是整除性，而这正是整数最基本的性质.

整除性所涉及的范畴主要有：加减乘除运算、奇数偶数、素数合数、算术基本定理、数的进位制、最大公因数、最小公倍数、勾股数组等，在书的最后介绍两个重量级的猜想——费马大定理、哥德巴赫猜想.

第 2 章　整数的加减乘除运算

可以说是数统治着整个量的世界，而算数的四则运算则可以看作是数学家的全部装备.

——麦克斯韦

　　运算是通过已知量的可能的组合，获得新的量的一种行为，其前提是量的相等，并依此进行代换，本质是集合之间的映射．如加法运算、减法运算、乘法运算、除法运算等等．

　　人类在很早以前就会计数．先会进行数的相加，很久以后学会了数的相乘和相减，为了要平均分配大量的果子或捕获的小动物，数的相除就应运而生了．这些运算统称为计算（Calculations）．"Calculation"一词起源于拉丁词 *Calculus*，意思是"小石子"，古罗马人在他们的计算板上就是用小石子来表示数的．

　　加减乘除四则运算是数学最基本的计算，主要是通过对一些已知量进行加、减、乘或除的组合，获得一个新的量——和、差、积或商．

　　这一章分别介绍整数的加减乘除法及其运算，其中对整数的除法及其运算进行重点介绍．

2.1　整数的加法及其运算

　　整数的加法是一切数字运算的基础．

　　在整数集 Z 中，若按一定次序取出两个元素，不妨假设依次取出 a 和 b，如果对于某一运算规则" + "在 Z 中有唯一确定的第三个元素，不妨假设 c 与之对应，则称在整数集 Z 中定义了一种运算，记作：

$$a + b = c.$$

其中 a 和 b 称为加法的加数，" + "称为加号，c 称为 a 与 b 相加的和．整数加法的数学本质如图 2.1 和图 2.2 所示．

　　整数的加法主要用于将两个及更多的整数进行合并，使之以一个整数的形式出现．

　　加法是同类事物的重复或累计，是数字运算的开始．种类不同的事物不能相加，如一个香蕉加一个苹果，结果只能等于两个水果，而这就存在分类与归类的关系，故一般认为不能相加．

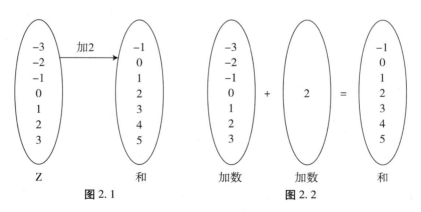

图 2.1　　　　　　　　　　　图 2.2

2.1.1　整数的加法运算规则

为避免表述上的来回返工，此处将不按数学史发展顺序，而根据读者认知实际进行介绍，即默认了已经存在零和负整数.

相反数指的是数值相反的两个数. 比如 2 的相反数是 -2，$-k$ 的相反数是 k 等.

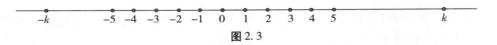

图 2.3

绝对值指的是某数在数轴上到原点的距离，是一个永恒的非负数. 比如 2 的绝对值是 2，-2 的绝对值也是 2，0 的绝对值是 0，而 $-k$ 的绝对值却未必是 k. 一般用双竖线表示绝对值，$|k| = k, k \geq 0$；$|k| = -k, k < 0$.

两个相同正负号的整数相加，绝对值相加，符号与加数相同；两个不同正负号的整数相加，绝对值相减（绝对值大的减绝对值小的），符号与绝对值大的加数相同. 也即：

正整数 + 正整数 = 正整数，

负整数 + 负整数 = 负整数，

正整数 + 负整数 = 正整数（若正整数大于负整数的相反数），

正整数 + 负整数 = 负整数（若正整数小于负整数的相反数）.

2.1.2 特殊的 "0"

在整数进行加法运算的时候出现了一种可能性，就是正整数和负整数相加时，若遇正整数刚好等于负整数的绝对值，此时必然出现一个整数来促使等式成立，于是就定义了零，来表示 "无". 即：

正整数 + 负整数 = 0（若正整数等于负整数的相反数）.

一般认为，代表零的原始符号发明于中南半岛或者印度地区，并且早在大约 1000 多年前就由玛雅人独立发明出来了.

一个整数加零，其和为这个整数. 也即：

正整数 $a + 0 = $ 正整数 a，

负整数 $b + 0 = $ 负整数 b.

特别的，零加零，其和为零，一个整数和它的相反数相加，其和为零. 也即：

$$0 + 0 = 0,$$

$$a + (-a) = 0.$$

2.1.3 整数的加法运算律

加法交换律：两个整数相加，互换加数的位置，其和不变. 也即：

$$a + b = b + a.$$

加法结合律：三个整数相加，可以将其中任意两个整数结合先行 "相加" 运算后再与第三个整数相加. 也即：

$$a + b + c = (a + b) + c = a + (b + c) = (a + c) + b.$$

例 2.1 兔子问题

如果有一对兔子，每一个月都生下一对小兔，而所生下的每一对小兔在出生后的第三个月也都生下一对小兔. 那么, 由一对兔子开始, 满一年时一共可以繁殖成多少对兔子?

按照规则, 开始时仅有 1 对兔子, 第一个月时会生下一对小兔, 故第一个月结束的时候应该有 $1+1=2$ 对兔子, 其中一对是大兔, 一对是初生小兔, 大兔子还会继续生产, 而小兔要到生下后的第三个月才会生产. 第二个月的时候大兔子还会再生一对小兔出来, 所以第二个月结束的时候应该有 $1+1+1=3$ 对兔子, 其中一对是大兔, 一对是满一月的小兔, 一对是初生小兔, 以此类推, 我们可以用列举法得到以下答案:

表 2.1

月份	1	2	3	4	5	6	7	8	9	10	11	12
兔子对数	2	3	5	8	13	21	34	55	89	144	233	377

故由一对兔子开始, 满一年时一共可繁殖成 377 对兔子.

意大利数学家斐波那契 (Leonardo Pisano, Fibonacci, Leonardo Bigollo, 1175—1250) 在其所著的《算盘全集》一书中提出了上述兔子繁殖问题, 随后发现后面一个月份的兔子总对数, 恰好等于前面两个月份兔子总对数的和.

$$2+3=5$$
$$3+5=8$$
$$5+8=13$$
$$8+13=21$$
$$13+21=34$$
$$21+34=55$$
$$34+55=89$$
$$55+89=144$$
$$89+144=233$$
$$144+233=377$$
$$\cdots\cdots$$

若将第一个月时候的兔子对数重复写一次, 将得到:

$$1, 1, 2, 3, 5, 8, 13, 21, 34, 55, 89, 144, 233, 377\cdots$$

这一串数字就被命名为斐波那契数列 (Fibonacci Sequence), 其中的数称为斐波那契数. 在我们周围存在许多诸如松果、凤梨、树叶的排列、某些花朵的花瓣数 (典型的有向日葵花瓣)、蜂巢、蜻蜓翅膀、超越数①e②（可以推出更多）、黄金分割③、黄金三角形④、黄金矩形、等角螺线、十二平均律⑤等都有斐波那契数的影子．在现代物理、准晶体结构、化学等领域，斐波那契数列都有直接的应用．

巧合的是，在东方，13 世纪南宋杰出的数学家和数学教育家杨辉⑥也发现了这一规律，并将其表示为一个"三角形"的形式，后人称为杨辉三角（见图 2.4），又称贾宪⑦（11 世纪前半叶中国北宋数学家）三角形、帕斯卡三角形（Pascal Triangle）．由贾宪、斐波那契、杨辉三人所生活的年代可以看出，他们几乎是在同一时期发现的这一特殊数学性质．事实上，在数学界，这样的事情屡见不鲜，比如牛顿和莱布尼兹几乎在同一时间不同国度同时进行了微积分的相关研究并做出了极大的贡献．

---

① 超越数是指不满足任何整系数（有理系数）多项式方程的实数，即不是代数数的数．它因欧拉说过的"它们超越代数方法所及的范围之内"（1748 年）而得名．几乎所有的实数都是超越数．1844 年，刘维尔（J. liouville，法，1809—1882）首先证明了超越数的存在性．厄米特与林德曼先后证明了 e 与 π 为超越数．

② 以 e 为底的对数叫作自然对数．在原子物理和地质学中考察放射性物质的衰变规律或考察地球年龄时便要用到 e，在用齐奥尔科夫斯基公式计算火箭速度时也会用到 e，在计算储蓄最优利息及生物繁殖问题时，也要用到 e．最有意思的是："将一个数分成若干等份，要使各等份乘积最大，怎么分？"要解决这个问题便要同 e 打交道．答案是：使等分的各份尽可能地接近 e 值．如把 10 分成 $10 \div e \approx 3.7$ 份，从另一个角度来看，$10 - 3.7 = 6.3$，这一数值很接近于黄金分割．

③ 黄金分割是指将整体一分为二，较大部分与整体部分的比值等于较小部分与较大部分的比值，其比值约为 0.618．这个比例被公认为是最能引起美感的比例，因此被称为黄金分割．1753 年，格拉斯哥大学的数学家西摩松（R. Simson）发现，随着数字的增大，斐波那契数列两数间的比值越来越接近黄金分割率，即随着 n 的无限增大，F_{n+1} 和 F_n 的比值越来越接近黄金比例．事实上，斐波那契数列的通项公式就是用黄金分割数表达的．

④ 黄金三角形是一个等腰三角形，其底与腰的长度比为黄金比值，正是因为其腰与边的比为 $(\sqrt{5} - 1) / 2$ 而被称为黄金三角形．黄金三角形是唯一一种可以用 5 个而不是 4 个与其本身全等的三角形来生成与其本身相似的三角形．由五角形的顶角是 36°可得出黄金分割的数值为 2sin18°［即 $2 \times \sin$（$\pi/$10）］．

⑤ 一种音乐定律方法，将一个纯八度平均分成十二等份，据此得到其比例．

⑥ 杨辉，字谦光，汉族，钱塘（今浙江杭州）人，南宋杰出的数学家和数学教育家，生平履历不详．曾担任过南宋地方行政官员，为政清廉，足迹遍及苏杭一带．他在总结民间乘除捷算法、"垛积术"、纵横图以及数学教育方面，均做出了重大的贡献．他是世界上第一个排出丰富的纵横图和讨论其构成规律的数学家．著有数学著作 5 种 21 卷，即《详解九章算法》12 卷（1261 年）、《日用算法》2 卷（1262 年）、《乘除通变本末》3 卷（1274 年）、《田亩比类乘除捷法》2 卷（1275 年）和《续古摘奇算法》2 卷（1275 年）（其中《详解》和《日用算法》已非完书），后三种合称为《杨辉算法》．朝鲜、日本等国均有译本出版，流传世界．杨辉还曾论证过弧矢公式，时人称为"辉术"．其与秦九韶、李冶、朱世杰并称"宋元数学四大家"．

⑦ 贾宪，北宋人，约于 1050 年左右完成《黄帝九章算经细草》，原书遗失，但其主要内容被杨辉（约 13 世纪中）著作所抄录，因能传世．杨辉《详解九章算法》（1261 年）载有"开方作法本源"图，注明"贾宪用此术"．这就是著名的"贾宪三角"，或称"杨辉三角"．《详解九章算法》同时录有贾宪进行高次幂开方的"增乘开方法"．

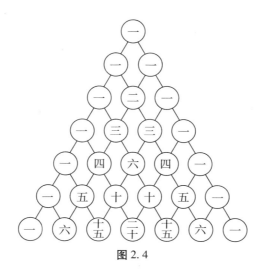

图 2.4

2.2　整数的减法及其运算

整数的减法是整数加法的逆运算.

在整数集 Z 中，若按一定次序取出两个元素，不妨假设依次取出 a 和 b，如果对于某一运算规则"$-$"在 Z 中有唯一确定的第三个元素，不妨假设 c 与之对应，则称在整数集 Z 中定义了一种运算，记作：

$$a - b = c.$$

其中 a 称为减法的被减数，b 称为减法的减数，"$-$"称为减号，c 称为 a 减 b 的差. 整数减法的数学本质如图 2.5 和图 2.6 所示.

$a - b = c$ 等价于 $a - c = b$，也等价于 $c + b = a$. 换句话说，在上述事实成立的情况下，已知 c, b，求 a 的运算就是加法运算，已知 a, b（或 a, c），求 c（或 b）的运算就是减法运算.

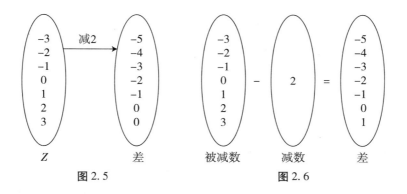

图 2.5　　　　　　　　　图 2.6

2.2.1 整数的减法运算规则

整数的减法运算可以将减数变为其相反数，然后再按加法的运算规则进行运算．即：
$$a - b = a + (-b).$$

一个整数减零，其差为这个整数．零减一个整数，其差为这个整数的相反数．如图2.7 所示，即：
$$a - 0 = a,$$
$$0 - a = -a.$$

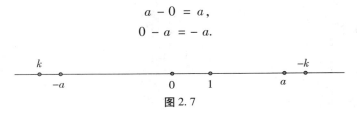

图2.7

2.2.2 整数减法的方法论意义

从数学贡献上讲，减法运算将自然数集扩充到了包含零、正整数（自然数）和负整数的整数环．在整数减法运算之前是没有零和负数的．事实上，人们一开始很容易接受自然数，却很难接受零和负数这两个概念，但由于进一步的认知了解到"既然一个自然数与另一个自然数可以相加，就会想到做相反的事情，也就出现了减法"．更进一步，"既然1与1相加等于2，那么1减1等于几呢？"一个现实的场景可能就是"我打猎得到一只兔子，交给了部落首领，我还有几只兔子？"于是就出现了 $1 - 1 = 0$．

更进一步地，当一个大的数（自然数）减一个小的数（自然数）时是没问题的，但是当情况相反了呢？当一个小的数（自然数）减一个大的数（自然数）就不够减了，这个时候需要假想出一个数来作为其差，于是产生了负数．

2.3 整数的乘法及其运算

整数的乘法是简化加法运算的必然程序．

在整数集 Z 中，若按一定次序取出两个元素，不妨假设依次取出 a 和 b，如果对于某一运算规则"×"在 Z 中有唯一确定的第三个元素，不妨假设 c 与之对应，则称在整数集 Z 中定义了一种运算，记作：
$$a \times b = c.$$

其中 a 和 b 称为乘法的乘数，有时也叫因数，"×"称为乘号，乘号在后续的学习中可以被简化成"·"或直接省略，也即 $c = a \cdot b = ab$，c 称为 a 与 b 相乘的积．整数乘法的数学本质如图2.8 和图2.9 所示．

整数的乘法主要用于将两个及更多的加数相加时进行简化，使之用较少步数的乘法运算完成较多步数的加法运算，或者说"乘法是相同加数进行加法的简便运算"．

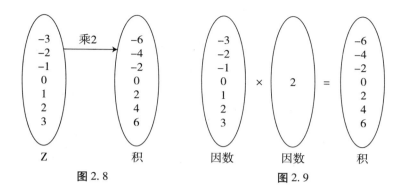

图 2.8　　　　　　　　　　图 2.9

例 2.2　$3+3+3+3+3$ 或 $5+5+5$ 可以用乘法表示成 3×5 或 5×3.

注意：像例 2.2 中这样，两个数相乘时，3×5 是不可以将乘号简化或省略的.

2.3.1　整数的乘法运算规则

两个相同正负号的整数相乘，绝对值相乘，符号为正；两个不同正负号的整数相乘，绝对值相乘，符号为负.也即：

$$正整数 \times 正整数 = 正整数,$$
$$负整数 \times 负整数 = 正整数,$$
$$正整数 \times 负整数 = 负整数,$$
$$负整数 \times 正整数 = 负整数.$$

2.3.2　特殊的"1"

由乘法的定义可知，会出现一种情况，就是相同加数的个数为 1，此时乘法就会变为：

$$a \times 1 = a.$$

一个整数与零相乘，其积为零.也即：

$$a \times 0 = 0.$$

2.3.3　整数的乘法运算律

乘法交换律：两个整数相乘，互换乘数的位置，其积不变.也即：

$$a \times b = b \times a.$$

乘法结合律：三个整数相乘，可以将其中任意两个整数结合先行"相乘"运算后再与第三个整数相乘.也即：

$$a \times b \times c = (a \times b) \times c = a \times (b \times c) = (a \times c) \times b.$$

分配律：两个整数的和与另一个整数相乘，可先将这两个整数分别与其相乘再求和.也即：

$$(a + b) \times c = a \times c + b \times c.$$

2.3.4　整数的乘方

若干个整数相乘：

$$\underbrace{a \times a \times \cdots \times g}_{n\text{个}} = a^n$$

称为 a 的 n 次方（次幂），a 称为幂底数，n 称为幂指数.

正整数的任何次方都是正整数，负整数的偶次方是正整数，负整数的奇次方是负整数，零的任何次方都是零.

特别的，规定非零数的零次方为 1，即 $a^0 = 1, a \neq 0.$

2.4 整数的除法及其运算

除法的本质属性是"平均分". 表示的是将一个整体（可以是一个数或一个量等）平均的分成若干份，如将一定数量的物品分成同样多少的份数，一件物品和一个接受者相匹配. 整数的除法可以理解为将一个整数平均分成若干（整数）份.

除法是乘法的逆运算：已知两个因数的积与其中一个因数，求另一个因数的运算，叫作除法. 用式子表示是：

$$被除数 \div 除数 = 商.$$

读作：被除数除以除数等于商，或除数除被除数等于商. 需要特别注意的是除数不能为 0.

例如：$6 \div 2 = 3$ 中，6 是被除数，2 是除数，3 是商. 如果在代数式中，$6 \div 2$ 还可以写成 $6:2$ 或者 $6/2$，此时读作 6 比 2.

在整数集 Z 中，若按一定次序取出两个元素，不妨假设依次取出 a 和 $b(b \neq 0)$，如果对于某一运算规则"$\div$"在 Z 中有唯一确定的第三个元素，不妨假设 c 与之对应，则称在整数集 Z 中定义了一种运算，记作：

$$a \div b = c.$$

其中 a 称为除法的被除数，b 称为除法的除数，"$\div$"称为除号，c 称为 a 除 b 的商. 整数乘法的数学本质如图 2.10 和图 2.11 所示.

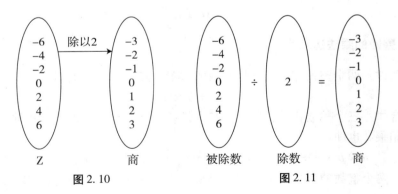

图 2.10　　　　　　　　　　　图 2.11

$a \div b = c$ 等价于 $c \times b = a.$

换句话说，在上述事实成立的情况下，已知 c, b，求 a 的运算就是乘法运算，已知 a，b（或 a, c），求 c（或 b）的运算就是除法运算.

2.4.1　整数除法的可能性

自然而然的，我们会追问，整数除以整数的情形．由小学数学知识抑或常识我们都知道，整数除以整数可能是整数也可能不是整数（除不尽，有余数）．所以整数的除法运算可能的结果是：

$$整数 \div 整数 = \begin{cases} 整数 & 除得尽. \\ 非整数 & 除不尽. \end{cases}$$

当整数除以整数结果刚好为整数的时候，我们就称之为整除．

那当整数除以整数结果不是整数，也就是除不尽的时候，还可以继续做除法吗？答案是肯定的，至于如何进行这样的除法运算将在"带余除法"那一节中进行详细介绍．

加减乘三种运算对于整数集来说是封闭的，而除法却未必，也就是说，一个整数除以一个整数，其结果未必是一个整数，甚至有时候是不可进行除法运算的．

2.4.2　与零有关的除法运算

零除以任何非零的数等于零．也即：

$$0 \div a = 0.$$

任何数不能除以零，也就是说零不能作为除数．

例 2.3　请查询相关资料，说说为什么零不能作为除数．

2.4.3　运算规则

两个相同正负号的整数相除，绝对值相除，符号为正；两个不同正负号的整数相除，绝对值相除，符号为负．也即：

$$正整数 \div 正整数 = 正整数,$$
$$负整数 \div 负整数 = 正整数,$$
$$正整数 \div 负整数 = 负整数,$$
$$负整数 \div 正整数 = 负整数.$$

特别的，任何数除以 1 等于它自身．也即：

$$a \div 1 = a.$$

2.4.4　整数除法的方法论意义

从对数学的贡献来讲，除法运算将整数扩充到了包含零、正有理数和负有理数的有理数域．在整数除法运算之前是没有有理数的，因为除法要将整数平均分，而当不能平均分成若干份的时候，也就是有剩余的时候，就需要对剩余部分进行描述，于是就出现了非整数的有理数．

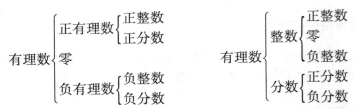

毕达哥拉斯学派认为"万物皆数"，其中的数特指整数和由整数进行加减乘除运算后得到的数，事实上，这就已经发展到了有理数. 所谓有理数有两种形式：一是分数，二是有限小数或无限循环小数.

2.5 典型问题

2.5.1 典型例题

例 2.4 求 a, b, c，使它们满足不等式：

$$a^2 + b^2 + c^2 + 3 < ab + 3b + 2c (a, b, c \in Z).$$

分析：利用整数的离散性及配方法.

解：由题意，整数 a, b, c 满足：

$$a^2 + b^2 + c^2 + 3 < ab + 3b + 2c,$$

即：

$$a^2 + b^2 + c^2 + 4 \leqslant ab + 3b + 2c.$$

配方得：

$$\left(a - \frac{b}{2}\right)^2 + 3\left(\frac{b}{2} - 1\right)^2 + (c - 1)^2 \leqslant 0.$$

由于：

$$\left(a - \frac{b}{2}\right)^2 \geqslant 0, \left(\frac{b}{2} - 1\right)^2 \geqslant 0, (c - 1)^2 \geqslant 0,$$

故有：$a - \dfrac{b}{2} = \dfrac{b}{2} - 1 = c - 1 = 0$，即：

$$a = 1, b = 2, c = 1.$$

例 2.5 设 $a, b \in Z$，证明：$4a - 1 \neq 4b + 1$.

证明：（反证法）假设：

$$4a - 1 = 4b + 1,$$

则 $4(a - b) = 2$，即：

$$a - b = \frac{1}{2} \notin Z,$$

而这与 Z 对减法运算封闭矛盾.

所以对任意 $a, b \in Z$，均有：

$$4a - 1 \neq 4b + 1.$$

例 2.6　试证：任何 ≥8 的正整数都能表示为若干个 3 和 5 的和.

分析：在这里主要涉及对"若干个"的理解. 所谓若干个，指的是 0 个或正数个，也就是平时所说的几个. 以此可以分为三种可能：首先是 0 个 3，几个 5；其次是 0 个 5，几个 3；最后一种可能是几个 3，几个 5.

证明：当 $n = 8$ 时，有 $8 = 3 + 5$，命题成立.

假设当 $n = k$（k 是正整数且 $k \geq 8$）时命题成立，即存在正整数 a, b 使得：

$k = 3a + 5b$；

或存在正整数 $a \geq 3$，使得 $k = 3a$；

或存在正整数 b，使得 $k = 5b$.

现在看 $n = k + 1$ 时命题成立情况.

按照分析中的三种可能分别进行讨论可以得到：

$k + 1 = 3a + 5b + 1 = 3(a + 2) + 5(b - 1)$

$k + 1 = 3a + 1 = 3(a - 3) + 5 \times 2$

$k + 1 = 5b + 1 = 3 \times 2 + 5(b - 1)$

由此可知这个命题当 $n = k + 1$ 时也成立.

所以，任何 ≥8 的正整数都能表示为若干个 3 和 5 的和.

例 2.7　有两堆棋子，数目相等，有两人玩耍，每人可以在任意一堆里任意取几颗，但不能同时在两堆里取，规定取最后一颗者胜. 试证：后取者必胜.

证明：设 n 是每一堆棋子的颗数.

当 $n = 1$ 时，先取者只能在一堆里取一颗，这样另一堆里留下的一颗就被后者取得，所以结论成立.

假设当 $1 \leq n \leq k$ 时结论成立.

现在我们来证明，当 $n = k + 1$ 时结论也成立.

在这种情况下，先取者可以在一堆棋子里取棋子 l（$1 \leq l \leq k$）颗.

这样，剩下的两堆棋子中，一堆有 $k + 1$ 颗，另一堆有 $k + 1 - l$ 颗，这是后取者可以在较多的一堆里取棋子 l 颗，使两堆棋子都有 $k + 1 - l$ 颗.

由归纳假设，可知后取者可以获胜.

根据第二数学归纳法，这个命题对所有正整数 n 来说，后取者必胜.

2.5.2　典型练习题

练习 2.1　试证：当 m, n 取遍全体正整数时

$$m + \frac{1}{2}(m + n - 2)(m + n - 1)$$

也取遍全体正整数.

练习 2.2　求证：对任意整数 m, n 必有 $6m - 1 \neq 6n + 1$.

练习 2.3　试分析正整数集对算术四则运算的封闭性.

练习 2.4 对任意 $m, n \in Z$，试说明 $7m + 8n$ 一定是整数.

练习 2.5 a, b, c 均为整数，且等式：

$$a + \frac{x}{b} = c$$

成立，试说明 x 也是整数.

练习 2.6 已知斐波那契数列 $\{f_n\}$ 满足：

$$f_1 = 1,$$
$$f_2 = 1,$$
$$f_n = f_{n-1} + f_{n-2} \,(n \geqslant 3),$$

试证：$f_n = \dfrac{1}{\sqrt{5}} \Big[\Big(\dfrac{1 + \sqrt{5}}{2} \Big)^n - \Big(\dfrac{1 - \sqrt{5}}{2} \Big)^n \Big].$

练习 2.7 十个男孩和 n 个女孩共买了 $n^2 + 8n + 2$ 本书，已知他们每人买的书的数量相同，且女孩的人数多于男孩的人数，问女孩人数是多少？

第 3 章 整除性

第一是数学，第二是数学，第三是数学.

——伦 琴

我们主要通过研究整除这一"特殊"情形进而推广到"一般"情形. 在整除的条件之下, 整数有一些很有价值的性质值得我们探究, 如最大公因数理论和算术基本定理等. 整除性是初等数论的基础.

图 3.1

整数的除法运算分为可以除得尽和除不尽两种可能. 除得尽指的是通过这一除法运算后其商也恰好是一个整数, 此时即为"整除". 除不尽指的是通过这一除法运算后其商不是一个整数, 此时即为"不整除". 整除是除法中的一种很特殊的情形, 是一种巧合, 而除不尽, 也就是不整除才是更加普遍的情形, 是除法的常态.

3.1　整　除

3.1.1　**整　除**

整除的数学定义: 设 $a, b \in Z, b \neq 0$, 若存在整数 q, 使得 $a = bq$, 则称 a 能被 b 整除, 或 b 整除 a, 记作: $b \mid a$. 否则称 a 不能被 b 整除, 或 b 不整除 a, 记作: $b \nmid a$.

例 3.1　整除: 6 和 3 都是整数, 且 3 不等于 0, 我们知道 $2 \times 3 = 6$, 也就是说存在整数 2 可以使得 $6 = 2 \times 3$, 于是称 6 能被 3 整除, 记作 $3 \mid 6$.

例 3.2　不整除: $5 \nmid 7$, 因为不存在整数 $n \in Z$ 可以使得 $5n = 7$, 事实上,
$$5 \times 1 + 2 = 7$$
而:
$$5 \times 2 - 3 = 7$$
也就是说, 如果存在某整数 $n \in Z$ 可以使得 $5n = 7$, 那么这个整数应该介于 1 和 2 之间, 而这是不可能的.

进一步, 若已知乘法算式:
$$x \cdot y = z,$$
其中的 x, y, z 都是整数且 x, y 不等于零, 那么就可以得出:
$$x \mid z, \quad y \mid z.$$
当 $b \mid a$ 时, a 是 b 的倍数, 而 b 是 a 的因数 (也称约数).

任意非零的整数 a 必然有因数 $\pm 1, \pm a$, 所以称这几个因数为平凡因数. 除了平凡因数, 有时还有其他一些因数.

当 $b \mid a$, 且 $b \neq \pm a, a \neq 0, b \neq \pm 1$ 时, 称 b 是 a 的真因数, 也称为非平凡因数、非显

然因数等.

例 3.3 讨论 35 的因数及其分类.

解：考虑到：

$35 = 1 \times 35$

$35 = 5 \times 7$

$35 = (-1) \times (-35)$

$35 = (-5) \times (-7)$

故 35 的因素有 ± 1，± 5，± 7，± 35.

真因数有 ± 5，± 7.

正因数有 1，5，7，35.

正的真因数有 5，7.

3.1.2　整除的方法论意义

在数学中，"一般"与"特殊"是极其重要的，是一种很重要的数学思想方法，可以由此引申出一连串的事物本质认识.所谓一般指的是具有一般性的事物，一般性就是共性，是所有对象都共同具有的.特殊指的是具有特殊性的事物，特殊性就是个性，是某一特殊对象所独自具有的.一般性包含在特殊性中，而特殊性是一般性的基础.

正因如此，数学学习往往会先对一般性进行一个简介，随后对特殊事物进行深入探讨，以此通过再次推理得出一般事物所具有的共性.在现实中，我们若能以这样的方法认识事物，其结论才更经得起推敲与实际的考验.

在数学中，有一种很常见的方法，就是为了证明某命题成立，我们往往需要假设这个命题的结论是成立的，然后通过原始概念、定义或特殊需要时候的约定进行反向推导，得出结论成立的基本条件.

以整除为例，我们知道所谓整除其实就是刚好除得尽，进行"平均分"后没有余数，通过其上一层级的概念，也就是父定义——乘法进行逆向推导，换句话说，在除得尽的时候，可以说除法完全就是乘法的逆运算.所以要说明除法，我们得考虑先从乘法入手.

第一步，为了进行"$a \div b$"这个运算，需要先从其逆运算着手，而在乘法中，涉及"$a \times b = c$"的三个数，按理把这三个数反推进入除法的时候将会变成"$c \div a = b$"（或者"$c \div b = a$"）.为了在除法中用我们更常用的"a,b"两个符号，不妨将其对调，将原始的乘法改为"$q \times b = a$"，这样并不会造成任何矛盾或不妥，于是得到的除法就是"$a \div b = q$".

第二步，前面的假设中的三个数都是整数，并且刚好存在"$q \times b = a$"这个结论，而我们需要考虑的是对于形如"$a \div b$"的除法的情形，也就是说，式子中的"q"是我们假设出来的，在整除性的表示中必须进行说明.于是"除得尽"的简单表述就会严谨化为"若存在整数 q".

第三步，考虑一个特殊情形，就是在乘法中永恒成立的关于零的特殊性的问题：一个整数与零相乘，其积为零.换句话说，即便前面所有的假设都成立，比如存在整数 $q = 3$，可以使得"$q \times 0 = 0$"，却无法得出"$0 \div 0 = q = 3$"的结论.否则，任意整数 q，比

$q = 5, 7, 100 \cdots \cdots$ 都可以使得 "$q \times 0 = 0$" 成立，这样的话将会导致 "$0 \div 0$" 的结论不唯一的麻烦，相应的还应考虑 "$3 \div 0$" "$5 \div 0$" 之类的除数为零的情形．于是为了得到唯一一个商，对于除法来说，我们还需要把除数为零的情形排除在外．至此，才能得到整除的数学定义．

意义：在往后的学习中，涉及证明整除性的时候，可以考虑将其商求出，通过该商是不是整数来判断其整除性．如果已经找到了这个身为整数的商，那自然可以通过定义证明其整除了．这就是数学中最常见的数学证明方式——用定义进行证明．

3.2　整除性

整数的整除具有很多性质，在此，我们只涉及一些很常见的性质，为了表述上的方便，以后若无特别说明，我们用字母 $a, b, c \cdots$ 表示整数．

3.2.1　整除性

性质 1　$1 \mid a, b \mid 0, a \mid a$.

考虑到阅读上的简便，我们将以 "说明" 而非 "严谨证明" 的方式继续．

首先，1 的特殊性：因为任何数乘以 1 都是该数本身，故任意一个整数 a 都可以记作 $a = a \times 1$，所以对任意整数 a 都有 $1 \mid a$.

其次，0 的特殊性：因为 0 除以任何数都为 0，故对任意一个整数 b 都有 $0 = b \times 0$ 成立，所以对任意整数 b 都有 $b \mid 0$.

最后，整除的自反性：对任意一个整数 a 都有 $a = a \times 1$ 成立，所以对任意整数 a 都有 $a \mid a$.

性质 2　若 $a \mid b$，则 $-a \mid b, a \mid -b, -a \mid -b, |a| \mid |b|$. 反之也成立．

整除性与整数的正负无关．

事实上，如果 $a \mid b$，那么涉及一个除法算式 $b \div a = c$，也就是说有那么一个整数 c 可以满足这个等式．既然如此，当把 a 或 b 的符号变掉，换成它的相反数之后，也会存在一个除法算式可以成立，只需要考虑它们的商 c 的正负号即可．

例 3.4　已知 $5 \mid 10$，这是因为 $10 \div 5 = 2$，那么要看所有 ± 5 和所有 ± 10 的整除关系的时候，只需要考虑：

$$10 \div 5 = 2,$$
$$10 \div (-5) = -2,$$
$$(-10) \div 5 = -2,$$
$$(-10) \div (-5) = 2,$$

因此，$5 \mid 10, -5 \mid 10, 5 \mid -10, -5 \mid -10, |5| \mid |10|$.

性质 3　$a \mid b \Leftrightarrow \pm a \mid \pm b$.

将性质 2 进行推广即可．

性质 4 若 a，b 是正整数，$a \mid b$，则 $a \leqslant b$.

除法算式中整数大小关系比较.

除法的本质是"平均分"，那么将一个正整数分成正整数份，比如将 10 个苹果分给 5 个小朋友，每人分得的苹果个数不可能超过总数 10.

性质 5 若 a 是 b 的真因数，则 $1 < |a| < |b|$.

继续性质 4 的讨论，只有当小朋友的个数恰好为一个的时候，分得的苹果数才刚好等于 10，其他时候都会少于总数 10.

所谓小朋友个数恰好为 1，指的是除法中一种必然出现的情况. 也就是说，1 必然是任意整数的因数，是平凡的. 除了 1 之外还有一种情况就是因数刚好等于整数本身，这个时候也是平凡的. 当然，加上正负号不影响这个结论.

除了上述的 ± 1，$\pm |b|$ 之外的因数统称为真因数，因此在性质 4 中排除了 $a = b$ 的情况之后即可得 $1 < |a| < |b|$.

性质 6 $a \mid b$，$b \mid a$，$a = \pm b$.

由性质 4 知道，若 a，b 都是正整数，则由 $a \mid b$ 可得 $a \leqslant b$，$b \mid a$ 可得 $b \leqslant a$，要 $a \leqslant b$ 和 $b \leqslant a$ 同时成立有且仅有一种可能：$a = b$.

如果 a，b 不一定是正整数，则由性质 2 和性质 3 可以知道若同为负整数则 $a = b$. 若一个是正整数一个是负整数则 $a = -b$.

性质 7 若 $b \mid a$，则 $a = 0$ 或 $|b| \leqslant |a|$.

一方面，对任意的整数 b 都有 $b \mid 0$，所以 $a = 0$.

另一方面，如果 $a \neq 0$，则由性质 4 知道，若 a，b 都是正整数，则由 $b \mid a$ 可得 $b \leqslant a$. 若 a，b 都是负整数，则 $a \leqslant b$，此时 $|b| = -b \leqslant -a = |a|$. 如果 a，b 异号，也就是一正一负，那由性质 3 可以归纳为同号进行判断.

性质 8 $b \mid a$，$a \neq 0 \Rightarrow |b| \leqslant |a|$.

由性质 7，排除 $a = 0$ 的可能即可.

性质 9 $b \mid a$ 且 $|a| \leqslant |b| \Rightarrow a = 0$.

由性质 7 知道 $a = 0$ 或 $|b| \leqslant |a|$. 而 $|a| \leqslant |b|$. 所以，只有一种可能就是 $a = 0$.

性质 10 $a \mid b$，$b \mid c \Rightarrow a \mid c$.

整除的传递性.

在数学或其他科学研究中，有好多事物都具有传递性，比如实数大小关系比较中有：$a < b$，$b < c \Rightarrow a < c$. 集合关系中有：$A \subseteq B$，$B \subseteq C \Rightarrow A \subseteq C$. 在所有这些关系中，整除的传递性可以说是开了个好头.

具体来说，$a \mid b$ 意味着有一个整数 q 可以使得 $b = qa$ 成立，同理 $b \mid c$ 意味着有一个整数 p 可以使得 $c = pb$ 成立，把这两个一定会成立的等式联立就可以得到 $c = pqa$，显然 pq 是整数，所以结论成立.

例 3.5 因为 $2 \mid 4$，$4 \mid 12$，所以 $2 \mid 12$.

事实上，还有进一步的结论就是：

$$12 \div 2 = (4 \div 2) \times (12 \div 4).$$

性质 11　$a \mid b$，$a \mid c$，则 $a \mid (b \pm c)$.

推广可得：x，y 为任意整数，$a \mid b$，$a \mid c$，则 $a \mid (bx + cy)$. 反之也成立.

进一步推广有：如果有 $b \mid a_i (i = 1，2，\cdots，k，k \geqslant 2)$，则 $b \mid (a_1 m_1 + a_2 m_2 + \cdots + a_k m_k)$，此处 $m_i (i = 1，2，\cdots，k)$ 是任意的整数.

这就是整除的线性可加性.

作为推论 11 的应用，我们知道，若 d 整除一个方程一端的所有项，则它也整除此方程的另一端. 因此，若 $a + b = c$，且 $d \mid a$，$d \mid c$，则 $d \mid b$. 又若：

$$3x + 81y + 6z + 363 = w.$$

则 $3 \mid w$，因为 3 整除该方程左端所有项. 类似的，若：

$$3x^2 + 15xy + 5y^2 = 0$$

则：

$$3 \mid 5y^2，5 \mid 3x^2.$$

性质 12　若 $m \neq 0$，则 $a \mid b \Leftrightarrow ma \mid mb$.

这就是整除的可乘性.

整数的整除不会因为在被除数和除数上同时乘以一个非零整数而改变，事实上，同时乘以任意非零整数之后，对除法的商是没有影响的，因此，原来的商是整数则变化后的商也是原整数.

例 3.6　求证：

（1）若一个数的末位数字能被 2 整除，则这个数能被 2 整除；

（2）若一个数的末两位数字组成的数能被 4 整除，则这个数能被 4 整除.

证明：（1）设 $a = 10b + c$（b 是整数，$c \in \{0,1,2,\cdots,9\}$）.

因为 $2 \mid 10$，故 $2 \mid 10b$.

又因为 $2 \mid c$，所以 $2 \mid a$.

（2）设 $a = 100b + \overline{cd}$（b 是整数，$c,d \in \{0,1,2,\cdots,9\}$）.

因为 $4 \mid 100$，故 $4 \mid 100b$.

又因为 $4 \mid \overline{cd}$，所以 $4 \mid a$.

3.3　带余除法

针对不整除，我们应当用怎样的办法来合理表达呢？

可以这样理解，所谓不整除，是因为被除数多了或者少了，如果能够加上或者减掉一个合适的数，所有的不整除将会变成整除. 比如：$5 \mid 20$，但是 $5 \nmid 21$，$5 \nmid 18$，若能将 21 或 18 通过加或减而得到 20，则整除成立.

$$\because 21 - 1 = 5 \times 4，\therefore 21 = 5 \times 4 + 1.$$

在除法算式中，也可以写成：

$$21 \div 5 = 4 \cdots 1$$

其中 4 称为 $21 \div 5$ 的商，1 称为 $21 \div 5$ 的余数. 像这样在结果中带着余数的除法就叫作带余除法．当涉及的除数和被除数较小的时候，可以通过简单计算甚至乘法口诀得到商和余数，否则就得考虑用计算器了，可如果遇到是一个用字母或其他形式表示的整数除法怎么办呢？

例 3.7　1270 除以某自然数，其商为 74，求除数与余数．

分析：设除数为 x，余数为 r，则有 $1270 = 74x + r, r = 1270 - 74x$，且 $0 < r < x$. 由此可求得 x 可能取的整数值．

解：令除数为 x，余数为 r，据题意有：
$$1270 - 74x > 0, 1270 - 75x < 0.$$

由此求得：
$$\frac{1270}{75} < x < \frac{1270}{74},$$

即：
$$16\frac{14}{15} < x < 17\frac{6}{37}.$$

因此，$x = 17$，$r = 1270 - 74x = 12$．

说明：若可能取几个整数，则应一一列举给出各种可能的答案．

显然，用上面的表述来作为"带余除法"的定义是不够严谨的．我们需要重新思考其严谨的数学概念．

3.3.1　带余除法

带余除法：设 $a, b \in Z$，且 $b \neq 0$，则存在唯一的 $q, r \in Z$，使得：
$$a = bq + r, 0 \leqslant r < |b|.$$

上式等价于：
$$a \div b = q \cdots r, 0 \leqslant r < |b|.$$

其中的 q 称为 b 除 a（a 除以 b）的不完全商（有时候也简称为商），r 是 b 除 a 的最小非负余数，简称余数．

证明：考虑整数序列
$$\cdots, -2|b|, -|b|, 0, |b|, 2|b|, \cdots.$$

则 a 必在上述序列的某相邻两项之间，不妨假定
$$q|b| \leqslant a < (q+1)|b|.$$

于是 $0 \leqslant a - q|b| < |b|$，

令 $r = a - q|b|$，

则有 $0 \leqslant r < |b|$，

因此，当 $b > 0$ 时，

有 $a = bq + r$；

当 $b < 0$ 时，有 $a = b(-q) + r$

这样，我们就证明了 q 和 r 的存在性．

下面证明 q, r 的唯一性．假设存在另外一组：

q'，$r' \in Z$，使得 $a = bq + r$，$0 \leqslant r < |b|$ 成立.

即：

$a = bq' + r'$，$0 \leqslant r' < |b|$，

则有：

$$-|b| < r - r' = b(q' - q) < |b|.$$

因此 $b(q' - q) = 0$，从而有 $r - r' = 0$，

即 $q' = q$，$r' = r$，所以唯一性成立.

例如，当 $b = 15$，则当 $a = 255$ 时

$a = 17b + 0$，$r = 0 < 15$，

而 $q = 17$；当 $a = 147$ 时，

$a = 27b + 12$，$0 < r = 12 < 15$，

而 $q = 27$；当 $a = -81$ 时，

$a = -6b + 9$，$0 < r = 9 < 15$，而 $q = -6$.

由带余除法的定义可知：$a = bq + r$，$0 \leqslant r < |b|$ 时，$b \mid a$ 的充要条件是 $r = 0$.

带余除法是初等数论证明中最基本、最直接也是最重要的工具.

例 3.8　求证：任意给出的五个数中，必有三个数之和能被 3 整除.

证明： 设这五个整数是 a_i，令：

$a_i = 3q_i + r_i (0 \leqslant r_i \leqslant 3, i = 1,2,3,4,5)$.

分别考虑一下两种情形：

（i）若在 r_1，r_2，$\cdots$，r_5 中数 0，1，2 都出现，不妨设：

$$r_1 = 0，r_2 = 1，r_3 = 2，$$

此时

$$a_1 + a_2 + a_3 = 3(q_1 + q_2 + q_3) + 3$$

能被 3 整除；

（ii）若在 r_1，r_2，$\cdots$，r_5 中数 0，1，2 至少有一个不出现. 则根据抽屉原理至少有三个 r_i 要取相同的值. 不妨设 $r_1 = r_2 = r_3 = r$（r 是 0，1，2 中的某一个），此时

$$3 \mid a_1 + a_2 + a_3 = 3(q_1 + q_2 + q_3) + 3r.$$

综合情形（i）和（ii）可知，所证结论成立. 证毕.

3.3.2　带余除法的方法论意义

在数学抑或是生活中，常常会出现特殊与一般关系的两个对象，整数除法中的整除和不整除就是一个典型例子，为了研究一般化的对象，在数学中往往先研究特殊对象，随后通过进行类比等得出所需要研究的一般化对象的结论. 比如在几何学习中的正方形、长方形乃至四边形等.

在整除中，需要做到的就是存在整数 q，使得"$a = bq$"成立. 在不整除中，出现的情形就是不存在能使得 $a = bq$ 成立的整数 q. 问题是存在只需要找出来一个即可，而不存在却不能用一两个整数来说明. 严谨地说，需要将所有的整数都进行试除一遍，这是不可

能实现的，于是需要找到一种方法，来用几个有限的整数施行一定的步骤就能代替对于无限个数的整数的一遍遍试除．为此，按照"整除的方法论意义"中的介绍，我们自然会想到先将不整除的算式假设出来．

所谓不整除，就是对于任意的整数 q，都有" $a \neq bq$ "成立．这是一个不等式，需要将其化为等式，一个最正常不过的想法就是"多了就减少一些，少了就加上一些"以让它们能够整除，这里的减少一些和加上一些自然地对应了数学中的加法和减法，于是我们可以将" $a \neq bq$ "假设为 $a + r = bq$ 或者 $a - k = bq$，也就是 $b \mid (a + r)$ 或 $b \mid (a - k)$．这样就将不整除转化成了整除，也就构建起了不整除和整除之间的核心桥梁．

下一步，考虑其中的 r 和 k，以 r 的一个具体数值为例进行讨论．

如 $a = 18$，$r = 2$，$b = 5$，此时因为可以找到一个整数 $q = 4$ 使得 $bq = a + r$ 成立，所以 $5 \mid 18 + 2$ 成立．

与此同时，将 r 换成 7，12，17……都可以得到相同的结果，唯一的不同之处只在于其商 q 的取值有所变化，于是需要将 r 唯一化．具体做法是让 $r < b$，考虑到整数还分正整数和负整数，优化为 $r < \mid b \mid$．k 的情形类似．

又因为 r 和 k 都是假设出来的，其本质都是整数，唯一区别只在于一个是需要加上的数一个是需要减去的数，所以为了表述上的方便，不妨统一为同一个 r，此时涉及一个变化的地方就是加减号，所以将 $r < \mid b \mid$ 进一步优化为 $0 < r < \mid b \mid$．当 $r = 0$ 时就是整除．

事实上，如果 $0 < b$，那么我们可以构造一个整数序列：

$$0, b, 2b, 3b, 4b, \cdots$$

对于任意给定的正整数 a，可能出现两种情况：a 恰巧和序列中的某一项，不妨设为 qb 相等，也即 $a = qb$，此时相当于当整除时候的 $r = 0$．另一种情形是 $a \neq qb$，这个时候需要找出序列中的两个数，使得：

$$qb < a < (q + 1)b = bq + b.$$

涉及负整数的地方请读者自行验证．

为了将整除的特殊情形也考虑进去，使之成为一个更一般的"除法"，只需要将 $r = 0$ 加入条件即可，于是可以得到带余除法的严谨定义．

3.4 典型问题

3.4.1 典型例题

例 3.9 已知一个两位数去除 251，得到的余数是 41，求这个两位数．

解： 由带余除法知：

$$a \div b = q \cdots r，0 \leqslant r < \mid b \mid．$$

将 251 和 41 代入得：

$$251 \div b = q \cdots 41，0 \leqslant 41 < \mid b \mid，$$

也即

$$251 - 41 = bq$$
$$210 = bq$$
$$210 = 2 \times 3 \times 5 \times 7$$

210 的两位数的因数有：10，14，15，21，30，35，42，70，因为所求除数需要大于余数 41，故所求的两位数为 42 或 70.

例 3.10　请在 503 后面添加一个三位数，使得所得到的这个六位数能同时被 7，9，11 整除.

分析： 要使得这个数能同时被 7，9，11 整除，只需该数能整除 $7 \times 9 \times 11 = 693$，故可考虑用 503000 或者 504000 来除 693，然后再运用带余除法就可以得到要添加的数字.

解： ∵ $7 \times 9 \times 11 = 693$，

而 $504000 = 693 \times 727 + 189$，

因此 $504000 - 189 = 503811 = 693 \times 727$，

$$503811 - 693 = 503118 = 693 \times 726,$$

也就是说 503811 和 503118 都能被 693 整除，于是，所需添加的数是 811 或 118.

例 3.11　若 $n > 1$，$(n-1) \mid (n+11)$，求 n.

解： 因为
$$n + 11 = (n - 1) + 12$$
且
$$(n - 1) \mid (n + 11)$$
所以
$$(n - 1) \mid 12$$
又因为 $n > 1$，所以 $n = 2$，3，4，5，7，13.

例 3.12　求证：$37 \mid 333^{777} + 777^{333}$

证明： 因为 $37 \times 3 = 111$，所以
$$333^{777} = (111 \times 3)^{777} = (37 \times 9)^{777} = 37 \times (37^{776} \times 9^{777})$$
那么
$$37 \mid 333^{777}$$
同理可证：$37 \mid 777^{333}$.

所以 $37 \mid 333^{777} + 777^{333}$.

例 3.13　设对所有的正整数 n，有 $10 \mid (3^m + 1)$，求正整数 m.

证明： 由 $3^{m+4n} + 1 = 3^{m+4n} - 3^m + 3^m + 1 = 3^m(81^n - 1) + (3^m + 1)$，$10 \mid (81^n - 1)$ 知，要使 $10 \mid (3^{m+4n} + 1)$，必须有 $10 \mid (3^m + 1)$.

因任一正整数被 4 除所得余数为 0，1，2 或 3 故：
$$m = 4q, m = 4q + 1, m = 4q + 2, m = 4q + 3.$$
若 $m = 4q$（q 为正整数），则 $3^{4q} + 1$ 的末尾字是 2；

若 $m = 4q + 1$（q 为非负整数），则 $3^{4q+1} + 1$ 的末尾数字是 4；

若 $m = 4q + 2$（q 为非负整数），则 $3^{4q+2} + 1$ 的末尾数字是 0；

若 $m = 4q + 3$（q 为非负整数），则 $3^{4q+3} + 1$ 的末尾数字是 8；

综上，当 $m = 4q + 2$（q 为非负整数）时，$10 \mid (3^m + 1)$，从而 $10 \mid (3^{m+4n} + 1)$

例 3.14 求所有的正整数 n，使得它能够被不大于 $\sqrt{n}$ 的所有正整数整除.

解：此题可用多种方法求解.

法一：对每个正整数 n，存在唯一的 $q \in N^*$，使得 $q^2 \leqslant n < (q+1)^2$，于是每个正整数 n 可以唯一地表示成 $n = q^2 + r, 0 \leqslant r \leqslant 2q$.

由于正整数 $q \leqslant \sqrt{n}$，则有题设 $q \mid n$，从而 $q \mid r$，于是 $r = 0, q, 2q$，即正整数 n 必有如下形式：

$$n = q^2, q^2 + q, q^2 + 2q,$$

其中正整数 $q \leqslant \sqrt{n}$.

若 $q = 1$，则 $n = 1, 2, 3$.

若 $q \geqslant 2$，则 $q - 1 \mid n$，以下按 n 的三种形式分类求解：

当 $n = q^2$ 时，有

$$q - 1 \mid q^2 \Leftrightarrow q - 1 \mid 2q - 1 \Leftrightarrow q - 1 \mid 1 \Leftrightarrow q = 2;$$

当 $n = q^2 + q$ 时，有

$$q - 1 \mid q^2 + q \Leftrightarrow q - 1 \mid 3q - 1 \Leftrightarrow q - 1 \mid 2 \Leftrightarrow q - 1 = 1, 2,$$

即 $q = 2, 3$；

当 $n = q^2 + 2q$ 时，有

$$q - 1 \mid q^2 + 2q \Leftrightarrow q - 1 \mid 4q - 1 \Leftrightarrow q - 1 \mid 3 \Leftrightarrow q - 1 = 1, 3,$$

即 $q = 2, 4$.

因此，必有

$$n \in \{1, 2, 3, 4, 6, 8, 12, 24\},$$

经检验，这些正整数都满足题设条件.

法二：取 $q = [\sqrt{n}]$.

若 $q \geqslant 2$，则 $q - 1 \mid n, q \mid n$.

据 $(q - 1, q) = 1$，得 $q(q - 1) \mid n$；

再由不等式 $(q - 1)q < q^2 \leqslant n$，得

$$2(q - 1)q \leqslant n = (\sqrt{n})^2 < (q + 1)^2, 2 \leqslant q \leqslant 4.$$

故 q 只能取 1，2，3，4 从而 $n \leqslant 25$.

对这些正整数逐一检验，得到符合题设的正整数 n 得 1，2，3，4，6，8，12，24.

例 3.15 求所有的正整数 n，使得存在唯一的正整数对 (x, y) 满足 $n = \dfrac{x^2 + y}{xy + 1}$.

解：当 $n = 1$ 时，有 $1 = \dfrac{x^2 + (x + 1)}{x(x + 1) + 1}$ 对一切正整数对 $(x, y) = (x, x + 1)$ 成立，表示

不唯一.

当 $n \geqslant 2$ 时，令 $n = \dfrac{x^2 + y}{xy + 1}$ ，则 $x \geqslant 2$ 以及 $y = \dfrac{x^2 - n}{nx - 1} \in N^*$ ，有

$$nx - 1 \mid x^2 - n,$$

这等价于

$$nx - 1 \mid n^2(x^2 - n)$$
$$\Leftrightarrow nx - 1 \mid (nx)^2 - 1 + 1 - n^3$$
$$\Leftrightarrow nx - 1 \mid n^3 - 1,$$

所以 $n^3 - 1 \geqslant nx - 1$ ，即 $2 \leqslant x \leqslant n^2$.

记 $d = \dfrac{n^3 - 1}{nx - 1}$ ，则 $xd = n^2 + \dfrac{d - 1}{n} \in N^*$ ，所以 $d = 1$ 或 $d \geqslant n + 1$.

如果是 $d \geqslant n + 1$ ，则 $\dfrac{n^3 - 1}{nx - 1} \geqslant n + 1$ ，有

$$x \leqslant \frac{n^2 + 1}{n + 1} < n, \text{ 从而}$$

$nx - 1 > x^2 - 1 > x^2 - n$ ，矛盾！故只能是

$\dfrac{n^3 - 1}{nx - 1} = d = 1, x = n^2, y = \dfrac{n^4 - n}{n^3 - 1} = n.$

所以，对任意的正整数 $n \geqslant 2$ ，存在唯一的正整数对

$$(x, y) = (n^2, n),$$

满足 $n = \dfrac{x^2 + y}{xy + 1}$.

综上所述，所求正整数 n 是大于 1 的一切正整数.

例 3.16　已知 $n \in N$ 且 $4 \nmid n$ ，求证：$5 \mid (1^n + 2^n + 3^n + 4^n)$.

证明：$\because 1^n + 2^n + 3^n + 4^n$

$= 1^n + 2^n + (5 - 2)^n + (5 - 1)^n$

$= 1^n + 2^n + 5^n + C_n^1 \cdot 5^{n-1} \cdot (-2) + C_n^2 \cdot 5^{n-2} \cdot (-2)^2 + \cdots + C_n^{n-1} \cdot 5 \cdot (-2)^{n-1} +$

$(-2)^n + 5^n + C_n^1 \cdot 5^{n-1} \cdot (-1) + C_n^2 \cdot 5^{n-2} \cdot (-1)^2 + \cdots + C_n^{n-1} \cdot 5 \cdot (-1)^{n-1} + (-1)^n$

$= 1^n + 2^n + (-2)^n + (-1)^n + 5t_1 + 5t_2$

（这里 $t_1 = 5^{n-1} + C_n^1 \cdot 5^{n-2} \cdot (-2) + C_n^2 \cdot 5^{n-3} \cdot (-2)^2 + \cdots + C_n^{n-1} \cdot (-2)^{n-1}$ ，

$t_2 = 5^{n-1} + C_n^1 \cdot 5^{n-2} \cdot (-1) + C_n^2 \cdot 5^{n-3} \cdot (-1)^2 + \cdots + C_n^{n-1} \cdot (-1)^{n-1}$ ）

$\because 4 \nmid n$ ，

$\therefore n = 4q + r, r = 1, 2, 3.$

当 $r = 1$ 时，$(-2)^n = (-2)^{4q+1} = (-2)^{4q}(-2) = -2 \cdot 2^{4q} = -2^{4q+1} = -2^n, (-1)^{4q+1}$

$= 1$ ，此时，$1^n + 2^n + (-2)^n + (-1)^n = 0$ ，结论成立.

当 $r = 2$ 时，$(-2)^n = (-2)^{4q+2} = 2^{4q+2}, (-1)^n = (-1)^{4q+2} = 1$ ，此时

$1^n + 2^n + (-2)^n + (-1)^n = 2 + 2 \cdot 2^{4q+2} = 2 + 2 \times 4^{2q+1}$ ，而

$4^{2q+1} = (5 - 1)^{2q+1}$

$$= 5^{2q+1} - C_{2q+1}^1 \cdot 5^{2q} + C_{2q+1}^2 5^{2q-1} - \cdots - 1,$$

$$\therefore 2 + 2 \times 4^{2q+1}$$

$$= 2 \times (5^{2q+1} - C_{2q+1}^1 5^{2q} + C_{2q+1}^2 5^{2q-1} - \cdots + C_{2q+1}^{2q} 5).$$

故当 $r = 2$ 时,$5 \mid [1^n + 2^n + (-2)^n + (-1)^n]$.

当 $r = 3$ 时,与 $r = 1$ 时类似,也有

$$5 \mid [1^n + 2^n + (-2)^n + (-1)^n]$$

综上所述,已知 $n \in N$ 且 $4 \nmid n$ 时,$5 \mid (1^n + 2^n + 3^n + 4^n)$.

例 3.17 证明 $7 \mid \overline{abcabc}$

分析: $\overline{abcabc}$ 是一类特殊的六位数,其意义是

$$\overline{abcabc} = a \times 10^5 + b \times 10^4 + c \times 10^3 + a \times 10^2 + b \times 10 + c \,(a \neq 0)$$
$$= \overline{abc} \times 10^3 + \overline{abc}$$
$$= \overline{abc} \times 1001$$

证明: 因为

$$\overline{abcabc} = \overline{abc} \times 10^3 + \overline{abc} = \overline{abc} \times 1001$$

所以:$1001 \mid \overline{abcabc}$.

又因为:

$$1001 = 7 \times 143$$

即:

$$7 \mid 1001$$

所以 $7 \mid \overline{abcabc}$.

例 3.18 若 $n \in N$,求证:$9 \mid [(3n+1) \times 7^{n-1}]$.

证明: 设 $f(n) = (3n+1) \times 7^{n-1}$,则

$$f(n-1) = (3n-2) \times 7^{n-1} - 1,$$

所以

$$f(n) - f(n-1) = 9(2n+1) \times 7^{n-1},$$

即 $f(n) = f(n-1) + 9(2n+1) \times 7^{n-1}$,从而

$$9 \mid f(n) \Leftrightarrow 9 \mid f(n-1).$$

因为 $9 \mid f(0) = 0$,所以 $9 \mid f(1)$,$9 \mid f(n-1)$,$\cdots$,$9 \mid f(n-1)$,$9 \mid f(n)$.

例 3.19 若 $N = 2^{2000} - 2^{1998} + 2^{1996} - 2^{1994} + 2^{1992} - 2^{1990}$,则 $9 \mid N$.

证明: $\because N = 2^{2000} - 2^{1998} + 2^{1996} - 2^{1994} + 2^{1992} - 2^{1990}$

$$= 2^{1990}(2^{10} - 2^8 + 2^6 - 2^4 + 2^2 - 1)$$
$$= 2^{1990}(2^2 - 1)(2^8 + 2^4 + 1)$$
$$= 2^{1990} \times 3 \times 273$$
$$= 9 \times 91 \times 2^{1990}$$

$\therefore 9 \mid N$.

例 3.20 m,n 都是整数，若 $3 \mid 2n - m$．求证：$9 \mid 8n^2 + 10mn - 7m^2$．

证明： 由题意有：

$9 \mid 8n^2 + 10mn - 7m^2$

$= 8n^2 - 4mn + 14mn - 7m^2$

$= 4n(2n - m) + 7m(2n - m)$

$= (2n - m)(4n + 7m)$

$= (2n - m)[4n - 2m + 9m]$

$= (2n - m)[1(2n - m) + 9m]$，

因为 $3 \mid 2n - m, 3 \mid 9m \Rightarrow 3 \mid [2(2n - m) + 9m]$，

所以 $9 \mid 8n^2 + 10mn - 7m^2$．

例 3.21 x,y,z 均为整数．若 $11 \mid (7x + 2y + 5z)$，求证：$11 \mid (3x - 2y + 12z)$．

证明： 因为

$4(3x - 7y + 12z) + 3(7x + 2y - 5z)$

$= (12x - 28y + 48z) + (21x + 6y - 15z)$

$= 11(3x - 2y + 3z)$，

而 $11 \mid 11(3x - 2y + 3z)$．

且 $11 \mid 7x + 2y - 5z$，

因此 $11 \mid 4(3x - 7y + 12z)$．

又 $(11,4) = 1$，

所以 $11 \mid (3x - 2y + 12z)$．

例 3.22 n 是自然数．证明 $13 \mid 9^{3n+1} + 3^{3n+1} + 1$．

证明： 用数学归纳法．

（1）当 $n = 1$ 时，有

$$9^4 + 3^4 + 1 = 6643 = 13 \times 511,$$

$$13 \mid 9^4 + 3^4 + 1.$$

（2）设 $n = k$ 时，命题成立，即

$$13 \mid 9^{3k+1} + 3^{3k+1} + 1.$$

现在证明，$n = k + 1$ 时，命题也成立．

事实上，

$9^{3(k+1)+1} + 3^{3(k+1)+1} + 1$

$= 9^{3k+4} + 3^{3k+4} + 1$

$= 729 \cdot 9^{3k+1} + 27 \cdot 3^{3k+1} + 1$

$= (27 \cdot 9^{3k+1} + 27 \cdot 3^{3k+1} + 27) + 702 \cdot 9^{3k+1} - 26$

$= 27(9^{3k+1} + 3^{3k+1} + 1) + 13(54 \cdot 9^{3k+1} - 2)$．

由归纳假设，$13 \mid (9^{3k+1} + 3^{3k+1} + 1)$，

又 $13 \mid 13(54 \cdot 9^{3k+1} - 2)$，

所以 $13 \mid [9^{3(k+1)+1} + 3^{3(k+1)+1} + 1]$.

即 $n = k + 1$ 是命题成立.

根据数学归纳原理可得, 对任意自然数 n 都有

$$13 \mid 9^{3n+1} + 3^{3n+1} + 1.$$

例 3.23 如果 $m = 3n + 1$, n 为自然数, 求证 $1 + 3^m + 9^m$ 能被 13 整除.

分析: 绝大多数整除问题是与自然数有关的, 因此数学归纳法也是处理这类问题的一种重要手段, 而选择归纳的对象是证明的关键.

证明: 用数学归纳法证明.

第一步: 当 $n = 1$ 时, 原式 $= 9^4 + 3^4 + 1 = 6643$ 是 13 的倍数.

第二步: 假设当 $n = k$ 时, 命题成立, 即 $1 + 3^{3k+1} + 9^{3k+1}$ 能被 13 整除.

第三步: 现在考虑 $n = k + 1$ 时的情况. 由于

$$1 + 3^{3(k+1)+1} + 9^{3(k+1)+1}$$
$$= 1 + 3^{3k+4} + 9^{3k+4}$$
$$= 1 + 27 \times 3^{3k+1} + 729 \times 9^{3k+1}$$
$$= 27 + 27 \times 3^{3k+1} + 27 \times 9^{3k+1} + 702 \times 9^{3k+1} - 26$$
$$= 27(1 + 3^{3k+1} + 9^{3k+1}) + 702 \times 9^{3k+1} - 26$$

显然上式是能被 13 整除的, 这就证明了 $n = k + 1$ 时成立.

所以原命题对任意的自然数都成立.

例 3.24 设自然数 a, b 的末位数字是 3 或 7, 试证对任意自然数 n 和 m, $a^{4n+2m} - b^{2m}$ 能够被 20 整除.

分析: 既然 a 的末位数字是 3, 那么它就可以写成 $10a_1 + 3$ 的形式, 再利用二项式定理进行分析.

证明: 不妨先设 $a = 10a_1 + 7, b = 10b_1 + 3$, 则

$$a^{4n+2m} - b^{2m}$$
$$= (10a_1 + 7)4n + 2m - (10b_1 + 3)2m$$
$$= [^10(10 + 14a_1 + 4) + 9]2n + m - [^10(10 + 6b_1) + 9]m$$
$$= [^10a_2 + 9]2n + m - [^10b_2 + 9]m,$$

其中 $a_2 = 10a_1^2 + 14a_1 + 4, b_2 = 10b_1^2 + 6b_1$ 均是偶数. 由二项式定理知道, 只要能证明 $9^{2n+m} - 9^m$ 能被 20 整除, 那么原数就一定是 20 的倍数了. 而 $9^{2n+m} - 9^m = 9^m(9^{2n} - 1) = 9^m(81^n - 1)$, 其中 $81^n - 1$ 是能被 20 整除的.

对于其他情况, 类似于上述讨论同样可以得到相同的结果. 故 $a^{4n+2m} - b^{2m}$ 能被 20 整除.

例 3.25 当 $n \in N_+$ 时, 求证: $23(5^{2n+1} + 2^{n+4} + 2^{n+1})$.

证明: 此题可用多种方法证明.

法一: 用数学归纳法证明.

当 $n = 1$ 时,

$$\because 5^{2n+1} + 2^{n+4} + 2^{n+1} = 5^3 + 2^5 + 2^2$$
$$= 125 + 32 + 4$$
$$= 161 = 23 \times 7.$$

$\therefore n = 1$ 时, 结论成立.

假设 $n = k$ 时结论成立, 即

$$23(5^{2k+1} + 2^{k+4} + 2^{k+1})$$

当 $n = k + 1$ 时,

$$5^{2n+1} + 2^{n+4} + 2^{n+1} = 5^{2(k+1)+1} + 2^{k+1+4} + 2^{k+1+1}$$
$$= 5^{2k+3} + 2^{k+5} + 2^{k+2}$$
$$= 25 \times 5^{2k+1} + 2 \times 2^{k+4} + 2 \times 2^{k+1}$$
$$= 23 \times 5^{2k+1} + 2 \times (5^{2k+1} + 2^{k+4} + 2^{k+1}).$$

$\because 2323 \times 5^{2k+1}, 23(5^{2k+1} + 2^{k+4} + 2^{k+1})$,

$\therefore n = k + 1$ 时, 结论也成立.

则 n 为任何整数时, $23(5^{2n+1} + 2^{n+4} + 2^{n+1})$.

法二: 设法拼拆出因子 23 即可.

$$f(n) = 5^{2n+1} + 2^{n+4} + 2^{n+1} f(n)$$
$$= 5^{2n+1} + 2^n(2^4 + 2)$$
$$= 5^{2n+1} + 18 \cdot 2^n$$
$$= 5 \cdot 25^n + 18 \cdot 2^n$$
$$= 23 \cdot 25^n - 18(25^n - 2^n).$$

因为 $(25 - 2) \mid 25^n - 2^n$, 即 $23 \mid 25^n - 2^n, 23 \mid 23 \cdot 25^n$,

所以 $23 \mid 5^{2n+1} + 2^{n+4} + 2^{n+1}$.

例 3.26 证明一个整数 a 既不能被 2 也不能被 3 整除, 则 $a^2 + 23$ 必能被 24 整除.

证明: 由条件得 $2 \nmid a$, 且 $3 \nmid a$, 因此, $8 \mid (a^2 - 1)$,

且 $3 \mid (a^2 - 1)$, 从而 $24 \mid (a^2 - 1)$,

故 $24 \mid (a^2 + 23) = (a^2 - 1) + 24$.

例 3.27 a, b 均为整数, 且 a, b 都不被 2, 3 整除. 求证: $24 \mid a^2 - b^2$.

分析: 要证 $24 \mid a^2 - b^2$, 只要判定 a^2, b^2 被 24 除时的余数相等即可.

证明: 考虑用 24 去除 a, 则有

$$a = 24q + r(0 \leqslant r < 24).$$

由于 a 不被 2 整除, 也不被 3 整除, 因此 r 只能取 1, 5, 7, 11, 13, 17, 19, 23 这八种值.

于是 a 可表为

$24q + 1, 24q + 5, 24q + 7, 24q + 11, 24q + 13, 24q + 17, 24q + 19, 24q + 23$,

不论哪一种情况, a^2 被 24 除后余数均为 1.

比如 $a = 24q + 17$, 则

$$\begin{aligned}
a^2 &= (24q + 17)2 = 24^2 q^2 + 2 \times 24q \times 17 + 289 \\
&= 24^2 q^2 + 2 \times 24q \times 17 + 12 \times 24 + 1 \\
&= 24(24q^2 + 2q \times 17 + 12) + 1.
\end{aligned}$$

可知，a^2 被 24 除余 1，其余同样可以判定.

同理可得 b^2 被 24 除余 1.

因此，$a^2 - b^2$ 被 24 除余 0.

即 $24 \mid a^2 - b^2$.

例 3.28 n 为非负整数，证明 $39 \mid 2^{12n+9} - 5^{4n+1}$.

证明：

$$\begin{aligned}
f(n) &= 2^{12n+9} - 5^{4n+1} \\
&= 2^{3(4n+3)} - 5^{4n+1} \\
&= 8^{4n+3} - 5 \cdot 5^{4n} \\
&= 512 \cdot 8^{4n} - 5 \cdot 5^{4n} \\
&= (13 \times 39 + 5)8^{4n} - 5 \cdot 5^{4n} \\
&= 13 \times 39 \times 8^{4n} + 5 \times 8^{4n} - 5 \cdot 5^{4n} \\
&= 13 \times 39 \times 8^{4n} + 5(64^{2n} - 25^{2n}).
\end{aligned}$$

因为 $64 - 25 = 39$，所以 $39 \mid 5(64^{2n} - 25^{2n})$.

因此 $39 \mid 2^{12n+9} - 5^{4n+1}$.

例 3.29 若 $n \in Z$，求证：$73 \mid f(n) = 8^{n+2} + 9^{2n+1}$.

证明： 因为 $f(n) = 64 \times 8^n + 9 \times 81^n$

$$\begin{aligned}
&= (64 \times 8^n + 9 \times 81^n) + (9 \times 8^n - 9 \times 8^n) \\
&= 73 \times 8^n + 9(81^n - 8^n),
\end{aligned}$$

因为 $73 \mid (81^n - 8^n)$，所以 $73 \mid f(n) = 8^{n+2} + 9^{2n+1}$.

例 3.30 设 $n \in N_+$，求证：$512 \mid 3^{2n} - 32n^2 + 24n - 1$.

分析： 设 $f(n) = 3^{2n} - 32n^2 + 24n - 1$，欲证 $512 \mid f(n)$，我们可运用递推思想转而证明 $512 \mid f(1)$，以及 $512 \mid f(n + 1) - f(n), n \in N_+$.

证明： 设 $f(n) = 3^{2n} - 32n^2 + 24n - 1$，

因为 $f(1) = 0$，则 $512 \mid f(1)$；

又 $f(n + 1) - f(n) = [3^{2n+2} - 32(n + 1)^2 + 24(n + 1) - 1] - (3^{2n} - 32n^2 + 24n - 1)$

$$= 8(3^{2n} - 8n - 1).$$

因为 $512 = 8 \times 64$，故只需证明 $64 \mid (3^{2n} - 8n - 1)$.

设 $g(n) = 3^{2n} - 8n - 1$，

因为 $g(1) = 0$，则 $64 \mid g(1)$.

又 $g(n + 1) - g(n) = [3^{2n+2} - 8(n + 1) - 1] - (3^{2n} - 8n - 1)$

$$= 8(3^{2n} - 1)$$

$$= 8(9^n - 1)$$
$$= 8(9 - 1) \cdot (9^{n-1} + 9^{n-2} + \cdots + 1)$$
$$= 64(9^{n-1} + 9^{n-2} + \cdots + 1).$$

可见，$64 \mid [g(n+1) - g(n)]$，

于是，$64 \mid g(n)$，因此 $512 \mid f(n)$.

例 3.31　证明 $10 \underbrace{\cdots}_{50个0} 01$ 能被 1001 整除.

证明：因为

$$10 \underbrace{\cdots}_{50个0} 01 = 10^{51} + 1$$
$$= (10^3)^{17} + 1$$
$$= (10^3 + 1) [(10^3)^{16} - (10^3)^{15} + \cdots - 10^3 + 1].$$

所以，$10^3 + 1 = 1001$ 整除 $10 \underbrace{\cdots}_{50个0} 01$. 证毕.

例 3.32　设 p 和 q 均为自然数，使得

$$\frac{p}{q} = 1 - \frac{1}{2} + \frac{1}{3} - \cdots - \frac{1}{1318} + \frac{1}{1319},$$

证明：p 可被 1979 整除. （第 21 届 IMO）

分析：关于有限项和的整除问题，在前面我们已有所接触，其中一个重要的解题途径是将有限个加项进行适当地调整，重新组合，以发现一些隐藏的规律.

证明：$\dfrac{p}{q} = 1 - \dfrac{1}{2} + \dfrac{1}{3} - \cdots - \dfrac{1}{1318} + \dfrac{1}{1319}$

$$= \left(1 + \frac{1}{2} + \frac{1}{3} + \cdots + \frac{1}{1319}\right) - 2\left(\frac{1}{2} + \frac{1}{4} + \cdots + \frac{1}{1318}\right)$$

$$= \left(1 + \frac{1}{2} + \cdots + \frac{1}{1319}\right) - \left(1 + \frac{1}{2} + \cdots + \frac{1}{659}\right)$$

$$= \frac{1}{660} + \frac{1}{661} + \cdots + \frac{1}{1318} + \frac{1}{1319}$$

$$= \left(\frac{1}{660} + \frac{1}{1319}\right) + \left(\frac{1}{661} + \frac{1}{1318}\right) + \cdots + \left(\frac{1}{989} + \frac{1}{990}\right)$$

$$= 1979 \times \left(\frac{1}{660 \times 1319} + \frac{1}{661 \times 1318} + \cdots + \frac{1}{989 \times 990}\right).$$

将等式两边同乘以 1319!，得

$$1319! \times \frac{p}{q} = 1979 \cdot N,$$

其中 N 是自然数. 由此可见 1979 整除 $1319! \times p$，但是注意到 1979 是一个素数，同时 1979 显然不能整除 1319!，所以 p 能被 1979 整除.

说明：题目中的 1979 只是为了与当年的年份一致而已. 一般地，可以将 1979 换成一个形如 $3k + 2$ 的素数，相应地将 1319 换成 $2k + 1$（k 为自然数）.

例3.33 设 $3 \mid (a^2 + b^2)$，证明 $3 \mid a$ 且 $3 \mid b$，其中 a, b 为任意整数．

证明： 令 $a = 3q_1 + r_1, b = 3q_2 + r_2, r_i = 0, 1, 2(i = 1, 2)$

则：

$$\begin{aligned}
a^2 + b^2 &= (3q_1 + r_1)2 + (3q_2 + r_2)^2 \\
&= 9q_1^2 + 6q_1 r + r_1^2 + 9q_2^2 + 6q_2 r + r_2^2 \\
&= (9q_1^2 + 6q_1 r + 9q_2^2 + 6q_2 r) + (r_1^2 + r_2^2) \\
&= 3(3q_1^2 + 2q_1 r + 3q_2^2 + 2q_2 r) + (r_1^2 + r_2^2) \\
&= 3q + r_1^2 + r_2^2 (q \in Z).
\end{aligned}$$

因为：

$$3 \mid (a^2 + b^2)$$

所以：

$$3 \mid (r_1^2 + r_2^2)$$

所以：

$$r_1 = r_2 = 0$$

所以：$3 \mid a$ 且 $3 \mid b$．

例3.34 a, b 均为整数，若 $7 \mid a^2 + b^2$，求证 $7 \mid a$ 且 $7 \mid b$．

证明： 设 $a = 7m + \alpha, b = 7n + \beta$．

其中 α, β 取值 $0, 1, 2, 3, 4, 5, 6$．

因为 $a^2 + b^2 = 49(m^2 + n^2) + 14(\alpha m + \beta n) + \alpha^2 + \beta^2$．$7 \mid a^2 + b^2 \Leftrightarrow 7 \mid \alpha^2 + \beta^2$．

但 α^2 和 β^2 只能取 $0, 1, 4, 9, 16, 25, 36$．

当且仅当 $\alpha = 0, \beta = 0$ 时才有 $\alpha^2 + \beta^2$ 被 7 整除．

此时 $a = 7m$ 且 $b = 7m$．

所以 $7 \mid a$ 且 $7 \mid b$ 成立．

例3.35 求证：$n^2 \mid [(n + 1)^n - 1]$．

证明： 因为

$$\begin{aligned}
(n + 1)^n &= n^n + C_n^1 n^{n-1} + C_n^2 n^{n-2} + \cdots + C_n^n n + 1 \\
&= n^n + C_n^1 n^{n-1} + C_n^2 n^{n-2} + \cdots + n^2 + 1
\end{aligned}$$

所以 $(n + 1)^n - 1 = n^n + C_n^1 n^{n-1} + C_n^2 n^{n-2} + \cdots + n^2$，于是 $n^2 \mid [(n + 1)^n - 1]$．

例3.36 设 $m > n \geq 0$，证明：$(2^{2^n} + 1) \mid (2^{2^m} - 1)$．

证明： 由于 $m > n \geq 0$，故 $m - n - 1 \geq 0$．

$$2^{2^m} - 1 = (2^{2^{n+1}})^{2^{m-n-1}} - 1 = (2^{2^{n+1}} - 1)[(2^{2^{n+1}})^{2^{m-n-1}-1} + \cdots + 2^{2^{n+1}} + 1]$$

所以 $(2^{2^{n+1}} - 1) \mid (2^{2^m} - 1)$．又 $2^{2^{n+1}} - 1 = (2^{2^n} + 1)(2^{2^n} - 1)$，

因此 $(2^{2^n} + 1) \mid (2^{2^m} - 1)$．

于是有 $(2^{2^n} + 1) \mid (2^{2^m} - 1)$．

例 3.37　如果 $a,b \in Z$，且 $a \neq b$，当 $n \in N_+$ 时，则 $(a-b)(a^n - b^n)$．

证明： 如果 $b = 0$，则 $a \neq 0$．而 aa^n，故有

$$(a-b)(a^n - b^n)$$

同样，如果 $a = 0$ 时结论仍然成立．

如果 $b \neq 0, a \neq 0$，构造下列等比数列：

$$b^{n-1}, ab^{n-2}, \cdots, a^{n-2}, a^{n-1}.$$

根据等比数列的求和公式得：

$$b^{n-1} + ab^{n-2} + \cdots + a^{n-2}b + a^{n-1} = \frac{a^n - b^n}{a - b},$$

即：

$$a^n - b^n = (a-b)(b^{n-1} + ab^{n-2} + \cdots + a^{n-2}b + a^{n-1}).$$

$a - b \neq 0$，$a^{n-1} + \cdots + b^{n-1}$ 是整数，

所以 $(a-b)(a^n - b^n)$，结论成立．

例 3.38　已知整数 m,n,p,g 适合 $(m-p) \mid (mq+np)$．证明 $(m-p) \mid (mq+np)$．

证明： 因为

$$mq + np = (mn + pq) - (m-p)(n-p)$$

而：

$$(m-p) \mid (mq+np)$$
$$(m-p) \mid (m-p)(n-p)$$

故可知结论成立．

例 3.39　若 $ax_o + by_o$ 是形如 $ax + by$（x,y 是任意整数，a，b 是两个不全为零的整数）的数中的最新正整数，试证；$(ax_o + by_o) \mid (ax+by)$．

证明： 由 $ax + by = (ax_o + by_o)q + r (0 \leqslant r < ax_o + by_o)$，

知 $r = ax + by - (ax_o + by_o)q = a(x - x_o q) + b(y - y_o q)$，即 r 也是形如 $ax + by$ 的数．

但因 $0 \leqslant r < ax_o + by_o, ax_o + by_o$ 是形如 $ax + by$ 的数中最小正整数，

故 $r = 0$．此时 $ax + by = (ax_o + by_o)q$，

即：

$$(ax_o + by_o) \mid (ax + by)$$

例 3.40　设 n,k 是正整数，证明：n^k 与 n^{k+4} 的个位数字相同．

证明： 记 $n = 10q + r (r = 0,1,\cdots,9)$．

则 $n^{k+4} - n^k$ 与 $r^{k+4} - r^k = r^k(r^4 - 1)$ 被 10 除的余数相同．

对 $r = 0,1,\cdots,9$ 进行验证即证得结论．

例 3.41　设 $a_1, a_2, \cdots, a_n$ 为不全为零的整数，以表示集合

$$A = \{y \mid y = a_1 x_1' + \cdots a_n x_n', x_i \in z, 1 \leqslant i \leqslant n\}$$

中的最小正数，则对于任何 $y \in A$，有 $y_0 \mid y$；特别地，有 $y_0 \mid a_i (1 \leqslant i \leqslant n)$．

证明： 设 $y_0 = a_1 x_1' + \cdots a_n x_n' \in A$.

对任意的 $y = a_1 x_1 + \cdots + a_n x_n \in A$ ，存在 $q, r \in z$. 使得

$$y = q y_0 + r (0 \leqslant r < y_0) .$$

因此

$$r = y - q y_0 = a_1 (x_1 - q x_1') + \cdots + a_n (x_n - q x_n') \in A$$

如果 $r \neq 0$ ，那么，因为 $0 < r < y_0$ ，所以 r 是 A 中比 y_0 还小的正整数. 这与 y_0 的定义矛盾. 所以 $r = 0$ ，即 $y_0 \mid y$.

显然 $a_i \in A (1 \leqslant i \leqslant n)$ ，所以上述结论得 $y_0 \mid a_i (1 \leqslant i \leqslant n)$.

例 3.42 设 m 和 n 为正整数，$m > 2$ ，证明 $(2^m - 1) \nmid (2^n + 1)$.

证明：（1）$n < m$ 时 $(m > 2)$ ，显然有 $2^m - 1 > 2^n + 1$ ，故此时 $2^m - 1 \nmid 2^n + 1$.

（2）当 $n = m$ 时，由于 $2^n + 1 < 2^{m+1} - 2$ ，故 $1 < \dfrac{2^n + 1}{2^m - 1} < 2$ ，所以此时 $2^m - 1 \nmid 2^n + 1$.

（3）当 $n > m$ 时，可以用带余除法转换到上述特殊情形.

设 $n = mq + r, 0 \leqslant r < m$ ，而 $q \geqslant 0$ ，由于

$$2^n + 1 = (2^{mq} - 1) 2^r + 2^r + 1 ,$$

而

$$(2^m - 1) \mid (2^{mq} - 1), (2^m - 1) \nmid (2^{r+} 1) \ (0 \leqslant r < m) ,$$

从而当 $n > m$ 时有 $2^m - 1 \nmid 2^n + 1$.

综合（1），（2），（3）得 $2^m - 1 \nmid 2^n + 1$.

例 3.43 证明：对于任何整数 n, m ，等式 $n^2 + (n+1)^2 = m^2 + 2$ 不可能成立.

证明： 对于任何整数 n, m ，考察等式

$$n^2 + (n+1)^2 = m^2 + 2$$

的左右两边被 4 除的余数.

左边除以 4 的余数只可能是 1，而右边除以 4 的余数有可能是 2 或 3，故可知等式不可能成立.

例 3.44 证明：若 a 被 9 除的余数是 3，4，5 或 6，则方程 $x^3 + y^3 = a$ 没有整数解.

证明： 对任意的整数 x , y ，记 $x = 3 q_1 + r_1$, $y = 3 q_2 + r_2$, $0 \leqslant r_1, r_2 < 3$. 则存在 $Q_1,$ $Q_2, R_1, R_2 \in Z$ 使得 $x^3 = 9 Q_1 + R_1$, $y^3 = 9 Q_2 + R_2$ ，其中 R_1 和 R_2 被 9 除的余数分别与 r_1^3 和 r_2^3 被 9 除的余数相同，即 $R_1 = 0, 1$ 或 $8, R_2 = 0, 1$ 或 8.

因此，$x^3 + y^3 = 9 (Q_1 + Q_2) + R_1 + R_2$ ，所以，$R_1 + R_2$ 被 9 除的余数只可能是 0，1，2，7 或 8，故 $x^3 + y^3$ 不可能等于 a .

例 3.45 设整数 $k \geqslant 1$ ，证明：

（1）若 $2^k \leqslant n < 2^{k+1}, 1 \leqslant a \leqslant n, a \neq 2^k$ ，则 $2^k \nmid a$ ；

（2）若 $3^k \leq 2n - 1 < 3^{k+1}, 1 \leq b \leq n, 2b - 1 \neq 3^k$，则 $3^k \nmid (2b - 1)$．

证明：反证法．

（1）假设 $2^k \mid a$，则存在整数 q，使得

$$a = 2^k q.$$

显然 q 只可能是 0 或 1，此时 $a = 0$ 或 2^k．

这都是不可能的，所以 $2^k \nmid a$．

（2）假设 $3^k \mid (2b - 1)$，则存在整数 q，使得

$$2b - 1 = 3^k q.$$

显然 q 只可能是 $0, 1$ 或 2．此时 $2b - 1 = 0, 3^k$ 或 $2 \cdot 3^k$．这都是不可能的．

所以 $3^k \nmid (2b - 1)$．

例 3.46　连续三个整数中必有一个数被 3 整除．

证明：设这三个连续整数为

$$a, a + 1, a + 2.$$

根据以 3 为标准对整数的分类，一个整数要么是 $3k$ 型，要么是 $3k + 1$ 型，要么是 $3k + 2$ 型．

如果 $a = 3k$，显然 a 就是被 3 整除的数；

如果 $a = 3k + 1$，

$$a + 2 = (3k + 1) + 2 = 3k + 3 = 3(k + 1),$$

此时 $a + 2$ 就是被 3 整除的数；

如果 $a = 3k + 2$，

$$a + 1 = (3k + 2) + 1 = 3k + 3 = 3(k + 1),$$

此时 $a + 1$ 就是被 3 整除的数．

综上可得，连续三个整数中必有一个数是 3 的倍数．

仿上述证明可得证：

连续 4 个整数中必有一个数是 4 的倍数；

连续 5 个整数中必有一个数是 5 的倍数；

……

一般地，可得：

定理：连续 n 个整数中必有一个数是 n 的倍数．

这个例题在证明整除问题时是很有用的．

例 3.47　设 $a > 0$，证明：相邻的 a 个整数有且仅有一个被 a 整除．

证明：设相邻的 a 个整数是 $m, m + 1, \cdots, m + a - 1$．

令 $m = aq + r_0 (0 \leq r_0 \leq a - 1)$，则

$$m + j = aq + (r_0 + j) + aq_j + r_j (0 \leq r_j < a; j = 0, 1, \cdots, a - 1).$$

若 $r_0 = 0$，则 $a \mid m$，且对于 $1 \leq j \leq a - 1$，

若 $r_0 \neq 0$，则当 $j + r_0 < a$ 时，$r_j = j + r_0 \geq a$ 时，$r_j = j + r_0 - a$．

此时取 $j_0 = a - r_0$ ，则 $r_{j_0} = 0$ ．

有 $a \mid (m + r_{j_0})$ ．

除此之外 $0 < r_j < a$ ，

均有 $a \nmid (m + j)(1 \leqslant j < a, j \neq j_0)$ ．

因而结论得证．

例 3.48　求证：（1）$6 \mid (n^3 - n)$ ；

（2）若 n 为奇数，则 $8 \mid (n^2 - 1)$ ．

证明：（1）因为

$$n^3 - n = n(n^2 - 1) = (n - 1)n(n + 1) ，$$

所以 $6 \mid (n^3 - n)$ ．

（2）设 $n = 2m + 1(m \in Z)$ ，

则 $n^2 - 1 = 4m(m + 1)$ ．

因为 $2 = 2! \mid m(m + 1)$ ，

所以 $8 \mid 4m(m + 1)$ ，

故 $8 \mid (n^2 - 1)$ ．

例 3.49　设 n 是自然数，证明：

$$n(n^2 - 1)(n^2 - 5n + 26)$$

可以被 120 整除．

分析： 通过观察发现，

$$n(n^2 - 1) = (n - 1)n(n + 1)$$

恰好是三个连续自然数的乘积．这启发我们想到二项式系数，由定理 6 知道，$C_n^k (0 \leqslant k \leqslant n)$ 是整数，这样其展开式的含义就变为：任意 k 个连续自然数的乘积一定能被 $k!$ 整除．而题目中的 120 恰好等于 5! ．

证明： 记 $f(n) = n(n^2 - 1)(n^2 - 5n + 26)$ ．

当 n 为最初几个自然数时，$f(n)$ 是能被 120 整除的，即 $f(1) = 0, f(2) = 120$ 和 $f(3) = 480$ ．

当 $n > 3$ 时，

$$f(n) = (n - 1)n(n + 1)(n - 2)(n - 3) + 20(n - 1) \cdot n \cdot (n + 1) ．$$

$f(n)$ 的第一部分是五个连续自然数的乘积，它能被 5! = 120 整除；

第二部分中，$(n - 1) \cdot n(n + 1)$ 是三个连续自然数的乘积，能被 3! = 6 整除．

所以 $20(n - 1)n(n + 1)$ 可被 $20 \times 6 = 120$ 整除．

从而 $f(n)$ 能被 120 整除．

例 3.50　若 a, b, c 为整数，且 $a^3 + b^3 + c^3$ 是 24 的倍数，求证：

$$(a^5 + b^5 + c^5) + 4(a + b + c)$$

是 120 的倍数．

证明： 考虑到 120 是 5! 的特殊性，而题中又涉及某数的五次幂，故应考虑将其转换

为连续五个整数的乘积. 由于

$$a^5 - 5a^3 + 4a = a(a^2 - 1)(a^2 - 4) = (a - 2)(a - 1)a(a + 1)(a + 2)$$

是五个连续整数之积. 因而它是 $5! = 120$ 的倍数.

同理

$$b^5 - 5b^3 + 4b = b(b^2 - 1)(b^2 - 4) = (b - 2)(b - 1)b(b + 1)(b + 2)$$

$$c^5 - 5c^3 + 4c = c(c^2 - 1)(c^2 - 4) = (c - 2)(c - 1)c(c + 1)(c + 2)$$

都是 120 的倍数.

因此, $(a^5 + b^5 + c^5) + 4(a + b + c) - 5(a^3 + b^3 + c^3)$ 是 120 的倍数.

又因为 $a^3 + b^3 + c^3$ 是 24 的倍数,

故 $5(a^3 + b^3 + c^3)$ 是 120 的倍数.

所以 $(a^5 + b^5 + c^5) + 4(a + b + c)$ 是 120 的倍数.

例 3.51　证明：若两个自然数 n 和 k 互素，则 C_n^k 可以被 n 整除.

分析：只要我们注意得到 C_n^k 和 C_{n-1}^{k-1} 之间的关系，就可以很快地解决这个问题.

证明：因为

$$\begin{aligned} C_n^k &= \frac{n(n - 1)\cdots(n - k + 1)}{k \times (k - 1) \times \cdots \times 2 \times 1} \\ &= \frac{n}{k} \cdot \frac{(n - 1)\cdots(n - k + 1)}{(k - 1) \times \cdots \times 2 \times 1} \\ &= \frac{n}{k} C_{n-1}^{k-1}. \end{aligned}$$

即 $k \cdot C_n^k = n \cdot C_{n-1}^{k-1}$. 由于 k 和 n 互素，而 n 整除 $k \cdot C_n^k$，所以 C_n^k 是 n 的倍数.

例 3.52　证明：对于任意给定的 n 个整数中，必可以从中找出若干个数作和，使得该和能被 n 整除.

证明：设 n 个整数为 $a_1, a_2, \cdots, a_n$.

作 $s_1 = a_1, s_2 = a_1 + a_2, \cdots, s_n = a_1 + a_2 + \cdots + a_n$.

如果 s_i 中有一个被 n 整除，则结论已证；

否则，存在 $s_i, s_j (i < j)$. 使得 s_i 与 s_j 被 n 除的余数相等.

于是 $n \mid (s_j - s_i) = a_{i+1} + \cdots + a_j$.

例 3.53　已知等差数列的各项均是整数，它的公差是不能被 5 整除的奇数，求证这个数列的任何连续十项中必有一项能被 10 整除.

分析：由于我们所接触的数列通常都有一些联系项与项之间的关系式，如等差数列和等比数列的通项公式，递推数列的递推关系式等等，因此在讨论时利用这些公式是非常重要的. 设 $a_1, a_2, \cdots, a_{10}$ 是这个等差数列的任意连续十项，我们只要能证明在这十个数中必有一个的个位数字是零就可以了. 因此设 $r_1, r_2, \cdots, r_{10}$ 是 $a_1, a_2, \cdots, a_{10}$ 被 10 除后的余数，显然 $0 \leqslant r_i \leqslant 9 (i = 1, 2, \cdots, 10)$. 最后借助等差数列的通项公式，便可说明 r_i 中必有一个是零.

证明：设此等差数列的公差为 d，为简单起见，就设 $a_1,a_2,\cdots,a_{10}$ 为任何连续的十项；于是有

$a_i = 10q_i + r_i$（其中 q_i 和 r_i 都是整数，$0 \leq r_i < 10, i = 1,2,\cdots,10$）.

当 $0 \leq j < k \leq 9$ 时，因为 $a_k - a_j = 10(q_k - q_j) + (r_k - r_j)$.

又因为 $a_k - a_j = (k-j)d$.

所以必然有

$$10(q_k - q_j) + (r_k - r_j) = (k-j)d.$$

也即

$$r_k - r_j = (k-j)d - 10(q_k - q_j).$$

因为 $0 \leq j < k \leq 9$，所以 $k-j$ 不是 10 的倍数；

又由已知 d 不是 5 的倍数而且是奇数，于是 2 和 5 都不是 d 的因数，从而 $(k-j)d$ 不是 10 的倍数.

但 $10(q_k - q_j)$ 是 10 的倍数，故 $r_k - r_j$ 不会是 10 的倍数.

而 $0 \leq r_i < 10$，则 $-10 < r_k - r_j < 10$，从而 $r_k \neq r_j$.

可见 $r_1,r_2,\cdots,r_{10}$ 是不相等的十个数，但它们只能取 $0,1,2,\cdots,9$ 这十个值，因此必有一个为零，故 a_i 中一定有一个能被 10 整除.

例 3.54　试证明当 n 为自然数时，$\dfrac{(3n)!}{6^n \cdot n!}$ 是整数.

分析：本题还可以换一种提法，即 $(3n)!$ 是 $6^n n!$ 的倍数. 与阶乘有关的整除问题比较多，处理这种问题的办法就是恰当的约分.

证明：分子是 $3n$ 个连续自然数的乘积，最大的是 $3n$.

而分母也是一些自然数的乘积，最大的要超过 $3n$，我们先从 $3n$ 考虑，

令 $N = 3^n \times n!$，则

$$N = 3^n \times 1 \times 2 \times 3 \times \cdots \times n = 3 \times 6 \times 9 \times \cdots \times (3n).$$

而 $(3n)! = 1 \times 2 \times 3 \times \cdots \times (3n-2)(3n-1)(3n)$，可知 N 是 $(3n)!$ 中的一部分数的乘积.

于是从 $(3n)!$ 中消去 N 后，必然还含有 n 个偶数，所以

$$\frac{(3n)!}{6^n \cdot n!} = \frac{(3n)!}{2^n \cdot N}$$

必是整数.

例 3.55　设 n 是大于 1 的自然数，证明

$$1 + \frac{1}{2} + \frac{1}{3} + \cdots + \frac{1}{n}$$

不是整数.

分析：令原数为 M，若 M 是整数，则对任意的整数 A，AM 也一定是整数. 于是我们就要想办法找到这样的一个 A，使得 $\dfrac{A}{i}(1 \leq i \leq n)$ 中只有一项不是整数，而其余的项都

是整数，这样就会导致矛盾．

证明：很显然，每一个自然数都可以表示成 $2^k \cdot l$ 的形式，其中 k 为非负整数，l 是奇数．

设 $1, 2, \cdots, n$ 分别等于 $2^{\lambda_i} \cdot l_i$，$\lambda_i \geq 0$，l_i 是奇数，$i = 1, 2, \cdots, n$．

由于 $n > 1$，所以 $1, \cdots, n$ 中至少有一个是偶数，即至少有 $\lambda_i > 0$．

设 λ 是 $\lambda_1, \lambda_2, \cdots, \lambda_n$ 中最大的数，我们先来证明，在 $\lambda_1, \lambda_2, \cdots, \lambda_n$ 中只有一个与 λ 相等．

否则设

$$1 \leq k < j \leq n，\lambda_k = \lambda_j = \lambda，k = 2^{\lambda_k} \cdot l_k，j = 2^{\lambda_j} l_j.$$

因为 $k < j$，所以 $l_k < l_j$．

于是就有偶数 h 在 l_k 和 l_j 之间，故在 k 和 j 之间有 $2^{\lambda} \cdot h$，而 $2 \mid h$，就有 $h = 2m$，$2^{\lambda} \cdot h = 2^{\lambda+1} \cdot m$．

这与 λ 是最大数矛盾．

令 $A = 2^{\lambda-1} l$，其中 $l = l_1 \cdot l_2 \cdots l_n$，这个 A 就是我们所要找的．此时

$$AM = \frac{L}{2} + N，$$

其中 $\dfrac{L}{2} = \dfrac{A}{2^{\lambda} \cdot l_0}$，$l_0$ 是奇数，是 $l_1, l_2, \cdots, l_n$ 中使 2 的指数为 λ 的那一个．

N 是整数．显然 $\dfrac{L}{2}$ 不会是整数，故原来设 M 是整数是错误的，即原数不是整数．

说明：本题所涉及的知识不多，但方法是比较特殊的，下面的题目也可以考虑用类似的办法．

问题 1. 设 m 和 n 是自然数，证明

$$\frac{1}{3} + \frac{1}{5} + \cdots + \frac{1}{2n+1}，$$

$$\frac{1}{n} + \frac{1}{n+1} + \cdots + \frac{1}{n+m}$$

都不是整数．

问题 2. 设 $m > n \geq 1$，$a_1 < a_2 < \cdots < a_s$ 是不超过 m 且与 n 互素的全部正整数，记

$$S_m^n = \frac{1}{a_1} + \cdots + \frac{1}{a_s}$$

则 S_m^n 不是整数．

由此可直接推出

$$S = 1 + \frac{1}{2} + \frac{1}{4} + \frac{1}{5} + \cdots + \frac{1}{3n+1} + \frac{1}{3n+2}(n \geq 0)$$

不是整数．

例 3.56　试证：当 x 是整数时，多项式

$$f(x) = \frac{1}{5}x^5 + \frac{1}{2}x^4 + \frac{1}{3}x^3 - \frac{1}{30}x$$

的值仍是整数（全苏数学竞赛题）.

证明： 由于是五次多项式，根据定理，只需检验六个整数 x，看 $f(x)$ 是否为整数即可.

为了使计算量尽可能的少，可检验当 $x = -2，-1,0,1,2,3$ 时多项式的值. 容易看到：
$$f(-2) = -1, f(-1) = f(0) = 0, f(1) = 1, f(2) = 17, f(3) = 98,$$
均为整数.

由此可知 $f(x)$ 是整值多项式.

例 3.57 证明 $f(n) = n^3 - \dfrac{3}{2}n^2 + \dfrac{1}{2}n - 1$ 对任何正整数 n 都是整数，并且用 3 除时余 2.（北京 1956 年竞赛题）

分析： 考虑一个新的多项式
$$F(n) = \frac{f(n) + 1}{3}.$$

如果能证明 $F(n)$ 是一个整值多项式，那么也就说明 $f(n)$ 是一个整值多项式，并且 $f(n)$ 的每一个值被 3 除时余 2. $F(n)$ 也是一个三次多项式.

证明： 令 $n = 1,2,3,4$，分别计算 $F(n)$ 的值，得
$$F(1) = 0,$$
$$F(2) = 1,$$
$$F(3) = 5,$$
$$F(4) = 14.$$

根据定理可知 $F(n)$ 为整值多项式.

由分析中所讨论的可知该命题成立.

例 3.58 设 $P(x)$ 是四次整系数多项式，已知对任何整数 x，$P(x)$ 能被 7 整除. 证明：$P(x)$ 的所有系数均能被 7 整除.

分析： 用待定系数法，讨论 $P(x)$ 在 $x = 0,1,2,-1,-2$ 五个整数点处的值.

证明： 设 $P(x)$ 可以表为下列形式：
$$P(x) = a_4 x^4 + a_3 x^3 + a_2 x^2 + a_1 x + a_0,$$
则根据已知条件，有
$$P(0) = a_0 \Rightarrow a_0 = 7t_0 （t_0 是整数），$$
$$P(1) - P(-1) = 2a_1 + 2a_3 = 7t_1 （t_1 是整数），$$
即
$$7 \mid 2(a_1 + a_3)，而 (7,2) = 1,$$
于是 $7 \mid (a_1 + a_3)$，令 $a_1 + a_3 = 7t_2 （t_2 是整数）.$
$$P(1) + P(-1) = 2a_0 + 2a_2 + 2a_4$$
$$= 2(a_2 + a_4) + 2 \times 7t_0,$$
有
$$a_2 + a_4 = 7t_3 （t_3 是整数），$$

$$P(2) - P(-2) = 4a_1 + 16a_3$$
$$= 4(a_1 + 4a_3)$$
$$= 7t_4 \ (t_4 \text{ 是整数})$$

从而

$$(a_1 + 4a_3) = 7t_5 , \ (t_5 \text{ 是整数})$$
$$P(2) + P(-2) = 2a_0 + 8(a_2 + 4a_4) ,$$

可推出

$$a_2 + 4a_4 = 7t_6 . \ (t_6 \text{ 是整数})$$

所以

$$\begin{cases} a_1 + a_3 = 7t_2 \\ a_1 + 4a_3 = 7t_5 \end{cases} \Rightarrow a_1 \text{ 和 } a_3 \text{ 是 7 的倍数};$$

$$\begin{cases} a_2 + a_4 = 7t_3 \\ a_2 + 4a_4 = 7t_6 \end{cases} \Rightarrow a \ a_2 \text{ 和 } a_4 \text{ 是 7 的倍数}.$$

命题得证.

例 3.59　求所有大于 3 的正整数 n ，使得 $1 + C_n^1 + C_n^2 + C_n^3 \mid 2^{2000}$.

解: $1 + C_n^1 + C_n^2 + C_n^3 = \dfrac{1}{6}(n + 1)(n^2 - n + 6)$ ，

由 $1 + C_n^1 + C_n^2 + C_n^3 \mid 2^{2000}$ 存在正整数 $k(4 \leqslant k \leqslant 2000)$ ，使得

$$\frac{1}{6}(n + 1)(n^2 - n + 6) = 2^k ,$$

即

$$(n + 1)(n^2 - n + 6) = 3 \times 2^{k+1}.$$

所以

$$(\text{A}) \begin{cases} n + 1 = 2^s \\ n^2 - n + 6 = 3 \times 2^t \\ s + t = k + 1 \end{cases}$$

或

$$(\text{B}) \begin{cases} n + 1 = 3 \times 2^s \\ n^2 - n + 6 = 2^t , \\ s + t = k + 1 \end{cases}$$

其中 $s, t \in N^*$ ，且 $s \geqslant 3$.

由（A）得

$$(2^s - 1)^2 - 2^s + 7 = 3 \cdot 2^t ,$$

即

$$2^{2s} - 3 \cdot 2^s + 8 = 3 \cdot 2^t .$$

如果 $s > 3$ ，则

$$3 \cdot 2^t = 2^{2s} - 3 \cdot 2^s + 8 \equiv 8 (\bmod 16), t = 3 ,$$

这时 $n^2 - n + 6 = 24$ 无整数解；所以 $s = 3, n = 7, t = 4, k = 6$，满足题设.

由（B），得 $2^t = (3 \times 2^s - 1)^2 - (3 \times 2^s - 1) + 6 = 9 \cdot 2^{2s} - 9 \cdot 2^s = 8$.

如果 $s \geq 4$，则

$$2^t = 9 \cdot 2^{2s} + 8 \equiv 8 \pmod{16}, t = 3,$$

这时由 $n^2 - n + 6 = 8$ 有 $n = 2$ 不合题意，所以

$$s = 3, n = 23, t = 9, k = 11.$$

综上所述，满足题设的正整数 $n = 7, 23$.

例 3.60 试证：$s = 1 + \dfrac{1}{2} + \cdots + \dfrac{1}{n} (n > 1)$ 不是整数.

证明： 设 k 是满足条件 $2^k \leq n$ 的最大整数，p 是所有不大于 n 的正奇数的乘积，则

$$2^{k-1} ps = 2^{k-1} p\left(1 + \frac{1}{2} + \cdots + \frac{1}{n}\right)$$

的展开式中，除 $2^{k-1} p \dfrac{1}{2^k}$ 外，其余各项均为整数，所以 $2^{k-1} ps$ 不是整数，因而 s 不是整数.

例 3.61 证明 $f(n) = \dfrac{1}{3} n^3 + \dfrac{1}{2} n^2 + \dfrac{1}{6} n$ 是整值函数.

分析： $f(n) = \dfrac{2n^3 + 3n^2 + n}{6}$，要证 $f(n)$ 是整值函数，只需证 $6 \mid 2n^3 + 3n^2 + n$ 即可.

证明： $f(n) = \dfrac{1}{3} n^3 + \dfrac{1}{2} n^2 + \dfrac{1}{6} n$

$$= \frac{2n^3 + 3n^2 + n}{6}$$

$$= \frac{n(2n^2 + 3n + 1)}{6}$$

$$= \frac{n(n+1)(2n)}{6}$$

$$= \frac{n(n+1)(n+2) + (n-1)n(n+1)}{6}.$$

由于 $n(n+1)(n+2)$ 以及 $(n-1)n(n+1)$ 都是 3 个连续整数的乘积，必有

$$6 \mid n(n-1)(n+2), 6 \mid (n-1)n(n+1).$$

所以 $6 \mid n(n-1)(n+2) + (n-1)n(n+1)$，

即 $f(n) = \dfrac{n(n-1)(n+2) + (n-1)n(n+1)}{6}$ 是整数，

所以 $f(n) = \dfrac{1}{3} n^3 + \dfrac{1}{2} n^2 + \dfrac{1}{6} n$ 是整值函数.

3.4.2 典型练习题

练习 3.1 若按被 2 除所得余数分类，所有整数可分成哪几类？除数是 $3, 4, m(m \in N^*)$ 呢？

练习 3.2　一个数除以 7 余 2，若被除数扩大 9 倍，余数是多少？（福州市 1988 年小学生"迎春杯数学竞赛题"）

练习 3.3　某数除以 3 余 2，除以 4 余 1，该数除以 12 余几？（首届"华罗庚金杯"少年数学邀请赛复赛题）

练习 3.4　若 a 除以 b 商 c 余 r，则 am 除以 bm 商 _____ 余 _____？

练习 3.5　按照下列规律排列的图案，若用 1，2，3，4……编号，第 79 号是什么图案？

※※○○○　○　□※※○○○　○　□※※○○○　○　□…

练习 3.6　有一种盒子能装 3 斤糖，另一种盒子能装 6 斤糖．假定每个盒子必须装满，试问：能用这两个盒子来装完 100 斤糖吗？

练习 3.7　a,b,c 均是整数，若 $a+b$ 被 c 整除，能断言 $c|a$ 且 $c|b$ 吗？为什么？

练习 3.8　a,b,c 均是整数，若 $a-b$ 被 c 整除，且 b 被 c 整除，那么 a 被 c 整除吗？为什么？

练习 3.9　两个非零整数 a,b 满足 $ab = a + b$，试确定 a,b 的值，为什么？

练习 3.10　证明：求一数除以 99 的余数，可将此数从右端起每两个数字分一段，将各段相加，求其和除以 99 的余数．

练习 3.11　从等式 $1001 = 7 \times 11 \times 13$ 推导下述法则：要求一数除以 7，11 或 13 的余数，可将该数从右端起分为每三个数字一段，算出奇数段的和，有必要的话加上 7，11 或 13 的一个倍数，再减去偶数段的和，再将差数除以 7，11 或 13 那么除得的数就是所求的余数．

练习 3.12　设 a,b 都不是 3 的倍数，试证：$a+b$ 和 $a-b$ 中有且仅有一个是 3 的倍数．

练习 3.13　设 a,b 两数不能以 3 整除，$a^6 - b^6$ 能以 3 整除．

练习 3.14　设一数的十位数字的两倍加上各位数字能被四整除，则此数能被 4 除．

练习 3.15　设 $a \geqslant b$，不论 a,b 为何数，$a,b,a+b,a-b,2a+b,2a-bc$ 各数中必有一个被 5 整除．

练习 3.16　设 a 不能被 5 整除，那么 $a^4 - 1$ 是 5 的倍数，

这种证明的方式是常用的，要点在于：要求除以 5 的余数时，$a^4 - 1$ 中的 a 可用 a 除以 5 的余数来代替，这个余数可能是 1，2，3，4，我们逐个考察这些情况，结果都是被 5 整除的．

练习 3.17　设一数的个位数字加上其他数字之和的 4 倍，所得的和能被 6 整除，则此数能被 6 整除．

练习 3.18　设 a 不能被 7 整除，那么 $a^6 - 1$ 被 7 整除．

练习 3.19　一个四位数 $\overline{abcd}$，满足 $\overline{abc} - 2d$ 被 7 整除，证明 $7|\overline{abcd}$．

练习 3.20　$\overline{abc}$ 是三位数，若 $7 | 2a + 3b + c$，求证 $7 | \overline{abc}$．

练习 3.21　设一数的百位数字的 4 倍加上十位数字的两倍，又加上个位数字，所得之和能被 8 整除，则此数能被 8 整除．

练习 3.22　某数跟他按反序书写的数，两者之差能被 9 整除．

练习 3.23　$\overline{abc}$ 是三位数，若 $9 | a + b + c$，求证 $9 | \overline{abc}$．

练习 3.24　设一数字的个数是偶数，则此数跟它按反序书写的数之和能被 11 整除.

练习 3.25　设 n 是正整数，试用数学归纳法证明：$11 \mid \left[3^{n+1} + 3^{n-1} + 6^{2(n-1)} \right]$.

练习 3.26　证明 $11 \mid \overline{abcabc}$.

练习 3.27　证明 $13 \mid \overline{abcabc}$.

练习 3.28　设 n 为奇数，试证：$16 \mid (n^4 + 4n^2 + 11)$.

练习 3.29　x, y 均是整数，若 $x \div y = 48$，则必有 $16 \mid x$，为什么？

练习 3.30　乘积 $ab(a^2 + b^2)(a^2 - b^2)$ 总能被 30 整除.

练习 3.31　数 $a^7 - a$ 总能被 42 整除.

练习 3.32　若 $2 \mid n, 5 \mid n$ 则 $70 \mid n$.

练习 3.33　若 $5 \mid n$ 且 $17 \mid n$，则 $85 \mid n$.

练习 3.34　求证：$143 \mid 2002^{3003} + 3003^{2002}$.

练习 3.35　证明：如果三个连续自然数中有一个是自然数的立方，那么它们的乘积能被 504 整除.（波兰 1958 年竞赛题）

练习 3.36　设数 a 以 4 收尾而又不能被 4 整除，那么乘积 $a(a^2 - 1)(a^2 - 4)$ 能被 480 整除.

练习 3.37　数 $a^{13} - a$ 总能被 546 整除.

练习 3.38　设 a, b 是两个给定的整数，并且存在整数 x, y 使 $ax + by = 1$，如果再有 $a \mid n$，$b \mid n$，则 $ab \mid n$.

练习 3.39　设 $n \neq 1$. 证明：$(n-1)^2 \mid n^k - 1$ 的充分必要条件是 $(n-1) \mid k$.

练习 3.40　设 $n \neq 1$，试证 $(n-1)^2 \mid (n^k - 1)$ 的充要条件是 $(n-1) \mid k$.

练习 3.41　设 l 是一个给定的正整数，若 $d \mid (a+b+c)$，$d \mid (a^l - b^l)$ 且 $d \mid (b^l - 1)$，试证：对任意的正整数 n，有 $d \mid (a^{nl+1} + b^{nl+1} + c)$.

练习 3.42　设 n 是奇数，试证：$2^{2n}(2^{2n+1} - 1)$ 的最后两位数为 28.

练习 3.43　设 n 是正整数，试证：存在唯一一对整数 k, l，使得 $n = \dfrac{k(k-1)}{2} + 1 (0 \leqslant l < k)$.

练习 3.44　设 $k \geqslant 2$ 是正整数，试证：任意整数 a 均可唯一表示成 $a = b_n k^n + b_{n-1} k^{n-1} + \cdots + b_1 k + b_0$ 的形式，这里 $0 < b_n < k, 0 \leqslant b_i < k (i = 0, 1, \cdots, n-1)$.

练习 3.45　判断以下方程有无整数根，若有整数根则求出所有这种根：

$(i) \ x^2 + x + 1 = 0$;　　　　　　　$(ii) \ x^2 - 5x - 4 = 0$;

$(iii) \ x^4 + 6x^3 - 3x^2 + 7x - 6 = 0$;　　　$(iv) \ x^3 - x^2 - 4x + 4 = 0$.

练习 3.46　若 $x^2 + ax + b = 0$ 有整数根 $x_0 \neq 0$，则 $x_0 \mid b$. 一般的，若 $x^n + a_{n-1}x^{n-1} + \cdots + a_0 = 0$ 有整数根 $x_0 \neq 0$，则 $x_0 \mid a_0$.

练习 3.47　试证：$s = \dfrac{1}{3} + \dfrac{1}{5} + \cdots + \dfrac{1}{2n+1} (n \geqslant 1)$ 不是整数.

练习 3.48　试证：对于正整数 n，多项式 $f(n) = \dfrac{1}{3}n^3 + \dfrac{1}{2}n^2 + \dfrac{1}{6}n$ 总取整数值.

第 4 章　奇数与偶数

罗秀才："小小麻雀莫逞能，三百条狗四下分．一少三多要单数，看你怎样分得清．"

刘三姐："九十九条打猎去，九十九条看羊来．九十九条守门口，还剩三条财主请来当奴才．"

——歌剧《刘三姐》

从简到繁考察一切除法运算．首先是被除数为零的，零除以任意一个整数商都为零．其次是除数为 1 的，任意一个整数除以 1 其商都为原整数不变．除数是 −1 的时候，任意一个整数除以 −1 其商为原整数的相反数．事实上，一切对负整数的整除性讨论都可以归结到正整数的讨论上进行．

进一步，考虑除数是 2 的情况，任意一个整数除以 2，可能得出整除或者不整除两种不同的结果，将此结果进行分类，就会得出奇数与偶数．

4.1　奇数偶数

4.1.1　奇数偶数

能被 2 整除的整数，如：

$\cdots -14,\ -12,\ -10,\ -8,\ -6,\ -4,\ -2,\ 0,\ 2,\ 4,\ 6,\ 8,\ 10,\ 12,\ 14\cdots$

称为偶数（even number），一般用 $2k(k \in Z)$ 表示．一般把大于零的偶数叫作正偶数，生活中也叫双数．

注意：0 也是偶数．

容易看出，偶数的个位数是 0，2，4，6，8．

不能被 2 整除的整数：

$\cdots -13,\ -11,\ -9,\ -7,\ -5,\ -3,\ -1,\ 1,\ 3,\ 5,\ 7,\ 9,\ 11,\ 13\cdots$

称为奇数（odd number），一般用 $2k+1$ 或 $2k-1(k \in Z)$ 表示．大于零的奇数叫作正奇数，生活中也叫单数．

正奇数：

1，3，5，7，9，13，15，17，19，21，23，25，27，29，31，33，35，37，39$\cdots$

容易看出，奇数的个位数是 1，3，5，7，9．

由上述定义易知，任意一个奇数和任意一个偶数不可能相等．

于是，按照能否被 2 整除，整数可以分为：

$$整数\begin{cases}奇数\begin{cases}正奇数 \\ 负奇数\end{cases} \\ 偶数\begin{cases}正偶数 \\ 负偶数\end{cases}\end{cases}$$

平方数指的是像 1，4，9，16，25，36，49，64，81，100，121，144，169$\cdots\cdots$这样的能够表示成一个数的平方的数．

$$1 = 1^2$$
$$4 = 2^2$$
$$9 = 3^2$$
$$16 = 4^2$$
$$25 = 5^2$$
$$36 = 6^2$$

······

请读者再次查阅第 1 章图 1.5. 以左图为例，第一个平方数就只有 1 个点. 第二个平方数是以单位 1 为边长构成的正方形，此时的平方数是构成这个正方形的点的个数 4. 第三个平方数是以 2 倍单位长为边长构成的正方形，此时的平方数是构成这个正方形的点的个数 9. 以此类推，可得一系列的平方数.

若以图 1.5 的右图为例，可以这样看：第一个平方数就是左下角以单位长为边长的 1 个小正方形，其个数为 1. 第二个平方数就是左下角边长为 2 的小正方形，其个数为 4，第三个平方数就是以边长为 3 的小正方形，其个数为 9. 以此类推，也可得到一系列的平方数.

但是以点为例和以正方形为例的几何意义稍有区别，不过在数的讨论中，这并没有什么影响.

偶数的平方被 4 整除. 奇数的平方被 8 除余 1.

例 4.1　若 n 是奇数，证明：$8 \mid (n^2 - 1)$.

证明：令 $n = 2k + 1$，则
$$n^2 = (2k + 1)^2 = 4k^2 + 4k + 1$$

所以：
$$n^2 - 1 = 4k^2 + 4k = 4k(k + 1)$$

显然，$2 \mid k(k + 1)$，而 $4 \mid 4$，

所以：$8 \mid n^2 - 1$.

事实上，由上面的讨论我们可以知道：任意奇数的平方除以 2，4，8 所得的余数都是 1.

例 4.2　证明：平方数的（正）因数的个数是奇数.

分析：抓住 n 的因数是成对的这一特点：有因数 d，就有因数 $\{1, 2, \cdots mn\}$.

解：每个自然数 n 的因数是成对出现的：如果 d 是 n 的因数，那么 $\dfrac{n}{d}$ 也是 n 的因数：

d 不同时，$\dfrac{n}{d}$ 也不相同.

当 $d \neq \sqrt{n}$ 时，d 与 $\dfrac{n}{d}$ 不相等.

只有当 n 为平方数时，$\sqrt{n}$ 是 n 的因数，与它配对的数就是 $\sqrt{n}$ 自身.

所以当且仅当 n 为平方数时，n 的因数个数为奇数.

在 d 是 n 的因数时，我们把 $\dfrac{n}{d}$ 称为 n 的共轭因数.

这样，n 为平方数时，它有一个共轭（自己和自己共轭）的因数 $\sqrt{n}$.

反过来，如果 n 有共轭的因数，那么它一定是平方数.

例 4.3　用 $\tau(n)$ 表示 n 的因数个数，试确定

$$\tau(1) + \tau(2) + \cdots + \tau(199)$$

的奇偶性.

解：由上例的讨论可知，非平方数的因数个数是偶数，平方数的因数个数是奇数.

因为 $45 > \sqrt{199} > 44$，所以 1 至 199 中有 44 个平方数，即所求式中有 44 项为奇数，于是由基本性质 1 得所求式是偶数.

例 4.4　能否将 $\lim\limits_{x\to\infty}\{1,2,\cdots,972\}$ 分为 12 个互不相交的子集，每个子集含 81 个元素，并且各个子集的元素的和相等？如果能，怎样分？

解：如果存在所述的分法，那么和 $1 + 2 + 3 + \cdots + 972$ 应该是 12 的倍数，可是

$$1 + 2 + \cdots + 972 = \frac{1 + 972}{2} \times 972 = 973 \times 81 \times 6$$

不是 12 的倍数，矛盾！

所以，无法将题中的集合分成 12 个互不相交的子集符合要求.

说明，一般地：

（1）设 $n > 1$，当 n 为奇数，m 为正偶数时，无法将集合 $\{1,2,\cdots,mn\}$ 分为 m 个互不相交的子集，并且各个子集的元素的和相等. 不难由

$$1 + 2 + \cdots + mn = (1 + mn) \times n \times \frac{m}{2}$$

不是 m 的倍数知道这个结论成立.

（2）当 $n > 1$ 及 m 均为奇数或偶数时，我们都可以将集合 $\{1,2,\cdots,mn\}$ 分为 m 个互不相交的子集，满足上面所述的要求.

例 4.5　在一条线段的内部任取 n 个点，将这些点及线段端点依次记为 $A_0,A_1,\cdots,$ A_{n+1}，并且将端点 A_0 染上红色，A_{n+1} 染上蓝色，其余各点染上红色或蓝色. 称两端颜色不同的线段 $A_iA_{i+1}(0 \leqslant i \leqslant n)$ 为"好线段". 证明，好线段的条数为奇数.

分析：记两种颜色的点为"$+1$"和"-1"，运用基本性质解决这个问题.

解：将红色的点记为 $+1$，蓝色的点记为 -1.

考虑每条线段 A_iA_{i+1} 的两端的数的乘积.

当且仅当 A_iA_{i+1} 是好线段时，乘积是 -1.

将上述 $n+1$ 个乘积乘起来.

这时 A_0,A_{n+1} 各出现一次，中间的点 $A_i(1 \leqslant i \leqslant n)$ 出现两次，于是 $n+1$ 个乘积的积为 $1 \times (-1) = -1$.

这表明 $n+1$ 个乘积中，乘积为 -1 的个数是奇数，即好线段的条数为奇数.

4.1.2 奇数偶数的方法论意义

第一，整数可以按照能不能被 2 整除分成奇数和偶数两类，那么能不能按照类似的方法将其分类呢？比如按照能不能被 3 整除分成能被 3 整除和不能被 3 整除两类，更进一步，用带余除法的余数进行分类，将整数按照被 3 除后余数是 0、余数是 1 和余数是 2 分成三类，余数是 0 也就是能被 3 整除，余数是 1 和 2 事实上是将不能被 3 整除分成了两类.

$$整数\begin{cases}= 3k:能被~3~整除,除以~3~余数为~0. \\ \neq 3k:不能被~3~整除.\begin{cases}= 3k + 1:除以~3~余数为~1. \\ = 3k + 2:除以~3~余数为~2.\end{cases}\end{cases}$$

以此类推，将能不能被 4 整除分成两类，或者用带余除法的余数分成分别以 0，1，2，3 作为其余数的四类.

$$整数\begin{cases}= 4k:能被~4~整除,除以~4~余数为~0. \\ \neq 4k:不能被~4~整除.\begin{cases}= 4k + 1:除以~4~余数为~1. \\ = 4k + 2:除以~4~余数为~2. \\ = 4k + 3:除以~4~余数为~3.\end{cases}\end{cases}$$

将能不能被 5 整除分成两类，或者用带余除法的余数分成分别以 0，1，2，3，4 作为其余数的五类.

$$整数\begin{cases}= 5k:能被~5~整除,除以~5~余数为~0. \\ \neq 5k:不能被~5~整除.\begin{cases}= 5k + 1:除以~5~余数为~1. \\ = 5k + 2:除以~5~余数为~2. \\ = 5k + 3:除以~5~余数为~3. \\ = 5k + 4:除以~5~余数为~4.\end{cases}\end{cases}$$

以此类推，我们可以将这一想法推广到所有自然数上，也就是将整数按照能不能被 m 整除分成两类，或者用带余除法的余数分成分别以 0，1，2，$\cdots$，$m - 1$ 作为其余数的 k 类. 事实上，这就是数论研究中同余思想的原始基础.

$$整数\begin{cases}= mk:能被~m~整除,除以~m~余数为~0. \\ \neq mk:不能被~m~整除.\begin{cases}= mk + 1:除以~m~余数为~1. \\ = mk + 2:除以~m~余数为~2. \\ \cdots\cdots \\ = mk + (m - 1):除以~m~余数为(m - 1).\end{cases}\end{cases}$$

第二，如果按照奇数偶数的分类方式进行分类，我们可以有这样一个默认的结果，那就是一个奇数增加或减少 1，就变成另一类，也就是偶数. 同样的，一个偶数增加或减少 1，就变成奇数. 换一个表达方式，连续两个整数，必然有一个是奇数而另一个是偶数.

继续前面的假设，如将整数分成被 3 除后余数分别是 0，1，2 的三类，那么，将任意一个给定整数增加 1，这个数将变成另一类，但这个另一类却依据其余数确定.

考虑到我们经常要研究整数的整除性，若给定两个连续整数，那么这两个整数必然属于余数为 0，1 或者 1，2 或者 2，0 三种情况中的一种.

假设给定三个连续整数，那么这三个整数除以 3 后其余数必然分别为 0，1，2，也就是说，其中必有一个能被 3 整除.

如果将这个假设继续下去呢？愿感兴趣的读者自己继续下去.

例 4.6 每个正整数是否恰好能表示成五个形式 $5q, 5q + 1, 5q + 2, 5q + 3, 5q + 4$ 中的一个（$q \geqslant 0$ 是某个整数）？如何判断一个特定的数（例如 6666）是这些形式中的哪一种？

表 4.1

0	1	2	3	4		100	101	102	103	104
5	6	7	8	9		105	106	107	108	109
10	11	12	13	14		110	111	112	113	114
15	16	17	18	19		115	116	117	118	119
20	21	22	23	24		120	121	122	123	124
25	26	27	28	29		125	126	127	128	129
30	31	32	33	34		130	131	132	133	134
35	36	37	38	39		135	136	137	138	139
40	41	42	43	44		140	141	142	143	144
45	46	47	48	49		145	146	147	148	149
50	51	52	53	54		150	151	152	153	154
55	56	57	58	59		155	156	157	158	159
60	61	62	63	64		160	161	162	163	164
65	66	67	68	69		165	166	167	168	169
70	71	72	73	74		170	171	172	173	174
75	76	77	78	79		175	176	177	178	179
80	81	82	83	84		180	181	182	183	184
85	86	87	88	89		185	186	187	188	189
90	91	92	93	94		190	191	192	193	194
95	96	97	98	99		195	196	197	198	199

解：本题的实质就是将整数按照能否被 5 整除进行分类，分为了被 5 除后余数为 0，1，2，3，4 的五类．于是我们很容易得到某给定的数除以 5 之后的余数，如因为：

$$5 \mid 6665$$

所以，6666 除以 5 的余数是 1．

例 4.7　找规律判定"300"位于下面哪个字母的下边．（美国 1989 年小学数学奥赛题）

A	B	C	D	E	F	G
01	02	03	04	05	06	07
08	09	10	11	12	13	14
15	16	17	18	19	20	21

分析：事实上，这个题要考虑的就是如果将所有数按照除以 7 所得分成余数分别为 1，2，3，4，5，6，0 的 7 类，所以对于所要求的"300"的位置，只需要用带余除法去试除，然后得到的余数就可以知道应该在哪个位置了．

解：观察可以发现，每7个数组成一组，故 $300 = 7 \times 42 + 6$ 与6同在 F 的下边．

4.2 奇数偶数的加减乘除

奇数偶数的加减乘除运算具有一些有意思的性质，这些性质在整数性质讨论中即是主体，又可以作为讨论其他非奇数偶数性质时候的一些基础性工具．

4.2.1 加减运算

从奇数出发，不妨设为奇数 $a = 2k + 1 (k \in Z)$，则 $a + 1 = 2k + 2 = 2(k + 1)$ 必然可被2整除，也即必为偶数．再进一步 $a + 2 = 2k + 3 = 2(k + 1) + 1$ 不可能被2整除，必为奇数．这说明：任意奇数加1为偶数，任意偶数加1为奇数．换句话说，任意连续的两个整数必然有一个是奇数，而另一个是偶数．

任意连续的两个整数一般表示为 n，$(n + 1)$ 或 $(n - 1)$，n，任意连续的三个整数一般表示为 n，$(n + 1)$，$(n + 2)$．

将上述 $a + 1$ 相关讨论中的"1"推广到任意奇数，将 $a + 2$ 中的"2"推广到任意偶数，可以得到：

$$偶数 \pm 偶数 = 偶数；$$
$$奇数 \pm 奇数 = 偶数；$$
$$偶数 \pm 奇数 = 奇数．$$

进一步推而广之，奇数个奇数的和是奇数，偶数个奇数的和是偶数，任意整数个偶数的和是偶数．

著名数学家毕达哥拉斯发现一个有趣的奇数现象：将奇数连续相加，每次的得数正好是平方数．这体现在奇数和平方数之间有着密切的重要联系．如：

$$1 = 1^2$$
$$1 + 3 = 2^2$$
$$1 + 3 + 5 = 3^2$$
$$1 + 3 + 5 + 7 = 4^2$$
$$1 + 3 + 5 + 7 + 9 = 5^2$$
$$1 + 3 + 5 + 7 + 9 + 11 = 6^2$$
$$1 + 3 + 5 + 7 + 9 + 11 + 13 = 7^2$$
$$1 + 3 + 5 + 7 + 9 + 11 + 13 + 15 = 8^2$$
$$1 + 3 + 5 + 7 + 9 + 11 + 13 + 15 + 17 = 9^2$$
$$1 + 3 + 5 + 7 + 9 + 11 + 13 + 15 + 17 + 19 = 10^2$$

据此我们可以得出这样一个结论：任意一个奇数都可以写成两个整数平方差的形式．任意两个奇数的平方差是2，4，8的倍数．

我们知道：

$$4 = 2 + 2$$
$$6 = 3 + 3$$

$$8 = 3 + 5$$
$$10 = 5 + 5$$
$$12 = 5 + 7$$
$$14 = 7 + 7$$
$$16 = 5 + 11$$
……

于是我们可以猜想，是不是每一个大于 2 的偶数都可以写成两个奇数的和，答案显而易见．进一步猜想：是不是每一个大于 2 的偶数都可以写成两个素数的和？这就是著名的哥德巴赫猜想．

4.2.2　乘法运算

令 $k \in Z$，如果 a 是偶数，那么 $1 \times a, 2a, 3a, 4a, \cdots, ka$ 也是偶数．

如果 a 是奇数，那么 $1 \times a, 3a, 5a, 7a, \cdots, (2k+1)a$ 也是奇数，而 $2a, 4a, 6a, \cdots, 2ka$ 是偶数．也即：

$$偶数 \times 偶数 = 偶数；$$
$$奇数 \times 奇数 = 奇数；$$
$$偶数 \times 奇数 = 偶数．$$

推而广之，任意个奇数的积是奇数，至少有一个乘数是偶数的积是偶数．特别的，奇数的平方是奇数，偶数的平方是偶数．

例 4.8　已知两个整数之和是奇数，求证两个整数之积是偶数．

证明：设两个整数为 a 与 b，已知 $a + b =$ 奇数，可知 a 与 b 的奇偶性相反（因为如若不然，a, b 同奇或同偶时，$a + b$ 为偶数，与 $a + b =$ 奇数不符），a 与 b 一奇一偶，所以 $a \times b$ 是一个偶数．

4.2.3　除法运算

偶数能被奇数整除，则商必为偶数．否则的话，假设偶数被奇数整除，其商为奇数，那么反过来可以得到一个等式——奇数 × 奇数 = 偶数，而由奇数偶数的乘法运算知道这是不可能的．

例 4.9　对任意 $n \in Z$，必有 $2 \mid n(n+1)$．

分析：题意是两个连续整数的积 $n(n+1)$ 必为偶数．由奇数偶数的加减法中的分析我们知道，$n(n+1)$ 中 n 和 $n+1$ 必有一个为奇数，一个为偶数，因为 2 必然整除偶数，故必然整除偶数和奇数的乘积．所以在证明中只需要讨论 n 的一切可能的取值即可．

证明：若 $n = 2k (k \in Z)$，则 $2 \mid n$，故 $2 \mid n(n+1)$．

若 $n = 2k + 1 (k \in Z)$，则

$$n + 1 = 2k + 2 = 2(k+1)，$$

于是有 $2 \mid n + 1$，

故 $2 \mid n(n+1)$．

结论得证.

4.3 "$3x+1$" 问题

"$3x+1$" 问题从一个叫作角谷猜想（又叫叙古拉猜想、冰雹猜想、克拉茨问题）的游戏开始，1976 年的一天，《华盛顿邮报》在头版头条上报道了一条数学新闻. 文中记叙了这样一个故事：

20 世纪 70 年代中期，美国各所名牌大学校园内，人们都像发了疯一般，夜以继日，废寝忘食地玩弄一个数学游戏. 这个游戏十分简单：任意写出一个自然数 x，并且按照以下的规律进行变换：

如果是奇数，则下一步变成 $3x+1$.

如果是偶数，则下一步变成 $\dfrac{x}{2}$.

不单单是学生，甚至连教师、研究员、教授与学究都纷纷加入. 为什么这个游戏这般引人入胜，其魅力又经久不衰？因为人们发现，无论 N 是怎样一个数字，最终都无法逃脱回到谷底 1. 准确地说，是无法逃出落入底部的 4，2，1 循环，永远也逃不出这样的宿命.

例如 $x=60$，可以一步一步得到 30，15，46，23，70，35，106，53，160，80，40，20，10，5，16，8，4，2，1. 任意再给出一个自然数进行试验，一样会得出最后的结果 4，2，1.

克拉茨（L. Collatz）在 1950 年召开的第十一届国际数学家大会[①]上提出：从任意一个自然数开始，经过有限次上述变换，能否最终得到 4，2，1？用数学的表示方式就是：

设 $n \in Z$，且

$$f(n) = \begin{cases} \dfrac{n}{2}, & (2 \mid n) \\ 3n+1, & (2 \nmid n) \end{cases}$$

令

$$f_1(n) = f(n), f_2(n) = f(f_1(n)), \cdots, f_k(n) = f(f_{k-1}(n))$$

① 国际数学家大会（International Congress of Mathematicians），是数学家们为了数学交流，展示、研讨数学的发展，会见老朋友、结交新朋友的国际性会议，是国际数学界的盛会. 四年举行一次，首届大会 1897 年在瑞士苏黎世举行，至 2010 年底共举行了 26 届. 1900 年巴黎大会之后，除两次世界大战期间外，未曾中断过，它已成为高水平的全球性数学科学学术会议. 出席大会的数学家的人数，最少的一次是 208 人，最多的一次是 4000 多人. 每次大会一般都邀请一批杰出数学家分别在大会上做一小时的学术报告和学科组的分组会上做 45 分钟学术报告，凡是出席大会的数学家都可以申请在分组会上做 10 分钟的学术报告，或将自己的论文在会上散发. 其中学科组一般分为 20 个左右. 国际数学联盟（IMU）于 1920 年 9 月 20 号在法国斯特拉斯堡（Strasbourg）成立，截至 2017 年共有 80 余个成员国. 它的主要任务是：（1）促进数学方面的国际交流；（2）组织召开国际数学家大会，以及两届大会之间各种数学方面的国际性专门会议；（3）颁发奖励，主要是菲尔兹奖（Fields Medal）、陈省身奖（Chern Medal）、高斯奖（Gauss Prize）、奈望林纳奖（Rolf Nevanlinna Prize）.

则存在有限正整数 $m \in N$，使得 $f_m(n) = 1$．

很明显，变换过程中一旦出现 2 的幂，问题就解决了，而 2 的幂有无穷多个，只要变换过程足够长，必定会碰到一个 2 的幂使问题以肯定形式得到解决．题意如此简单，可数十年来，尽管已经有无数数学家和数学爱好者尝试过，其中不乏天才和世界上第一流的数学家，但他们都没有成功．

二十年前，有人向数论学家保尔·厄尔多斯（Paul Erdos）介绍了这个问题，并且问他怎么看待现代数学对这个问题无能为力的现象．他回答说：数学还没有准备好回答这样的问题．这个猜想至今无人证明，也无人推翻．都猜想这个问题的答案是肯定的，但却无法做出普遍的证明，大家不妨一试．

4.4　典型问题

4.4.1　典型例题

例 4.10　设 $a_1, a_2, \cdots, a_n$ 是 $1, 2, \cdots, n$ 的任一排列，n 为正奇数．求证：$(a_1 - 1)(a_2 - 2)\cdots(a_n - n)$ 为偶数．

证明： 因为

$$(a_1 - 1) + (a_2 - 2) + \cdots + (a_n - n)$$
$$= (a_1 + a_2 + \cdots + a_n) - (1 + 2 + \cdots + n)$$
$$= 0,$$

这说明奇数个整数 $(a_1 - 1), (a_2 - 2), \cdots (a_n - n)$ 之和为偶数，那么 $a_1 - 1, a_2 - 2, \cdots, a_n - n$ 中至少有一个为偶数，所以 $(a_1 - 1)(a_2 - 2)\cdots(a_n - n)$ 为偶数．

例 4.11　已知 $x, y \in Z$，且 $x^2 - 4y = 1$，试讨论 x, y 的奇偶性．

解：（1）若 x 是奇数，则 $x = 2k + 1$．因为：
$x^2 - 4y = 1$，
所以：

$$y = \frac{1}{4}(x^2 - 1)$$
$$= \frac{1}{4}(x - 1)(x + 1)$$
$$= \frac{1}{4} \cdot 2k \cdot (2k + 2)$$
$$= k \cdot (k + 1).$$

又有 $2 \mid k(k + 1)$，
所以 y 是偶数．

（2）若 x 是偶数，则 $x = 2k$，此时

$$y = \frac{1}{4}(x^2 - 1)$$

$$= \frac{1}{4}(4k^2 - 1)$$

$$= k^2 - \frac{1}{4} \notin Z$$

矛盾．故 x 不能是偶数．

由（1）（2）知，x 是奇数，y 是偶数．

例 4.12 试证：双数个奇数的和是偶数．

证明： 设双数个奇数分别为 $2k_i + 1(i = 1, 2, \cdots, 2n, n \geqslant 1)$．因为

$$\sum_{i=1}^{2n}(2k_i + 1) = 2\sum_{i=1}^{2n}k_i + \sum_{i=1}^{2n}1 = 2\sum_{i=1}^{2n}k_i + 2n = 2\left(\sum_{i=1}^{2n}k_i + n\right)$$

所以 $2\sum\limits_{i=1}^{2n}(2k_i + 1)$．

结论成立．

例 4.13 若 a 是整数，且 $2 \mid a^2$，求证 $2 \mid a$．

证明： 假设 $2 \nmid a$，则整数 a 为奇数，所以 a^2 为奇数，这时有 $2 \nmid a^2$，与 $2 \nmid a^2$ 的条件矛盾，所以 $2 \nmid a$ 成立．

例 4.14 设数列 $1, 9, 8, 3, 4, 3, \cdots$．其中 a_{n+4} 为 $a_n + a_{n+3}(n = 1, 2, 3, \cdots)$ 的个位数字．试证

$$a_{1998}^2 + a_{1999}^2 + a_{2000}^2 + 3$$

是 4 的倍数．

分析： 只要能推断出 a_{1998}，a_{1999} 和 a_{2000} 的奇偶性即可．这是因为奇数的平方不能被 4 整除，而偶数的平方一定能被 4 整除，而如果这三个数中有两个偶数和一个奇数问题就解决了．

证明： 为了讨论原数列的奇偶性，我们构造一个新数列．当原数列中 a_n 为奇、偶数时，分别记为 1，0，则得到数列：

$\{b_n\}: 1, 1, 0, 1, 0, 1, 1, 0, 0, 1, 0, 0, 0, 1, 1, \cdots$，显然有下面的两个结论：（1）$a_n$ 与 b_n 的奇偶性相同；（2）$b_n = b_{n+15t}$（其中 t 是任意正整数）．

而 $2000 = 15 \times 133 + 5$，所以 $b_{2000} = b_5 = 0; b_{1999} = b_4 = 1; b_{1998} = b_3 = 0$．所以 a_{1998}，a_{1999}, a_{2000} 分别为偶数、奇数、偶数．

但偶数的平方被 4 整除，奇数的平方被 4 除余 1，所以 $a_{1998}^2 + a_{1999}^2 + a_{2000}^2 + 3$ 必是 4 的倍数．

说明： 采用相同的办法还可以证明 $a_{1985}^2 + a_{1986}^2 + \cdots + a_{1999}^2 + a_{2000}^2$ 也是 4 的倍数．

例 4.15　求证：若 n 为奇数，则 $8 \mid (n^2 - 1)$.

证明： 设 $n = 2k + 1 (k \in Z)$，则

$$n^2 - 1 = (2k + 1)^2 - 1 = 4k(k + 1).$$

由于 k 和 $k + 1$ 中必有一个是偶数，所以 $8 \mid (n^2 - 1)$.

证毕.

例 4.16　设 $a_1, a_2, \cdots a_n$ 是整数，且 $a_1 + a_2 + \cdots + a_n = 0$，$a_1 a_2 \cdots a_n = n$. 证明：$n$ 是 4 的倍数.

证明： 如果 $2 \nmid n$，则 $n, a_1, a_2, \cdots, a_n$ 都是奇数.

于是 $a_1 + a_2, \cdots, + a_n$ 是奇数个奇数之和，不可能等于零，这与题设矛盾.

所以 $2 \mid n$，即在 $a_1, a_2, \cdots, a_n$ 中至少有一个偶数.

如果只有一个偶数，不妨设为 a_1.

那么 $2 \nmid a_i (2 \leqslant i \leqslant n)$.

此时有等式 $a_2 + \cdots + a_n = -a_1$.

左端是一个 $(n - 1)$ 个奇数之和，其和为奇数.

右端是偶数，这是不可能的.

因此，在 $a_1, a_2, \cdots, a_n$ 中至少有两个偶数，即 $4 \mid n$.

例 4.17　设 $a_1, a_2, \cdots, a_n$ 是一组数，它们中的每一个都取 1 或 -1，而且

$$a_1 a_2 a_3 a_4 + a_2 a_3 a_4 a_5 + \cdots + a_n a_1 a_2 a_3 = 0,$$

证明 n 必须是 4 的倍数.

分析： 本题只需借助奇偶性的分析.

证明： 由于每个 a_i 均为 1 或 -1，从而每一个 $a_i a_{i+1} a_{i+2} a_{i+3}$ 也只能取 1 或 -1，而这 n 个数的和等于零，故取 $+1$ 和 -1 的个数是相等的，因此 n 必是偶数. 设 $n = 2m$，再考虑这 n 项的乘积，有

$$(a_1 a_2 \cdots a_n)^4 = 1,$$

于是这 n 项中取 -1 的项（m 项）也一定是偶数，即 $m = 2k$，从而 n 是 4 的倍数.

说明： 本题可以进一步推广为：设 $a_1, a_2, \cdots, a_n$ 是一组仅取 $+1$ 或 -1 的数，而且 $a_1 a_2 \cdots a_{2m} + a_2 a_3 \cdots a_{2m} a_{2m+1} + \cdots + a_n a_1 \cdots a_{2m-1} = 0 (2m < n)$，证明 n 是 4 的倍数.

例 4.18　设 $M = 5^{2003} + 7^{2004} + 9^{2005} + 11^{2006}$，求证：$8 \mid M$. 将此题进行推广并证明你的结论.

证明： 利用"任何奇数的平方减 1 都能被 8 整除"这一结论易证 $8 \mid M$.

类似可证得推广结论：若 $M = a^e + a^f + a^g + a^h$. 其中 $a, b, c, d (a < b < c < d)$ 是任意四个连续奇数. $e, f, g, h (e < f < g < h)$ 是任意四个连续自然数. 且 $a = 2m + 1 (m \in Z)$.

则当 m 是偶数、e 是奇数（或 e 是偶数、m 是奇数）时，$8 \mid M$.

例 4.19　设 n 是奇数，证明：$16 \mid (n^4 + 4n^2 + 11)$.

证明：因为 $n^4 + 4n^2 + 11 = (n^2 - 1)(n^2 + 5) + 16$.

由 n 是奇数，有 $8 \mid (n^2 - 1)$.

又因为 $2 \mid (n^2 + 5)$，

所以 $16 \mid (n^2 - 1)(n^2 + 5)$.

从而 $16 \mid (n^4 + 4n^2 + 11)$.

例 4.20 设 n 是正奇数，试证明

$$1^n + 2^n + \cdots + 9^n - 3(1^n + 6^n + 8^n)$$

能被 18 整除.

分析：注意到 $18 = 2 \times 9$，而 $(2, 9) = 1$，所以只要分别能证明原式既被 2 整除，同时也被 9 整除即可. 在证明原式是 9 的倍数时，要借助因式分解公式.

证明：显然

$$1^n + 2^n + \cdots + 9^n \text{ 和 } 3(1^n + 6^n + 8^n)$$

均是奇数，从而其差必为偶数，可原式是 2 的倍数，由于 n 是奇数，则有

$$1^n + 8^n, 2^n + 7^n, 3^n + 6^n \text{ 和 } 4^n + 5^n$$

都是 9 的倍数. 故原式可以变形为

$$9^n + (1^n + 8^n) + (2^n + 7^n) + (3^n + 6^n) + (4^n + 5^n) - 3(1^n + 8^n) - 3 \times 6^n$$

可见每一项都是 2 的倍数.

命题成立.

例 4.21 当 n 为正奇数时，求证：$60 \mid f(n) = 6^n - 3^n - 2^n - 1$.

证明：$f(n) = 6^n - 3^n - 2^n - 1$

$$= (6^n - 3^n) - (2^n + 1).$$

因为 n 为正奇数，所以

$3 \mid (6^n - 3^n), 3 \mid (2^n + 1)$，

从而 $3 \mid f(n)$.

同理 $4 \mid f(n), 5 \mid f(n)$.

因为 $3, 4, 5$ 两两互质，所以 $60 \mid f(n) = 6^n - 3^n - 2^n - 1$.

例 4.22 设 n 为奇数，证明：$n \mid \left(1 + \dfrac{1}{2} + \cdots + \dfrac{1}{n-1}\right) \cdot (n-1)!$.

分析：由于 $1 + \dfrac{1}{n-1} = \dfrac{n}{n-1}, \dfrac{1}{2} + \dfrac{1}{n-2} = \dfrac{n}{2(n-2)}, \cdots$，这些数的分子都是 n，分母都小于 $(n-1)!$，故而想到用配方法证明.

证明：由于 $\left(\dfrac{1}{k} + \dfrac{1}{n-k}\right) \cdot (n-1)! = \dfrac{n}{k(n-k)} \cdot (n-1)!, 1 \leqslant k \leqslant n-1$，

其中 $k(n-k) \mid (n-1)!$，

故有 $n \mid \dfrac{n}{k(n-k)} \cdot (n-1)!$.

所以设 $n = 2m + 1, m \geqslant 0$ ，则：

$$\left(1 + \frac{1}{2} + \cdots + \frac{1}{n-1}\right) \cdot (n-1)! = \left(1 + \frac{1}{2} + \cdots + \frac{1}{2m}\right) \cdot (n-1)!$$

$$= \left[\left(1 + \frac{1}{2m}\right) + \left(\frac{1}{2} + \frac{1}{2m-1}\right) + \cdots + \left(\frac{1}{m} + \frac{1}{m+1}\right)\right] \cdot (n-1)!$$

$$= \left[\frac{2m+1}{2m} + \frac{2m+1}{2(2m-1)} + \cdots + \frac{2m+1}{m(m+1)}\right] \cdot (n-1)!$$

$$= \left[\frac{n}{(n-1)} + \frac{n}{2(n-2)} + \cdots + \frac{n}{\left(n - \frac{n-1}{2}\right)\left(n - \frac{n-3}{2}\right)}\right] \cdot (n-1)!$$

显然，$(n-1) \mid (n-1)!, 2(n-2) \mid (n-1)!, \cdots, \left(n - \frac{n-1}{2}\right)\left(n - \frac{n-3}{2}\right) \mid (n-1)!$，

所以，$n \mid \left(1 + \frac{1}{2} + \frac{1}{3} + \cdots + \frac{1}{n-1}\right) \cdot (n-1)!.$

例 4.23 设 k 为正奇数，证明：$\dfrac{n(n+1)}{2} \mid (1^k + 2^k + \cdots + n^k).$

分析：因为 $(n, n+1) = 1$ ，故可考虑分别证明

$$n \mid 2(1^k + 2^k + \cdots + n^k), (n+1) \mid 2(1^k + 2^k + \cdots + n^k).$$

证明：问题等价于证明 $n(n+1) \mid 2(1^k + 2^k + \cdots + n^k).$

对于正整数 x, y ，有

$$x^k + y^k = (x+y)\left[x^{k-1} - x^{k-2}y + \cdots + (-1)^{k-2}xy^{k-2} + (-1)^{k-1}y^{k-1}\right],$$

则 $(x+y) \mid (x^k + y^k)$ ，

所以由 $2(1^k + 2^k + \cdots + n^k)$

$$= \left[1^k + (n-1)^k\right] + \left[2^k + (n-2)^k\right] + \cdots + \left[(n-1)^k + 1^k\right] + 2n^k,$$

可知 $n \mid 2(1^k + 2^k + \cdots + n^k).$

同样，由 $2(1^k + 2^k + \cdots + n^k) = (1^k + n^k) + \left[2^k + (n-1)^k\right] + \cdots + (n^k + 1^k)$ ，

可知 $(n+1) \mid 2(1^k + 2^k + \cdots + n^k)$ ，

又 $(n, n+1) = 1$ ，则 $n(n+1) \mid 2(1^k + 2^k + \cdots + n^k)$ ，

即 $\dfrac{n(n+1)}{2} \mid (1^k + 2^k + \cdots + n^k).$

例 4.24 设 r 是正奇数，证明：对任意正整数 n ，有 $(n+2) \nmid (1^r + 2^r + \cdots + n^r).$

证明：当 $n = 1$ 时，结论显然成立.

设 $n \geqslant 2$ ，令 $s = 1^r + 2^r + \cdots + n^r$ ，则

$$2s = 2 + (2^r + n^r) + \left[3^r + (n-1)^r\right] + (n^r + 2^r)$$

因为 r 是正奇数，上式右边中除第一项外，每一加项 $i^r + (n+2-i)^r$ 都能被 $i + (n+2-i) = n + 2 (2 \leqslant i \leqslant n)$ 整除，

因此 $2s = 2 + (n+2)Q_1$ ，其中 Q_1 是整数.

显然，$2s$ 被 $n+2$ 除得的余数是 2，

由于 $n + 2 > 2$，所以 $(n + 2) \nmid (1^r + 2^r + \cdots + n^r)$.

例 4.25　设 n 和 k 均是正整数，k 是奇数，记

$$S_k = 1^k + 2^k + \cdots + n^k,$$

求证：$S_1 \mid S_k$.

分析：本题首先要注意到 $S_1 = 1 + 2 + \cdots + n = \dfrac{1}{2}n(n + 1)$，要证明 $S_1 \mid S_k$，只要证明 $n(n + 1) \mid 2S_k$ 即可. 而 $(n, n + 1) = 1$，所以只需分别证明 $n \mid 2S_k$ 和 $(n + 1) \mid 2S_k$ 即可. 此外还要用到因式分解公式.

证明：由于

$$
\begin{aligned}
2S_k &= 1^k + 2^k + \cdots + n^k + n^k + (n - 1)k + \cdots + 1^k \\
&= \left[1^k + n^k \right] + \left[2^k + (n - 1)k \right] + \cdots + \left[n^k + 1^k \right],
\end{aligned}
$$

而 k 是奇数，所以由因式分解公式可知 $2S_k$ 是 $n + 1$ 的倍数.

同理

$$
\begin{aligned}
2S_k &= 0^k + 1^k + 2^k + \cdots + n^k + n^k + (n - 1)k + \cdots + 0^k \\
&= \left[0^k + n^k \right] + \left[1^k + (n - 1)k \right] + \cdots + \left[n^k + 0^k \right],
\end{aligned}
$$

所以 $2S_k$ 也是 n 的倍数，从而 $S_1 \mid S_k$.

例 4.26　已知一个奇数 β，使得整系数二次三项式 $ax^2 + bx + c$ 的值 $a\beta^2 + b\beta + c$ 也是奇数，其中 c 是奇数. 求证：方程 $ax^2 + bx + c = 0$ 没有奇数根.

证明：因为 c 是奇数，并且 $a\beta^2 + b\beta + c$ 是奇数，因此

$$a\beta^2 + b\beta = \beta(a\beta + b)$$

是偶数. 由于 β 是奇数，因此 $a\beta + b$ 是偶数，又由于 β 是奇数，因此 a 与 b 必须同为奇数或同为偶数.

要证明二次方程 $ax^2 + bx + c = 0$ 没有奇数根，仅需证明：对任意一个奇数 β'，都有

$$a\beta'^2 + b\beta' + c \neq 0$$

即可.

其实，由于 a 与 b 同奇或同偶，对任意一个奇数 β'，总有 $a\beta' + b$ 是偶数. 由此

$$
\begin{aligned}
a\beta'^2 + b\beta' + c &= \beta'(a\beta' + b) + c \\
&= \text{偶数} + \text{奇数} \\
&\neq 0.
\end{aligned}
$$

所以方程 $ax^2 + bx + c = 0$ 没有奇数根.

说明：要证明方程 $ax^2 + bx + c = 0$ 没有奇（偶）数根，仅需证明对任意一个奇（偶）数 β，总有 $a\beta^2 + b\beta + c \neq 0$ 即可. 运用数的奇偶性可以解决这一类问题，例如，从方程 $2x^2 - 5y^2 = 7$ 推出 y 是奇数，令 $y = 2k - 1, k \in Z^+$，从而证明原方程没有整数解.

例 4.27　设 $f(x) = ax^2 + bx + c, a, b$ 为整数，c 为奇数. 若存在奇数 m，使 $f(m)$ 为奇数，则方程 $f(x) = 0$ 无奇数根.

证明：由 c 与 $f(m) = am^2 + bm + c$ 均为奇数，知

$$f(m) - c = am^2 + bm = m(am + b)$$

为偶数. 因为 m 为奇数, 所以 $(am + b)$ 必为偶数, 且 a,b 必同为奇数或同为偶数.

对任一奇数 k, 有

$$f(k) = ak^2 + bk + c = k(ak + b) + c.$$

易知 $(ak + b)$ 必为偶数, 所以 $k(ak + b)$ 为偶数. 因为 c 为奇数, 所以 $f(k)$ 为奇数, 那么 $f(k) \neq 0$, 故方程 $f(x) = 0$ 无奇数根.

例 4.28 设 a, b, c 都是奇数, 证明方程

$$ax^2 + bx + c = 0$$

没有有理根.

证明: 假设 $ax^2 + bx + c = 0$ 有解 $\dfrac{r}{s} \in Q, r, s$ 不全为偶数 (否则可以约分简化), 即 r, s 全为奇数或恰有一个奇数.

如果 r, s 全是奇数, 那么由

$$a\left(\frac{r}{s}\right)^2 + b\left(\frac{r}{s}\right) + c = 0$$

得:

$$ar^2 + brs + cs^2 = 0$$

但上式左边各项均为奇数, 且为三项, 所以, 它们的和为奇数. 而右边为偶数 0, 矛盾! 如果 r, s 中恰有一个奇数, 那么 $ar^2 + brs + cs^2 = 0$ 式左边有两项是偶数, 而其余一项是奇数. 于是它们的和为奇数, 不等于偶数 0, 矛盾!

因此, 原方程无有理根.

例 4.29 试证: 当 n 顺次取正整数时, $\left[(1 + \sqrt{2})^n\right]$ 轮流取偶数、奇数.

证: 注意到 $\alpha = 1 + \sqrt{2}, \beta = 1 - \sqrt{2}$ 是方程 $x^2 - 2x - 1 = 0$ 的两个根. 记 $U_n = a^n + \beta^n (n \geq 1)$, 从

$$a^{n+2} - 2a^{n+1} - a^n = 0 \text{ 及 } \beta^{n+2} - 2\beta^{n+1} - \beta^n = 0,$$

得 $U_{n+2} = 2U_{n+1} + U_n (n \geq 1)$. 由归纳法可知, U_n 都是偶数.

又 $-1 < \beta < 0$, 当 n 为奇数时, $-1 < \beta^n < 0$, 即 $-1 < U_n - a^n < 0$, 所以 $U_n < a^n < U_n + 1$.

由此得 $\left[a^n\right] = U_n$ 是偶数.

同理可证, 当 n 为偶数时, $\left[a^n\right] = U_n - 1$ 是奇数.

例 4.30 已知 n 是奇数, $a_1, a_2, a_3, \cdots, a_n$ 是 $1, 2, \cdots, n$ 的一个排列. 证明 $(a_1 - 1)(a_2 - 2) \cdots (a_n - n)$ 是偶数.

证明 1: 因为

$$(a_1 - 1) + (a_2 - 2) + \cdots + (a_n - n) = (a_1 + a_2 + \cdots + a_n) - (1 + 2 + \cdots + n) = 0$$

是偶数且 n 为奇数, 所以 $a_1 - 1, a_2 - 2, \cdots, a_n - n$ 中至少有一个是偶数 ($a_1 - 1, a_2 - 2, \cdots,$

$a_n - n$ 全是奇数，则这个奇数个数的和是奇数，与它们的和为 0 矛盾）．

因此，这 n 个数的积一定是偶数．

证明 2：因为 n 是奇数，所以 $1,2,\cdots,n$（即 $a_1,a_2,\cdots,a_n$）中奇数比偶数多 1 个．从而在 $(a_1-1),(a_2-2),\cdots,(a_n-n)$ 这 n 个数对中，至少有一个数对的两个数都是奇数．它们的差是偶数．故

$$(a_1-1)(a_2-2)\cdots(a_n-n)$$

是偶数．

例 4.31 求最小的正整数 n，使得 n^3+2n^2 是一个奇数的平方．

解：由 $n^3+2n^2=n^2(n+2)$ 为奇数，知 n 为奇数．

要使 $n^2(n+2)$ 为完全平方数，则 $n+2$ 为完全平方数．

所以，最小的正整数 $n=7$．

例 4.32 求证：任意两个奇数的平方和不是完全平方数．

证明：设两个奇数分别为

$a=2n+1(n\in Z)$,

$b=2m+1(m\in Z)$.

因为 a^2,b^2 均为奇数，所以 $k=a^2+b^2$ 为偶数．

假设 k 为完全平方数，则只能是一个正偶数的平方．

设 $k=(2q)^2(q$ 为 $N^*)$，则 $k=4q^2$，故 $4\mid k$．

另一方面，由 $k=4(n^2+m^2+n+m)+2$，知 $4\nmid k$，矛盾．

所以任意两个奇数的平方和不是完全平方数．

例 4.33 试证：当 $n\in N_+$ 时，$\dfrac{1}{2}+\dfrac{1}{3}+\dfrac{1}{4}+\dfrac{1}{5}+\cdots+\dfrac{1}{n}$ 无论 n 取何值，其结果都不是整数．

证明：因为对于任一 $n\in N_+$，一定存在非负整数 α，使 $2^\alpha\leqslant n\leqslant 2^{\alpha+1}$．

将和式 $\dfrac{1}{2}+\dfrac{1}{3}+\dfrac{1}{4}+\dfrac{1}{5}+\cdots+\dfrac{1}{n}$ 通分，通分后的公分母为 $2^\alpha k$，k 是单数，这时上述和式的 n 个分数，除分数 $\dfrac{1}{2^\alpha}$ 通分后其分子变成奇数 k 之外，其余各分数的分子都至少乘了一个 2，故均为偶数，故通分后和式的分子为奇数．

因为通分后和式的公分母 $2^\alpha k$ 是偶数，分子是奇数，而任一奇数均不能被偶数整除，所以其结果不是整数．

所以结论成立．

例 4.34 能否将两个 1，两个 2，$\cdots$，两个 1990 排成一列，使得两个 $i(1\leqslant i\leqslant 1990)$ 之间恰好有 i 个数？

解：假设能满足题中要求，这些数可以从左至右编上号码

$$1 + 2 + \cdots + 2 \times 1990 = \frac{1 + 2 \times 1990}{2} \times 2 \times 1990 = 1990 \times (1 + 2 \times 1990)$$

为偶数.

但是另一方面，每两个数 i 中间恰有 i 个数.

所以，在 i 为奇数时，这两个 i 的号码有相同的奇偶性，号码的和为偶数；在 i 为偶数时，两个 i 的号码的和为奇数.

又由于 1 至 1990 中有 995（奇数）个偶数，所以 2 个 1，2 个 2，……，2 个 1990 中共有 995 对 i 的号码的和为奇数.

于是号码的总和为奇数. 两方面的结论矛盾.

因此，不可能将 2 个 1，2 个 2，…2 个 1990 排成一列满足所述要求. 当且仅当 m 除以 4 余 0（即被 4 整除）或余 3 时，有满足所述要求的排法.

例 4.35　平面上有 15 个点，任意三点都不在一条直线上. 能不能从每个点都引三条线段，且仅引三条线段和其余的某三点相连？

证明：如果某点 A 与某点 B 连线段，那么同时点 B 也与点 A 连这同一线段. 平面上共有 15 个点（任意三点不共线），总共所连线段数是

$$\frac{15 \times 3}{2} = \frac{奇数}{2}.$$

这是不可能的. 因此不可能从每个点都引且仅引三条线段与其余三条相连.

说明：此题可以转化为一类具有相同解题思路的问题. 例如，求证：17 个同学聚会，不可能每人和且仅和其余三人握手.

例 4.36　在黑板上写出 3 个自然数，然后擦去一个换成其他两个数的和减 1，这样继续做下去，最后得到 17，1967，1983. 问原来 3 个数能否为 2，2，2？

基本思路：考虑各个数的奇偶性.

解：假设原来 3 个数是偶数，那么操作一次得到 2 个偶数 1 个奇数.

接下去的一次操作：如果擦去一个偶数，那么得到的新数仍然是偶数；如果擦去一个奇数，那么得到的新数仍然是奇数. 于是，这一次操作得到的仍是 2 个偶数 1 个奇数. 所以原来的 3 个数不可能全是偶数 2，2，2.

例 4.37　甲先乙后轮流在黑板上任意擦去 2，3，4，……，n 这 $n - 1$ 个数中的一个. 游戏规定最后剩下二个数互质，则甲胜；反之，则乙胜. 试问当 (1) $n = 1988$；(2) $n = 1989$ 时，谁必胜？获胜的诀窍是什么？

解：共有 $n - 1$ 个数，当 n 是偶数时，如 $n = 1988$ 时黑板上共有奇数个数（1987 个），甲擦去一个任意一个数，如擦去 1988. 心里把剩余各数按如下分组：

$$(2, 3), (4, 5), (6, 7), \cdots, (1986, 1987).$$

这样，无论乙擦哪一个数，甲都擦那个数所在组的另一个数，最终所剩的二个数必是相邻自然数，它们是互质的，因此甲必胜.

(2) 当 $n = 1989$ 时，黑板上共有偶数个数，甲擦去一个，还剩下奇数（1988）个数.

轮流擦下去，最后该乙擦时还有三个数.

甲为了获胜要始终擦偶数，最后为乙留下一偶二奇，这时，乙在前面擦数时只要注意留下不互质数的二个奇数如 3，9 不擦，最后留下一个偶数及 3，9，乙擦去偶数，获胜.

如果甲某一步擦了奇数，那么乙 3，9 都不用留，随便擦奇数，最后剩下的三个数中必然至少有两个偶数. 乙为自己留下两个偶数不擦，则乙获胜.

总之，当 $n = 1989$ 时，采取上述方法，乙必胜.

说明：本题中 $n = 1988$，$n = 1989$ 两个数是可以任意改动的，其实只要给出 n 是奇数，或 n 是偶数即可. 证明中用到的对偶分组也是一个重要的方法.

例 4.38 圆周上有 1989 个点，给每一个点染两次颜色：或红、蓝，或全红，或全蓝. 最后统计知：染红色 1989 次，染蓝色 1989 次. 试证：至少有一点被染上红、蓝两种颜色.

证明：假设没有一点被染上红、蓝两种颜色，即第一次染红（蓝），第二次仍染红（蓝）. 不妨令第一次有 m 个点（$0 \leqslant m \leqslant 1989$）染红，第二次仍有且仅有这 m 个点染红，即有 $2m$ 个红点，但是

$$2m \neq 1989$$

所以，至少能有一点被染上红、蓝两种颜色.

例 4.39 线段 AB 的两个端点，一个标以红色，一个标以蓝色. 在线段中间插入 n 个分点，每个分点随意标上红色或蓝色，这样分得 $n+1$ 个不重叠的小线段，如果把两端点颜色不同的线段叫作标准线段，试证：标准线段的个数是奇数.

证明：当线段 AB 中插入一个点时，无论这点是红色还是蓝色，标准线段的条数仍是 1，即标准线段增加零条，以后插入的点共分两种情况：

如插入标准线段中，则标准线段的条数不增加；

如插入非标准线段中，则插入点与端点同色时，不增加标准线段条数，插入点与端点不同色时，增加两条标准线段.

这样，每新插入一个点时标准线段的条数或增加零条，或增加两条.

因此，如果假设使标准线段增加 2 条的点有 m 个，那么总的标准线段的条数

$$1 + 2m + (n - m) \cdot 0 = 2m + 1$$

是奇数.

例 4.40 设有 n 盏亮着的拉线开关灯，规定每次必须拉动 $n-1$ 个拉线开关. 试问：能否把所有的灯都关闭？证明你的结论或给出一种关灯的办法.

分析：先从简单情况想起：当 $n=1$ 时，显然不行；当 $n=2$ 时，1 号灯不动、2 号关上，或 2 号灯不动、1 号关上，可行. 当 $n=3$ 时，每盏灯拉动奇数次才能关上，3 个奇数的和仍是奇数，而 $n-1=2$，按规定总拉开关次数是偶数，因此不能把灯全部关闭. 由此猜想当 n 为偶数时可以，当 n 为奇数时不行.

证明：（1）当 n 为奇数时，每盏灯需拉动开关奇数次才能关闭，因此要把全部灯关闭，总拉动开关次数应是奇数个奇数的和，是奇数，但 $n-1$ 是偶数，按规定只能拉动任

意的偶数次开关，故无论如何都不能把全部亮着的灯光上；

（2）当 n 为偶数时，把 n 盏灯编序为 $1,2,3,\cdots,n$，仅需如下操作：

第一次，1 号灯不动，拉动其余开关；

第二次，2 号灯不动，拉动其余开关；

······

第 n 次，n 号灯不动，拉动其余开关．

这样每盏灯拉动 $n-1$ 次，即奇数次，因此可以用这种办法把全部的灯关闭．

说明： 类似地，"把倒扣的酒杯翻过来"等问题都可以采用本题的解题思路去解决．

例 4.41　如果两人互相握手，则每人都记握手一次．求证：握手是奇数次的人的总数一定是偶数．

分析： 初看，似乎这问题还缺一个条件，因为并没有给出有多少人在握手．由于问题只涉及数的奇偶性质，因此握手的人数是无关紧要的．

证明： 无论多少人握手，握手次数总和一定是偶数．

由于握手是偶数次的人的握手次数总和一定是偶数，因此握手是奇数次的人的握手总和也一定是偶数．

把握手次数是奇数的人分为两类：

一类是偶数个人握相同奇数次，如 4 人握 3 次等，其总和是偶数，记为偶$_总$；

一类是奇数个人握相同奇数次，如 7 人握 5 次等，其总合计为奇$_总$．

由于握手奇数次的人的握手次数总和是偶数，且

$$总和 = 偶_总 + 奇_总，$$

即有

$$偶数 = 偶_总 + 奇_总$$

因此，奇$_总$应该是偶数．只有偶数个奇数的和才等于偶数，所以奇$_总$应有偶数项．这就是说握相同奇数次的奇数个人应有偶数组，因此它的人数总和是偶数．

由于两类的人数都是偶数，因此握手次数是奇数次的人总和一定是偶数．

例 4.42　7 个茶杯，杯口全朝上，每次同时翻转 4 个茶杯称为一次运动．可否经若干次运动，使杯口朝下？

解： 容易知道，一个茶杯由口朝上翻转为口朝下，需经奇数次翻转．

设经 k 次运动可使杯口全朝下．此时，每个茶杯翻转的次数依次为：

$$a_1,a_2,a_3,a_4,a_5,a_6,a_7.$$

因为杯口全朝下，所以 $a_1,a_2,a_3,a_4,a_5,a_6,a_7$ 均为奇数，

于是 7 个茶杯翻转的总次数

$$a_1 + a_2 + a_3 + a_4 + a_5 + a + a_7 = s$$

必为奇数．

另一方面，由每次同时翻转 4 个茶杯为一次运动，得 $s = 4k$，这与 s 为奇数矛盾．

所以，不可能经过若干次运动使杯口全朝下．

4.4.2　典型练习题

练习4.1　设 4 个自然数之和为 1989，求：它们的立方和不是偶数．

练习4.2　求证：17 个同学聚会，不可能每人恰好握手 3 次．

练习4.3　圆周上有 1999 个点，给每一个点染两次颜色，每次染红色和蓝色，共染红色 1999 次，蓝色 1999 次，证明：至少有一个点两次染的颜色不同．

练习4.4　若 $m+n+23$ 是偶数，试判定 $(m-1)(n-1)+2003$ 是奇数还是偶数．

练习4.5　设为 n 奇数，试证：$16\mid(n^4+4n^2+11)$．

练习4.6　设 n 是奇数，试证：$2^{2n}(2^{2n+1}-1)$ 的最后两位数为 28.

练习4.7　证明：当 n 是偶数时，$2\mid 3^n+1$；当 n 是奇数时，$2^2\mid 3^n+1$；但无论 n 是偶数还是奇数，对任意整数 $\alpha>2$，都有 $2^\alpha\mid 3^n+1$

练习4.8　试证：对于任何正整数 $n\geqslant 3$，总存在奇数 x,y，使得 $2^n=7x^2+y^2$．

练习4.9　若 19^2+2^{19} 与 $1999\times p$ 之和是个偶数，则 $2\mid p-1999$．

练习4.10　若 n 是整数，求证 $\dfrac{n^{19}-n}{2}$ 是整数．

练习4.11　已知 p 为偶数，q 为奇数．方程组
$$\begin{cases} x-18y=p, \\ 99+3y=q \end{cases}$$
的解是整数，那么（　　）

A. x 是奇数，y 是偶数．

B. x 是偶数，y 是奇数．

C. x 是偶数，y 是偶数．

D. x 是奇数，y 是奇数．

练习4.12　已知一个偶数 β，使得整系数二次三项式 ax^2+bx+c 的值 $a\beta^2+b\beta+c$ 是奇数．求证：方程
$$ax^2+bx+c=0$$
没有偶数根．

练习4.13　由于是 β 偶数，且 $a\beta^2+b\beta+c$ 是奇数，所以 c 是奇数．由此，任一偶数 β' 都不能使得 $a\beta'^2+b\beta'+c=0$ 成立．

练习4.14　试证：1987 拆分为四个自然数之和，必有"一奇三偶"或"三偶一奇"．

练习4.15　把 $1,2,3,\cdots,1987$ 任意颠倒顺序后构成一个新的排列 $a_1,a_2\cdots,a_n$．求证：
$$(a_1-1)(a_2-2)\cdots(a_{1987}-1987)$$
永远是偶数．

练习4.16　试证：找不到两个自然数，使其差与和的乘积等于 1990．

练习4.17　有一个自身相交的闭折线，已知：它每一节相交一次．求证：这个闭折线的节数是偶数．

练习4.18　证明：不论 a,b,n 为何数，a^n-b^n 能被 $a-b$ 整除．

设 n 为偶数，则 a^n-b^n 能被 $a+b$ 整除．

设 n 为奇数, 则 $a^n + b^n$ 能被 $a + b$ 整除.

练习 4.19　（第 32 届美国数学奥林匹克）证明：对每个正整数 n, 存在一个可以被 5^n 整除的 n 位正整数, 它的每一位上的数字都是奇数.

练习 4.20　设 k 是正奇数, 证明：$(1 + 2 + \cdots + n) \mid (1^k + 2^k + \cdots + n^k)$.

第 5 章　素数与合数

　　简单的事情考虑得很复杂，可以发现新领域；把复杂的现象看得很简单，可以发现新定律.

　　　　　　　　　　　　　　　　　　　　——牛　顿

自然数集中，最简单的是 1. 一方面，它是第一个自然数. 另一方面，在加法运算的意义下，每一个自然数都能够表示为若干个 1 相加的形式，也就是说任何一个自然数都能被 1 用加法表示. 那么在乘法的意义下呢？

在乘法的意义下，素数是正整数集中组成自然数的基本"元素""原子". 除 1 外的每个正整数都可由若干个素数相乘得到. 也正因如此，人们对素数研究越多对素数越感兴趣，由素数引发的许多问题一直是人类智力挑战的极限. 比如素数的个数是有限还是无限？素数如何进行计数？乃至后面会单独进行介绍的诸如孪生素数猜想、哥德巴赫猜想等等.

5.1　素数合数

将整数按照能不能被其他数整除，也就是有没有除了 1 和这个数本身以外的因数进行分类可以分成两类. 如果除了自身以外没有别的因数，那么这个数就是素数. 否则的话这个数是可以由其他因数复合而成的，称为合数.

5.1.1　素数合数

整数 1 只能被它自身整除，换句话说，1 只有自身一个正因数.

整数 2 只能被 1 和 2 整除，换句话说，2 只有 1 和 2 两个正因数. 同样的，整数 3 也只能被 1 和 3 整除，换句话说，3 只有 1 和 3 两个正因数. 像这样的数还有很多，于是我们将其定义为素数：

素数指的是像：

2，3，5，7，11，13，17，29，23，29，31，37，41，43，47，53，59，61…

一样大于 1 且除了 1 和自身外，没有其他的正因数的整数，素数也叫质数或不可约数（准确说是整数集中的不可约数）.

由素数定义可以看出，整数 2 是最小的素数，事实上，2 也是素数中唯一的偶数，因为除了 2 以外的正偶数必然能被 2 整除.

整数 4 除了能被 1，4 整除外，还能被 2 整除，换句话说，4 除了 1 和 4 之外，还有 2 这个正因数. 同样的，整数 6 除了能被 1 和 6 整除之外，还能被 2 和 3 整除. 像这样的数我们将其定义为合数：

如果一个整数除了 1 和自身外，还有其他的正因数，则称该数为合数，也叫复合数. 一个大于 1 的整数，如果不是素数，那么就是合数.

每个合数必具有一个不等于 1 的素因数. 如果一个合数的某个因数是素数，那么称其为素因数，也叫质因数或质约数等. 如整数 6，除了 1 和 6 之外，还有 2 和 3 两个正因数，所以 6 是合数，而 2 和 3 都是 6 的素因数. 事实上，如果存在满足这个条件的正因数，那么他们总是一对一对出现的.

因为我们讨论是素数还是合数的时候是从 2 开始的，故不对 1 进行界定，也即 1 既不是素数，也不是合数. 所以如果要按照能不能被其他正整数整除来将正整数进行分类的话，一共可以分为三类：

第一类：只有一个整数 1.

第二类：全体素数.

第三类：全体合数.

也即：

$$\text{整数}\begin{cases}\text{正整数}\\0\\\text{负整数}\end{cases}\qquad\text{正整数}\begin{cases}1\\\text{素数}\\\text{合数}\end{cases}$$

5.1.2　素数合数的方法论意义

研究一个问题，自然希望能从最简单的开始着手，并将复杂的问题与简单的问题联系后进行考虑，这就是研究素数的出发点.

注意到如 $10 = 2 \times 5$，$12 = 3 \times 4 = 2 \times 2 \times 3$ 等是可以被表示成几个相对更小的数的乘积的形式的人，自然地会追问，整数能不能被拆分或合成？或者说，整数能不能按照可否被拆分或合成进行分类呢？

在数学研究初期，人们就发现了这么一个事实：有些数可以像这样分解，有些则不能. 更进一步，可以分解的这些数往往可以一步一步地分解下去，直到分解为不能再次分解. 这时候，人们觉得有必要将这样的分类记录下来，于是将整数按能否被分解为更小的数分成了两类，也就是素数与合数.

要对整数进行分类，首要的就是要确定对哪些整数进行这样的界定. 由简单的自然数知识可以知道，对于 1 这个特殊的数难以讨论，因为 1 的所有因数只有 1 本身，其所有可能的表示方式就是 $1 = 1 \times 1$，也就是说不能将 1 分解为更小的数. 按理说，这样就应该分在素数一类了. 可问题随之而来，1 这个"素数"的性质和其他的"素数"完全不相同. 如其他的数进行分解的时候，分解出来的数必然是一对一对的，比如 $5 = 1 \times 5$，$4 = 2 \times 2$，而 1 却有且仅有自己作为分解时候的因数. 于是，人们开始考虑将 1 排除在外.

随着社会的发展和数学研究的深入，出现了负整数，而一切负整数都小于 0 和任意正整数，所以，需要将"分解为更小的数"进行修正.

为了更深入地进行整数分解相关性质的研究，逐渐地将是素数还是合数的讨论范围缩小到大于 1 的整数. 换句话说，整数可以分为负整数、零、1、素数、合数. 要讨论一个整数是素数还是合数，需要满足的基本条件就是这个数要大于 1.

对于 1 和 0 的讨论相对简单，这里不再多说. 对于负整数的讨论，根据负数乘法的相关性质我们知道，只需要用 -1 去乘这个负整数就会得到其相反数，这个相反数必然是正整数，再对其相反数进行素数合数的相关研究即可.

例 5.1　判断 2003 是不是素数.

解：因 $\sqrt{2003} < \sqrt{2025} = 45$，而小于 45 的素数有

$$2，3，5，7，11，13，17，19，23，29，31，37，41，43$$

一一验证，可知 2003 不能被上述素数中的任一个整除，因此 2003 是素数.

5.2 厄拉多塞筛法

为了把素数从自然数中筛选出来, 古希腊的数学家厄拉多塞设计了一种筛法, 后人为了纪念他, 将这一方法命名为厄拉多塞筛法.

5.2.1 找出素数

找素数一直是数论爱好者的永恒主题.

问题是要判断一个较小的数能否分解甚至分解后的结果都是相对容易的, 只需要一一验证比这个给定的数小的整数是不是它的因素就可以了. 比如对于 5 来说, 只需要分别验证 2, 3, 4 都不是它的因数, 所以它是素数. 对于 6 来说, 第一个大于 1 的数 2 就是它的因数, 所以它是合数.

也就是说, 对于大于 1 的整数 n, 我们需要从 2 开始一个一个地进行验证. 在此涉及一个时间效率的问题, 如果这个数像 6 一样, 那一次验证即可, 但这只是一个巧合. 有许多数比如 101 却需要一直如此验证 100 次.

所以理性地说, 我们至多做 $n-1$ 次这样的验证可以得出整数 n 是素数还是合数的结论. 这种方法的最大好处是运算简单, 可以用计算机将其程序化, 交给计算机来进行计算, 只不过这样做的效率实在太低下, 我们需要找到更科学合理的方法来优化它.

我们往往是从一些特殊的例子着手考虑数学中的一些较为抽象或深奥的道理的. 在这里我们首推的也是最简单的方法就是试除法.

试除法就是用一个一个特殊的数去除给定的数, 看是否能被其整除.

比如第一个特殊的数 2, 我们知道, 能被 2 整除的数具有一个共同的 "外貌", 那就是其末尾数是 0, 2, 4, 6, 8 这五个能被 2 整除的偶数.

于是要判断一个整数是不是素数, 可以先看看其末尾数, 因为只有除了 2 本身是素数外, 其他的所有偶数都是合数.

以此类推, 考虑 5, 能被 5 整除的数的共同的 "外貌" 是末尾数为 0 或者 5. 于是凡是末尾数出现 0 和 5 的数都可以判断为合数, 当然, 这里要除去 5 这个素数本身. 用上面这种方法可以非常直观地判断出一些数是素数还是合数, 可并非所有需要判断的数都具有那样直观显著的 "外貌", 那就需要我们再进一步考虑其他一些像 2、5 这样可以作为其因数的简单数的倍数所具有的共同特征了.

比如对于 3, 其倍数的共同特征是各个位数上的数值之和能被 3 整除. 比如 4, 其倍数的共同特征是末尾两位数能被 4 整除. 比如 6, 其特征就是既能被 2 整除又能被 3 整除, 也就是同时具有 2 的倍数和 3 的倍数的特征……

于是, 要判断一个数是不是素数, 我们只需要用小于这个数的素数去试除即可, 因为比这个数小的合数也是可以由更小的素数相乘得到的.

用试除法固然是很好的, 对于有些特殊的数可以简单地判断出其归属. 但是这样的试除法在没有其他保证的情况之下, 依然可能需要做 $n-1$ 次才可以判断给定数 n 的归属. 于是有了下面的思考:

由于一个整数分解为更小的数之后，如 $a = bc$，其中 $b < a, c < a$，于是，我们想到是不是不需要真正去验算 $a - 1$ 次．比如 12，按理说，我们需要分别验证 2，3，4，5，6，7，8，9，10，11 这样的十个数是不是它的因数，而我们注意到，我们进行验证甚至找出它的所有因数的时候并不需要将其——验证，而只需要验证一半，也就是验证 2，3，4，5，6 即可．

这是不是一个特例呢？请感兴趣的读者再找几个整数进行验证试试看．

于是得到一个结论，要判断一个整数是素数还是合数，甚至要找出一个数的所有的正因数 a，只需要进行 $\dfrac{a}{2}$ 次，当然，当 a 是奇数的时候，就需要验证 $\dfrac{a+1}{2}$ 次．

有句俗语说：懒人创造了世界．在数学界，懒人的作用可不小呢！我们觉得要验证一半的次数还是太多，比如对于 101，我们需要判断 51 次，太多了，能不能再进一步优化呢？毕竟真正需要我们判断的数往往会很大．

我们从乘法入手进行思考会得到这样一个事实：将两个数 b, c 的乘积进行开平方 $\sqrt{bc}$，如果 $b = c$，那么恰巧可以进行开平方，这个数就是平方数．

比如 $3 \times 3 = 9$，那么 $\sqrt{9} = 3$，此时 $\sqrt{bc} = c$．

可巧合并没有那么多，更多数情况下 $b \neq c$，不妨假设 $b < c$，那么可以得到一个结论是 $b < \sqrt{bc} < c$．

比如 $2 \times 3 = 6$，那么 $2 < \sqrt{6} < 3$．

结合两种情况可以得到一个结论：

$$b < \sqrt{bc} < c.$$

于是回到除法之后，我们可以得到，只需要试除至多 $\sqrt{a}$ 次就可以．这时候，对于 101，我们至多只需要判断 11 次．据此还可以得出数学定理：

如果 a 是合数，q 是它的除了 1 以外的最小正因数，那么 $q \leqslant \sqrt{a}$．

这时候如果还需要继续优化，那么我们就会想到，要判断一个正整数 a 是不是素数必须将不大于 $\sqrt{a}$ 的数全部验证完？

我们来考察 a 的非 1 最小正因数：

如果 a 是素数，那么它的非 1 正因数只有它本身，所以它的非 1 最小正因数必然是素数．

如果 a 是合数，那么除了 1 和 a 之外肯定还有其他的正因数．

不妨将其中最小的记作 q，如果 q 是合数，那么 q 一定还有大于 1 且不等于 q 的正因数 q_1，于是我们有 $q_1 \mid q, q \mid a$，于是 $q_1 \mid a$，也就是说 q_1 是 a 的非 1 正因数，明显的 $q_1 < q$，而这和假设矛盾了．也就是说：每个大于 1 的整数，它的非 1 最小正因数一定是素数．

于是我们可以得到一个判断大于 1 的整数 a 是素数的充要条件：对于任意素数 $q \leqslant \sqrt{a}$，q 都不是它的因数，也即 $q \nmid a$．

综上所述，结合用素数试除和用小于 $\sqrt{a}$ 的素数试除，我们可以得到，对于 101，我们只需要判断 2，3，5，7 是不是其因数即可，这样只需要验证 4 次，最大幅度地减少了计算量．

5.2.2　厄拉多塞筛法

早在公元 250 年左右，古希腊数学家厄拉多塞（Eratosthenes）就在 5.1.1 的思考下发明了一种求出所有不大于数 a 的素数的方法，被后人称为厄拉多塞（Eratosthenes）筛法．具体方法以 $a = 100$ 为例介绍如下：

第一步：制作一个 1~100 的数字表格．

第二步：在表格中将 1 用斜线画出（不考虑素数合数，将其筛出），这一步一共画出 1 个数．

第三步：从 2 开始进行试除，因 2 是素数，故对大于 2 的整数进行，凡是能被 2 整除的数都用斜线画出．这一步画出的数分别是 4，6，8，10，12，14，16，18，20，22，24，26，28，30，32，34，36，38，40，42，44，46，48，50，52，54，56，58，60，62，64，66，68，70，72，74，76，78，80，82，84，86，88，90，92，94，96，98，100 一共 49 个偶数．考虑到画出的时机及原因不同，在表格中用了不同方向的斜线加以区分，相信读者一看表格（见表 5.1）便能明了．

第四步：从 3 开始进行试除，因 3 是素数，故对未画出且大于 3 的整数进行，凡是能被 3 整除的数都用斜线画出．这一步画出的数分别是 15，21，27，33，39，45，51，57，63，69，75，81，87，93，99 一共 15 个数．

第五步：大于 3 的下一个数应该是 4，但是因为 4 是 2 的倍数已经从表格中画出了，故从下一个素数 5 开始进行试除，因 5 是素数，故对未划出且大于 5 的整数进行，凡是能被 5 整除的数都用斜线画出．这一步画出的数分别是 25，35，55，65，85，95 一共 6 个数．

第六步：从下一个素数 7 开始进行试除，因 7 是素数，故对未画出且大于 7 的整数进行试除，凡是能被 7 整除的数都用斜线画出．这一步画出的数分别是 49，77，91 一共 3 个数．

第七步：因为 $\sqrt{100} = 10$，而大于 7 的下一个素数是 11，故不需要再次进行试除．这样就得到了一个被各种斜线画过后的数字表格，其中凡是没有画出的数字就是素数．

这个方法虽然简单，但效率低下．对于较小的数，比如不超过 10 位的数，用厄拉多塞筛法还可以在计算机上正常工作．对于较大的数，要得出结果估计一辈子都不可能了．

19 世纪，一位名叫库利克（Kulik）的奥地利天文学家造出 10^8 以内的素数表，断断续续花去了他 20 年的时间，却未得到人们的重视，连他的手稿也被图书馆遗失掉一部分，其中包括 12642600 至 22852800 之间的素数．

目前，最完善的素数表是查基尔（Zagier）做的，他把不超过 5×10^7 的素数全部列出来了．借助于电子计算机，不超过 104395301 的素数也已经有表可查．

表 5.1

1	2	3	4	5	6	7	8	9	10
11	12	13	14	15	16	17	18	19	20
21	22	23	24	25	26	27	28	29	30
31	32	33	34	35	36	37	38	39	40
41	42	43	44	45	46	47	48	49	50
51	52	53	54	55	56	57	58	59	60
61	62	63	64	65	66	67	68	69	70
71	72	73	74	75	76	77	78	79	80
81	82	83	84	85	86	87	88	89	90
91	92	93	94	95	96	97	98	99	100

5.3　素数的分布

为了更好地研究素数的分布，我们首先要找出一些一定范围内的素数作为特殊对象进行讨论，然后得出一些更为一般化的结论．将找出的素数整理出来单独放入一个集合，就是素数集．100 以内的素数集如图 5.1 所示．

2，3，5，7，11，13，17，19，23，29，31，
37，41，43，47，53，59，61，67，71，
73，79，83，89，97

图 5.1

一般地，我们用 $\pi(x)$ 来表示不大于某个实数 x 的素数的个数．比如：
$$\pi(5) = \pi(5.3) = 3.$$

为了让读者能够有机会自己发现一些更加有趣的规律，在本书的附录中，我们特意列出了一千万以内的素数表．

根据素数表，我们发现了素数分布有这样的一些基本情况：
$$\pi(100) = 25.$$
$$\pi(1000) = 168.$$
$$\pi(2000) = 303.$$
$$\pi(3000) = 430.$$
$$\pi(4000) = 550.$$
$$\pi(5000) = 669.$$
$$\pi(10000) = 1229.$$

于是，我们通过简单计算 $\pi(2000) - \pi(1000) = 135$，$\pi(3000) - \pi(2000) = 127$，$\pi(4000) - \pi(3000) = 120$，$\pi(5000) - \pi(4000) = 119$，$\pi(10000) - \pi(5000) = 560$. 这说明：

在 1 到 100 之间有 25 个素数.

在 1 到 1000 之间有 168 个素数.

在 1000 到 2000 之间有 135 个素数.

在 2000 到 3000 之间有 127 个素数.

在 3000 到 4000 之间有 120 个素数.

在 4000 到 5000 之间有 119 个素数.

在 5000 到 10000 之间有 560 个素数.

可以得出的一个结论就是以一定数目为间距，在这样的间距之间的素数会逐渐减少（注意：并不是严格单调减少）. 换句话说，素数会随着整数的变大而越来越稀少，那素数会不会是有限个呢？

5.4　关于素数的一些探索

5.4.1　素数的个数

定理：素数有无穷多个.

分析：按照最初的想法，要知道素数的个数，需要将所有的素数找出来，然后数一数就行了，可问题是整数的个数是无限的，我们可以用厄拉多塞筛法找出一定范围内的素数，但是要将所有的素数找出来，这个却是很费劲甚至是不可能实现的. 这就需要我们找到一种数学证明方法，来讨论这个问题.

在数学中，还真有这么一种证明方法——反证法. 简单地说，就是要证明某命题的真假性，我们就先假设要证明的命题的反面成立，然后通过逻辑推理得出一个与题设矛盾或者与一些已知事实、定理、真理等不相符的结论，通过这种方式来说明假设是错误的，以此来得到原先要证明的结论.

证明：（用反证法证明）

假设素数的个数是有限的，不妨将这有限个素数假设为 $p_1, p_2, \cdots p_s$.

构造一个数

$$p = p_1 p_2 \cdots p_s + 1,$$

由于素数有且仅有 $p_1, p_2, \cdots p_s$ 这 s 个，所以这个数 p 一定是一个合数.

于是由合数的概念可知，这个数可以表示成若干个素数的乘积的形式. 或者换句话说，一定存在至少一个素数 p' 可以使得 $p' \mid p$.

很显然，这个素数 p' 必定不是 $p_1, p_2, \cdots p_s$ 里的任何一个，否则的话 $p' \mid p_1 p_2 \cdots p_s$，而这将会导致 $p' \mid 1$ 这个不可能的结.

故矛盾推出素数 p' 必定不是 $p_1, p_2, \cdots p_s$ 里的任何一个.

也就是说，除了 $p_1, p_2, \cdots p_s$ 这 s 个素数之外至少还应该有一个素数 p'，这与假设是矛盾的，所以说素数的个数是无限的．

说明：（1）这个证明与欧几里得在《几何原本》第 IX 卷命题（20）中的证明大致相同．

（2）这个证明的基本思路是，在假设只有有限个质数的情形下，设法找一个新的与已有 $p_1, p_2, \cdots p_s$ 都不同的质数，但质数不容易找到，转而找一个合数不被 $p_1, p_2, \cdots p_s$ 整除．这样的思路常被用于证明某数的无限性．

（3）我们还可以用费马数证明质数是无限的．虽然费马数不全是质数，但费马数是无限的，因此，如果知道每两个费马数互素，那么每一个费马数的质因数都与其他的质因数不同，从而可知，质数的个数是无限．

（4）我们还可以对小于等于 x 的质数个数进行估计，从而不仅得到定性的结论：质数有无穷多个，而且知道质数在某个界限下的大概个数，这也是数论中常用的方法．

5.4.2　素数的表达式

不单单素数的个数是无限的，对于有些特殊形式的素数往往也有无限多个，所以人们对素数的表达式就开始了无尽的追求．

考察 100 以内的素数集 $\{2，3，5，7，11，13，17，19，23，29，31，37，41，43，47，53，59，61，67，71，73，79，83，89，97\}$，这里面的数除了 2 以外，所有的数要么可以表示成 $4n-1$，要么可以表示成 $4n+1$ 的形式．

事实上，由带余除法可以知道，一切大于 2 的整数都可以按照除以 4 后的余数分成下面的四类：

$$4n, 4n+1, 4n+2, 4n+3 (n \in N).$$

显然，$4n$ 和 $4n+2$ 是偶数，而大于 2 的素数都是奇数，而 $4n+3 = 4(n+1)-1$，也就是说大于 2 的素数不是形如 $4n-1$ 的，就是形如 $4n+1$ 的．

现在考察形如 $4n-1$ 的素数的个数．

如果将 100 以内的 $\{4n-1\}$ 用列举法表示出来的话那就是：$\{3，7，11，15，19，23，27，31，35，39，43，47，51，55，59，63，67，71，75，79，83，87，91，95，99\}$，于是我们可以求得这个集合与 100 以内的素数集的交集如图 5.2 所示．

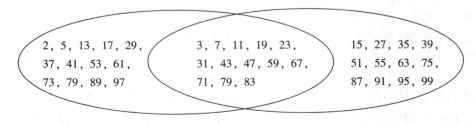

图 5.2

将两个集合的交集 $\{3，7，11，19，23，31，43，47，59，67，71，79，83\}$ 所涉及中的 $4n-1$ 和 n 对应如表 5.2 所示．

表 5.2

$4n-1$	3	7	11	19	23	31	43	47	59	67	71	79	83
n	1	2	3	5	6	8	11	12	15	17	18	20	21

也就是说，有许多素数可以表示成 $4n-1$ 的形式，事实上，还有以下这个加强版的定理：

定理：存在无限多个形如 $4n-1$ 的素数.

证明：（用反证法证明）

假设形如 $4n-1$ 的素数只有有限个，不妨把这有限个素数中最大的一个记作 p，构造一个整数 $x=4n_1-1$，其中 n_1 表示不超过 p 的所有奇素数的乘积. 显然，$x>p$. 又因为 x 的表达式也是 $4n-1$ 的形式，故而 x 应该是一个合数，那么就一定会有某个素数 $p_i\mid x$. 考察素数 p_i 的大小，如果 $p_i\le p$，那么 $p_i\mid n_1$，而这将导致 $p_i\mid 1$，这是不可能的，所以必然有 $p_i>p$，由于 x 是奇数，所以 p_i 只能是形如 $4n-1$ 或者 $4n+1$ 的素数，也就是说存在大于 p 的其他的形如 $4n-1$ 的素数 p_i，这与假设矛盾，故定理得证.

按照形如 $4n-1$ 的推理方法，考虑形如 $5n+1$ 的素数如表 5.3 所示：

表 5.3

$5n+1$	11	31	41	61	71	101	131	151	181	191	211	241	……
n	2	6	8	12	14	20	26	30	36	38	42	48	……

按照形如 $4n-1$ 的推理方法，考虑形如 $5n+2$ 的素数如表 5.4 所示：

表 5.4

$5n+2$	2	7	17	37	47	67	97	107	127	137	157	167	197	……
n	0	1	3	7	9	13	19	21	25	27	31	33	39	……

按照形如 $4n-1$ 的推理方法，考虑形如 $5n+3$ 的素数如表 5.5 所示：

表 5.5

$5n+3$	3	13	23	43	53	73	83	103	113	163	173	193	223	……
n	0	2	4	8	10	14	16	20	22	32	34	38	44	……

按照形如 $4n-1$ 的推理方法，考虑形如 $5n+4$ 的素数如表 5.6 所示：

表 5.6

$5n+4$	19	29	59	79	89	109	139	149	179	199	229	239	……
n	3	5	11	15	17	21	27	29	35	39	45	47	……

事实上，我们可以按照 $4n-1$ 的推理方法，继续对不同的形如 $an+b$ 的素数讨论下去，每一类都有许多素数，于是可以猜测每一类都包含无限多个素数.

当然，并非要把所有的整数都这样一次一次地试验完，对于形如 $an+b$ 的素数的讨论首先就要排除 a,b 有相同因数的情形，比如，一开始我们已经排除了形如 $4n+2$ 的情况，后面再次讨论的时候也要继续排除有如 $6n+2,6n+3,9n+3\cdots$ 的情形.

关于素数的形式，还有一个很有名的定理是德国数学家，解析数论的创始人狄利克雷（Johann Peter Gustav Lejeune Dirichlet）在 1837 年证明的，至今没有较为简单的证明，故在此只介绍定理，不对定理进行证明.

定理： 设正整数 k,l 互素，也即除了 1 之外没有其他的公因数，则形如 $kn+l$ 的素数有无限多个.

这个定理给出了关于某类特定形状的素数及其个数的无限性，是形如 $4n-1$ 等的一个延伸与归纳，对认识素数有极大的帮助. 据此，我们可以随意写出许多素数的基本表达式，并且知道这样的素数的个数是无限多个的.

素数定理： 设 $x \in R$，则

$$\lim_{x \to \infty} \frac{\pi(x)}{x/lnx} = 1$$

素数定理不仅仅是关于素数个数的一个理论结果，它在数学和计算机科学中也有很大的应用价值. 比如说，我们利用素数定理可以完成以下工作：

估计出随机选取的一个整数 x 是素数的概率为 $1/lnx$. 比如：一个不超过 10^{1000} 的数是素数的概率大约是 $1/ln10^{1000} \approx \dfrac{1}{2302}$.

近代数学奠基者之一，历史上最重要的数学家之一，享有"数学王子"之称的德国著名数学家高斯（Johann Carl Friedrich Gauss，1777—1855）在 1792 年就猜测到了素数定理的结论，甚至在此之前，法国数学家勒让德（Adrien Marie Legendre）也猜测出了该结果的另一个表达形式，但是他们都未能给出定理的证明.

直到 1850 年左右，俄罗斯数学家切比雪夫（Пафну́тий Льво́вич Чебышёв）证明了对充分大的 x 有下面的不等式：

$$0.92129 \frac{x}{lnx} < \pi(x) < 1.1056 \frac{x}{lnx}$$

1896 年，法国数学家哈达马（Jacques Solomon Hadamard，1865—1963）和比利时数学家德拉威力伯桑（C. dela Vallée – Poussin）分别独立地利用黎曼 ζ - 函数给出了证明.

1949 年，挪威数学家（A. Selberg）和匈牙利数学家保罗·厄尔多斯（P. Erdos）也分别给出了素数定理的初等证明，不过这些所谓的初等证明也是极其复杂的，我们在此只说明其结果和关键用途：当 x 趋近于正无穷大的时候，$\pi(x)$ 与 x/lnx 的比值趋近于 1. 如果将定理的趋势用表格（表 5.7）体现出来，就会发现这对确定一定范围内素数个数的帮助了.

估计利用试除法进行素性测试时需要的计算步骤. 在测试 x 是不是素数的时候，试除法中最多需要试除 $\pi(\sqrt{x})$ 个素数. 对于充分大的整数 x，我们有：

$$\pi(\sqrt{x}) \approx \frac{\sqrt{x}}{ln\sqrt{x}} = \frac{2\sqrt{x}}{lnx}.$$

如果小于 $\sqrt{x}$ 的素数表是已知的，且假定一台计算机执行一次除法需要 $\frac{lnx}{10^6}$ 秒，那么这台计算机检查 x 是否是素数大约需要：

$$\frac{2\sqrt{x}}{lnx} \times \frac{lnx}{10^6} = \frac{2\sqrt{x}}{10^6} \text{ 秒}.$$

这表明如果要直接使用这种方法验证一个 30 位数的素数将需要 63 年多的时间.

表 5.7

x	$\pi(x)$	x/lnx	$\dfrac{\pi(x)}{x/lnx}$
10^3	168	144.8	1.160
10^4	1229	1085.7	1.132
10^5	9592	8685.9	1.104
10^6	78498	72383.4	1.085
10^7	664579	620420.7	1.071
10^8	5761455	5428681.0	1.061
10^9	50847534	48254942.4	1.054
10^{10}	455052512	434294481.9	1.048
10^{11}	4118054813	3948131663.7	1.043
10^{12}	37607912018	36191206825.3	1.039
10^{13}	346065536839	334072678387.1	1.036
10^{14}	3204941750802	3102103442166.0	1.033

5.4.3 费马数

被誉为业余数学家之王的法国数学家皮埃尔·德·费马（Pierre de Fermat）在研究素数的形态的时候发现形如 $F_n = 2^{2^n} + 1$ 的数具有奇特的性质，并对 $n = 0,1,2,3,4$ 时进行了如下的验证：

$$F_0 = 3$$
$$F_1 = 5$$
$$F_2 = 17$$
$$F_3 = 257$$
$$F_4 = 65537$$

发现 F_n 都是素数. 于是他猜测当 $n \in N$ 时，F_n 都是素数.

后人为了纪念费马对此做出的贡献，称形如 $F_n = 2^{2^n} + 1$ 的数都为费马数. 由于计算量太大，在当时并没能直接算出第六个费马数是不是素数.

直到 1732 年，瑞士数学家欧拉（Leonhard Euler，1707—1783）算出了了第五个费马数：

$$F_5 = 641 \times 6700417$$

才说明了费马数不一定都是素数.

$F_6 = 274177 \times 67280421310721$

$F_7 = 59649589127497217 \times 5704689200685129054721$

$F_8 = 1238926361552897 \times 93461639715357977769163558199606896658405123754163818$
8580280321

$F_9 = 2424833 \times 7455602825647884208337395736200454918783366342657 \times 7416400626$
$27530801524787141901937474059940781097519023905821316144415759504705008092818711$
693940737

以及在大型计算机支持下计算出来的 $F_{10}, F_{11}, F_{12}, F_{13}$ 都是合数，并且现在还没有找到一个 $n \geqslant 5$，使得 F_n 是素数. 当然，人们也没有证出：当 $n \geqslant 5$ 时，F_n 都是合数. 在对费马数的研究中还发现，任意两个费马数都互素.

例 5.2 证明任意两个费马数 $(F_m, F_n) = 1 (m \neq n)$.

证明： 不妨设 $m > n$. 则当 $m > n \geqslant 0$ 时，费马数满足 $F_n \mid (F_m - 2)$.

即存在整数 t，使得 $F_m = 2 + tF_n$. 设 $d = (F_m, F_n)$. 则

$$d = (2 + tF_n, F_n) = (2, F_n) = 2.$$

故 $d = 1$ 或 $d = 2$. 但 F_n 显然是奇数.

故必有 $d = 1$，即费马数是两两互素的. 证毕.

5.4.4 梅森数

考察形如 $a^n - 1 (n \geqslant 2)$ 的素数. 比如 $31 = 2^5 - 1$ 这样的数，为了更好地看出规律，我们制作了表 5.8：

表 5.8

$2^2 - 1 = 3$	$2^3 - 1 = 7$	$2^4 - 1 = 3 \times 5$	$2^5 - 1 = 31$
$3^2 - 1 = 2^3$	$3^3 - 1 = 2 \times 13$	$3^4 - 1 = 2^4 \times 5$	$3^5 - 1 = 2 \times 11^2$
$4^2 - 1 = 3 \times 5$	$4^3 - 1 = 3^2 \times 7$	$4^4 - 1 = 3 \times 5 \times 17$	$4^5 - 1 = 3 \times 11 \times 31$
$5^2 - 1 = 2^3 \times 3$	$5^3 - 1 = 2^2 \times 31$	$5^4 - 1 = 2^4 \times 3 \times 13$	$5^5 - 1 = 2^2 \times 11 \times 71$
$6^2 - 1 = 5 \times 7$	$6^3 - 1 = 5 \times 43$	$6^4 - 1 = 5 \times 7 \times 37$	$6^5 - 1 = 5^2 \times 311$
$7^2 - 1 = 2^4 \times 3$	$7^3 - 1 = 2 \times 3^2 \times 19$	$7^4 - 1 = 2^5 \times 3 \times 5^2$	$7^5 - 1 = 2 \times 3 \times 2801$
$8^2 - 1 = 3^2 \times 7$	$8^3 - 1 = 7 \times 73$	$8^4 - 1 = 3^2 \times 5 \times 7 \times 13$	$8^5 - 1 = 7 \times 31 \times 151$

由表很容易看出，a 是奇数时，$a^n - 1$ 是偶数，从而不可能是素数. 同时，$a^n - 1$ 总是被 $a - 1$ 整除. 当然，读者可以利用自己学过的几何级数求和公式

$$x^n - 1 = (x - 1)(x^{n-1} + x^{n-2} + \cdots + x^2 + x + 1)$$

进行验证，使用 $x = a$ 的几何级数公式可得. 所以 $a^n - 1$ 是合数，除非 $a - 1 = 1$，即 $a = 2$. 当 $a = 2$ 时，$2^n - 1$ 也常常是合数.

考察下面表 5.9 中的这组数据：

表 5.9

n	2	3	4	5	6	7	8	9	10
$2^n - 1$	3	7	3×5	31	$3^2 \times 7$	127	$3 \times 5 \times 17$	7×73	$3 \times 11 \times 31$

从表中可以看出：

当 n 是偶数时，$2^n - 1$ 被 $3 = 2^2 - 1$ 整除.

当 n 被 3 整除时，$2^n - 1$ 被 $7 = 2^3 - 1$ 整除.

当 n 被 5 整除时，$2^n - 1$ 被 $31 = 2^5 - 1$ 整除.

这就有理由去猜测：如果 n 被 m 整除时，$2^n - 1$ 被 $2^m - 1$ 整除. 事实上，在整除那一章我们已经证明过这个结论了. 于是，我们可以得到这样一个结论，如果 n 是合数，则 $2^n - 1$ 也是合数.

定理：如果对整数 $a \geq 2$ 与 $n \geq 2$，$a^n - 1$ 是素数，则 a 必等于 2 且 n 一定是素数.

这个定理解决了形如 $a^n - 1 (n \geq 2)$ 的素数的情形，只需要考虑 $a = 2$ 且 n 是素数的情形. 早在 1644 年，神父梅森（Marin Mersenne，1588—1648）断言，当 $p = 2, 3, 5, 7, 13, 17, 19, 31, 67, 127, 257$ 时，形如 $M_p = 2^p - 1$ 的数必然是素数，并且这些是使得 $2^p - 1$ 为素数的仅有的小于 258 的素数. 值得一提的是，没人知道梅森当初是如何发现这些"事实"的，尤其是从最后看出他的列表并不正确，实际上梅森也犯了 5 个错误，他丢掉了 3 个数：61，89，107；又搞错了 2 个数：67，257. 当然，这错误反而更好地告诉我们一个事实：错误并不可怕，敢于猜想才能发现真理！

后来人们经过验证，认可了形如 $M_p = 2^p - 1$ 的素数并将其称为梅森素数. 前几个梅森素数是：

$$2^2 - 1 = 3, 2^3 - 1 = 7, 2^5 - 1 = 31, 2^7 - 1 = 127, 2^{13} - 1 = 8191.$$

当然，并不是 p 为素数时，数 M_p 都是素数，比如：

$$2^{11} - 1 = 2047 = 23 \times 89, 2^{29} - 1 = 536870911 = 233 \times 1103 \times 2089.$$

还有当 $p = 2, 3, 5, 7, 13, 17, 19, 31, 61, 89, 107, 127, 521, 607, 1279, 2203, 2281, 3217, 4253, 4423, 9689, 9941$ 时，$2^p - 1$ 都是梅森素数，目前人们只找到了 50 个梅森素数.

计算机技术的发展使得计算速度成几何级数增长，也使得验证数百位的素数情况成为可能. 在 1876 年，卢卡斯（E. Lucas）才证明了 $2^{127} - 1$ 是素数，并保持这一记录直到 20 世纪 50 年代才被突破，再往后发现梅森素数的速度也就越来越快了.

表 5.10 列举一些使用计算机发现的梅森素数及其发现者的姓名：

表 5.10

发现者	p 值	时间（年）
Robinson	521，607，1279，2203，2281	1952
Riesel	3217	1957
Hurwitz	4253，4423	1961
Gillies	9689，9941，11213	1963
Tuckerman	19937	1971
Noll，Nichel	21707	1978
Noll	23209	1979
Noll，Slowinski	44497	1979
Slowinski	86243	1982
Colquitt，Welsch	110503	1988
Slowinski	132049	1983
Slowinski	216091	1985
Slowinski，Gage	756839	1992
Slowinski，Gage	859433	1994
Slowinski，Gage	1257787	1996
Armengaud	1398269	1996
Spence	2976221	1997
Clarkson	3021377	1998
Hajratwala	6972593	1999
Cameron	13466917	2001
Shafer	20996011	2003
Findley	24036583	2004
M. Nowak	25964951	2005
C. Cooper	30402457	2005
S. Boone	32582657	2006

　　后面的大多数梅森素数几乎都是依赖于大型计算机的帮助验证得出的，比如 Findley 在 2004 年就是依靠 Woltman 的大型因特网梅森素数搜索（GIMPS）的软件发现的. 在这样越来越快的计算速度支撑之下，发现更大的梅森素数成为可能和必然，只要读者愿意，也可以寻找到已知的最大梅森素数.

　　当然，还有一个更有意思的工作留给读者，那就是著名的梅森素数猜想——存在无限多个素数 p，使得 M_p 是素数，至今为止，还无人给出完整证明.

5.4.5　孪生素数猜想

在素数表中容易看出这样一个事实：有许多相邻的两个素数的差是 2. 比如下面这些成双成对如孪生儿一样的素数：

$$3,5;\qquad 5,7;\qquad 11,13;\qquad 17,19;\qquad 29,31;$$
$$41,43;\qquad 59,61;\qquad 71,73;\qquad 101,103.$$

这样的素数被称作孪生素数.

1849 年，阿尔方·德·波利尼亚克（Alphonse de Polignac）提出了一般的猜想：对所有自然数 p，存在无穷多个素数对（$p,p+2$），$p=1$ 的情况就是孪生素数猜想.

素数有无限多个，那么孪生素数的对数是不是也是无限多个呢？希尔伯特在 1900 年国际数学家大会的报告上第 8 个问题中提出孪生素数猜想：存在无穷多个素数 p，使得 $p+2$ 是素数.

素数定理说明了素数在趋于无穷大时变得稀少的趋势. 而孪生素数，与素数一样，也有相同的趋势，并且这种趋势比素数更为明显.

1921 年，英国数学家戈弗雷·哈代和约翰·李特尔伍德提出一个与波利尼亚克猜想类似的猜想，通常称为"哈代 – 李特尔伍德猜想"或"强孪生素数猜想"（即孪生素数猜想的强化版）. 这一猜想不仅提出孪生素数有无穷多对，而且还给出其渐近分布形式.

2013 年 5 月，我国数学家张益唐在孪生素数研究方面所取得的突破性进展，他证明了孪生素数猜想的一个弱化形式. 在最新研究中，张益唐在不依赖未经证明推论的前提下，发现存在无穷多个之差小于 7000 万的素数对，从而在孪生素数猜想这个重要问题的道路上前进了一大步.

张益唐的论文在 5 月 14 号在网络上公开，5 月 21 日正式发表[①].

5 月 28 号，这个常数下降到了 6000 万.

5 月 31 号，下降到了 4200 万.

6 月 2 号，则是 1300 万.

6 月 3 号，500 万.

6 月 5 号，40 万.

在英国数学家 Tim Gowers 等人发起的"Polymath"计划中，孪生素数问题成为一个在全球数学工作者中利用网络进行合作的一个典型. 人们不断地改进张益唐的证明，进一步拉近了与最终解决孪生素数猜想的距离. 在 2014 年 2 月，张益唐的七千万已经被缩小到 246.

例 5.3　设 $p>3$ 为素数，n 为正整数，试证：对任意一对孪生素数 p 与 $p+2$，或者 $p+(p+2)=36n$，或者存在一对孪生素数 p 与 $p+2$，使得

$$[p+(p+2)]+[q+(q+2)]=36n.$$

证：因 p 与 $p+2$ 都大于 3 的素数，故必存在正整数 m，使得

---

① Zhang，Yitang. Bounded gaps between primes：Annals of Mathematics（Princeton University and the Institute for Advanced Study）.，May 21，2013.

$$p = 6m - 1, p + 2 = 6m + 1.$$

于是

$$p + (p + 2) = 12m.$$

若 $3 \mid m$，可令 $m = 3n$，这里 n 为正整数，此时 $p + (p + 2) = 36n$；

若 $3 \nmid m$，可令 $m = 3n_1 + r$，这里 n_1 为正整数，$r = 1$ 或 2.

当 $r = 1$ 时，令 $q = 11$，则 $q + 2 = 13$，此时

$$
\begin{aligned}
\left[p + (p + 2) \right] + \left[q(q + 2) \right] &= 12(3n_1 + 1) + 11 + 13 \\
&= 36(n_1 + 1) \\
&= 36n
\end{aligned}
$$

当 $r = 2$ 时，令 $q = 5$，则 $q + 2 = 7$，此时

$$
\begin{aligned}
\left[p + (p + 2) \right] + \left[q + (q + 2) \right] &= 12(3n_1 + 2) + 5 + 7 \\
&= 36(n_1 + 1) \\
&= 36n
\end{aligned}
$$

综上，命题得证.

5.4.6　哥德巴赫猜想

关于素数的研究还有很多，比如哥德巴赫猜想：每个大于 2 的偶数都能写成两个素数的和. 关于哥德巴赫猜想，我们将单列一章进行介绍，在此不再赘述.

例 5.4　每一个大于 11 的自然数都是两个合数的和.

解：每一个大于 6 的偶数都可以写成 $4 + 2k$ 的形式，其中 $k > 1$.

每一个大于 11 的奇数可以写成 $9 + 2k$ 的形式，其中 $k > 1$.

5.5　典型问题

5.5.1　典型例题

例 5.4　判别 32993 是否为质数.

解：因为 $182 > \sqrt{32993} > 181$

所以用不超过 181 的所有质数去除 32993，即用 2，3，5，7，11，13，17，19，23，29，31，37，41，43，47，53，59，61，67，71，73，79，83，89，97，101，103，107，109，113，127，131，137，139，149，151，157，163，179 去除 32993，看看能否整除.

由于这些数都不能整除 32993，所以 32993 是质数.

例 5.5　若素数 p 除以 30 后的余数 $r \neq 1$，试证 r 一定是素数.

证明：设 $p = 30k + r(r \neq 1)$. 因 p 是素数，故 r 不能是 2，3，5 的倍数，且 $1 < r < 30$.

于是可知若 r 是合数，则 r 必有一个不大于 $\sqrt{r} < \sqrt{30} < 6$ 的素因数，它们就是 2，3，5，

这就导致了矛盾. 故 r 只能是素数, 即

$$r = 7, 11, 13, 17, 19, 23, 29.$$

例 5.6　证明 $F_5 = 2^{32} + 1$ 不是素数.

证明: 因为

$$
\begin{aligned}
F_5 &= 2^{32} + 1 = 2^4 \times (2^7)4 + 1 \\
&= (1 + 2^7 \times 5 - 1)(2^7)4 + 1 \\
&= (1 + 2^7 \times 5)(2^7)4 + 1 - (2^7)4 \\
&= (1 + 2^7 \times 5)\left[(2^7)4 + (1 - 5 \times 2^7)(1 + 25 \times 2^{14}) \right] \\
&= 641 \times 6700417.
\end{aligned}
$$

所以 $F_5 = 2^{32} + 1$ 不是素数.

例 5.7　对于任意给定的自然数 n, 证明: 必有无穷多个自然数 a, 使 $n^4 + a$ 为合数.

基本思路: 配方并应用公式 $a^2 - b^2$.

证明: 取 $a = 4m^2$, 则

$$
\begin{aligned}
n^4 + 4m^4 &= n^4 + 4m^2 n^2 + 4m^4 - 4m^2 n^2 \\
&= (2m^2 + n^2)^2 - 4m^2 n^2 \\
&= (2m^2 + n^2 - 2mn)(2m^2 + n^2 + 2mn).
\end{aligned}
$$

当 $m > 1$ 时

$$2m^2 + n^2 - 2mn = (m - n)^2 + m^2 > 1$$

因此 $2m^2 - 2mn + n^2$ 是 $n^4 + a$ 的真因数, 即 $n^4 + a$ 为合数.

由 m 的任意性可知结论成立.

例 5.8　设 $n > 2$ 且 $a^n - 1$ 为素数, 试证: $a = 2$ 且 n 为素数.

证明: 假设 $a > 2$. 因 $n > 1$ 故 $1 < a - 1 < a^n - 1$, 且

$$a^n - 1 = (a - 1)(a^{n-1} + a^{n-2} + \cdots + a + 1),$$

从而 $a^n - 1$ 有真因数 $a - 1$, 即 $a^n - 1$ 不是素数, 矛盾.

因此 $a = 2$.

假设 n 是合数, 即 $n = kl$, 这里 $1 < k < n$, 那么

$$1 < 2^k - 1 < 2^n - 1$$

且

$$(2^k - 1) \mid (2^n - 1),$$

即 $2^n - 1$ 也不是素数.

因此, 要使 $a^n - 1$ 是素数, 必须 $a = 2$ 且 n 为素数.

例 5.9　试证: $F_5 = 2^{25} + 1$ 是合数.

证明: 设 $a = 2^7, b = 5$ 则

$$a - b^3 = 3,$$

$$1 + ab - b^4 = 1 + (a - b^3)b = 1 + 3b = 2^4,$$

于是

$$\begin{aligned}
F_5 = 2^{2^5} + 1 &= (2a)4 + 1 = 2^4 \cdot a^4 + 1 \\
&= (1 + ab - b^4)a^4 + 1 \\
&= (1 + ab)a^4 - a^4 b^4 \\
&= (1 + ab)(1 - ab + a^2 b^2 - a^3 b^3 + a^4)
\end{aligned}$$

即

$$(1 + ab) \mid F_5$$

故 F_5 是合数.

例 5.10 设 $n \geq 0, F_n = 2^{2^n} + 1$，试证：

(1) 若 $m \neq n, d > 1$，且 $d \mid F_n$，则 $d \nmid F_m$ 由此推出素数有无穷多个;

(2) $F_{n+1} = F_n \cdots F_0 + 2$.

证明： (1) 不妨设 $m > n$，且 $m = n + a (a \geq 1)$，则

$$\begin{aligned}
F_m = 2^{2^m} + 1 = 2^{2^{n+a}} + 1 &= (2^{2^n})2^a + 1 \\
&= (F_n - 1)2^a + 1 \\
&= qF_n + 2,
\end{aligned}$$

这里，是某一正整数.

假设 $d \mid F_m$. 因 $d \mid F_n$，则 $d \mid 2$，而 $d > 1$，故 $d = 2$.

但 F_n 是奇数，故 $d \mid F_n$，矛盾.

因此，若 $m \neq n, d > 1$，且 $d \mid F_n$，则 $d \nmid F_m$.

这说明当 $m \neq n$ 时，F_m 中的素因数与 F_n 中的素因数完全不同，而费马素数有无穷多个，故不同的素因数有无穷多个，由此推出素数有无穷多个.

(2) 本题可以用数学归纳法进行证明. 这里我们提供另一种证法.

在 $x^{2^{n+1}} - 1 = (x - 1)(x + 1)(x^2 + 1)(x^4 + 1)(x^8 + 1)\cdots(x^{2^n} + 1)$ 中，

令 $x = 2$，

得

$$2^{2^{n+1}} - 1 = (2^{2^0} + 1)(2^{2^1} + 1)(2^{2^2} + 1)(2^{2^3} + 1)\cdots(2^{2^n} + 1)$$

两边同时加上 2，即得：

$$F_{n+1} = F_n \cdots F_0 + 2.$$

例 5.11 证明有无穷多个 n，使 $n^2 + n + 41$ 表示合数，为 43 整除.

证明： 因为 43 是素数，所以当 $n = 1, 2, \cdots, 40$ 时

$$n^2 + n + 41$$

表示质数，而在 $n = 41k$ 时显然为合数.

当 $n = 43m + 1$ 时，

$$n^2 + n + 41 = (1 + 43m)^2 + 41 = 43[m(2 + m) + m + 1].$$

显然，此时 $43 \mid n^2 + n + 41$.

例 5.12　求所有的素数 p，使得 $43 \mid 7^p - 6^p - 1$.

分析：求高次幂对模的余数问题，通常应用"指数".

解：记 $f(p) = 7^p - 6^p - 1$，则 $f(2) = 7^2 - 6^2 - 1 = 12$ 不是 43 的倍数；

$f(3) = 7^3 - 6^3 - 1 = 7 \times 49 - 216 - 1 \equiv 7 \times 6 - 1 - 1 = 40 (\mathrm{mod}\,43)$ 不是 43 的倍数.

由 $6^3 = 216 \equiv 1 (\mathrm{mod}\,43)$ 以及 $7^6 \equiv 42^2 \equiv 1 (\mathrm{mod}\,43)$，对素数 $p > 3$ 按模 6 分类，必有 $p = 6l + 1 (l \in N^*)$ 或 $p = 6l + 5 (l \in N)$.

若素数 $p = 6l + 1 (l \in N^*)$，则

$f(p) = 7^{6l+1} - 6^{6l+1} - 1 \equiv 7 - 6 - 1 = 0 (\mathrm{mod}\,43)$，

即 $43 \mid f(p)$.

综上所述，所求素数 p 是一切不小于 5 的素数：5，7，11，13…

例 5.13　设 m 为正整数，试证：当 $\sigma(m)$ 为大于 3 的素数时，m 可表为 p^{2k}，这里 p 为素数，k 为正整数，且 $n - 1$ 不能被 $2k + 1$ 整除.

证明：（ i ）对任何素数 m，$\sigma(m) = 1 + m$ 为合数.

（ ii ）若 $m = p_1 p_2 \cdots p_s$，且 $p_1, p_2, \cdots p_s$ 为不同的素数，则

$$\sigma(m) = \sigma(p_1) \cdot \sigma(p_2) \cdots \sigma(p_s)$$

为合数.

（ iii ）若 $m = p_1^{l_1} p_2^{l_2} \cdots p_s^{l_s}$，且 $p_1, p_2, \cdots p_s$ 为不同的素数，$l_1, l_2, \cdots l_s$ 为非负整数，则 $\sigma(m) = \sigma(p_1^{l_1}) \sigma(p_2^{l_2}) \cdots \sigma(p_s^{l_s})$ 为合数.

（ iv ）对任何素数 p 及正整数 k，$\sigma(p^{2k+1})$ 的值均为合数，因为

$$\sigma(p^{2k+1}) = \frac{p^{2k+2} - 1}{p - 1} = \frac{(p^{k+1} - 1)(p^{k+1} + 1)}{p - 1}$$
$$= (p^k + p^{k-1} + \cdots + p + 1)(p^{k+1} + 1).$$

（ v ）若 p 为素数，k 为正整数，且 $(2k + 1) \mid (p - 1)$，则 $\sigma(p^{2k})$ 的值为合数.

事实上，令 $p = (2k + 1)u + 1$（u 为某正整数），则

$$\sigma(p^{2k}) = \sum_{i=0}^{2k} p^i = \sum_{i=0}^{2k} \left[(2k + 1)u + 1\right]^i$$

为 $2k + 1$ 的倍数，所以 $(2k + 1) \mid \sigma(p^{2k})$. 又 k 为正整数，故 $2k + 1 > 1$.

此外，当 $i > 0$ 时，总有 $p^i > 1$，故 $\sigma(p^{2k}) > 2k + 1$，因此 $\sigma(p^{2k})$ 为合数.

综上，结论成立.

例 5.14　设 p 是素数，而 m 是正整数，试证：$m^p - m$ 是 p 的倍数.

证明：令 $m = lp$（l 是正整数），则 $(lp)^p - lp$ 是 p 的倍数，即有无穷多个正整数 $lp (l = 1, 2, \cdots)$，使得 $m^p - m$ 是 p 的倍数.

假设 $m = k + 1$ 时，$(k + 1)^p - (k + 1)$ 是 p 的倍数，则由

$$(k + 1)^p - (k + 1) = (k^p - k) + c_p^1 k^{p-1} + c_p^2 k^{p-2} + \cdots\cdots + c_p^{p-1} k$$

$c_p^i = \dfrac{p(p - 1) \cdots (p - i + 1)}{i!} (1 \leqslant i \leqslant p - 1)$ 是 p 的倍数，知 $k^p - k$ 是 p 的倍数. 从而根据反向归纳法，对任意正整数 m，$m^p - m$ 都是 p 的倍数.

例 5.15 已知当 $x = \alpha, \beta, \gamma$ 时 $ax^2 + bx + c$，都是质数 p 的倍数，且 p 不能整除 $\alpha - \beta$，$\beta - \gamma$，或 $\gamma - \alpha$．求证：当 x 取任意整数时，总有

$$p \mid (ax^2 + bx + c).$$

分析：据题意有

$$a\alpha^2 + b\alpha + c; a\beta^2 + b\beta + c; a\gamma^2 + b\gamma + c$$

都被 p 整除，其差也被 p 整除，有 $(\alpha - \beta)[a(\alpha + \beta) + b]$ 被 p 整除．同理还可得到 $(\beta - \gamma)[a(\beta + \gamma) + b], (\gamma - \alpha)[a(\gamma + \alpha) + b]$ 都被 p 整除．

由此如能证明 a, b, c 都是 p 的倍数，则命题得证．

证明：据题意，有

$$p \mid (a\alpha^2 + b\alpha + c), p \mid (a\beta^2 + b\beta + c);$$
$$p \mid (a\gamma^2 + b\gamma + c).$$

由此可得

$$p \mid (\alpha - \beta)[a(\alpha + \beta) + b],$$
$$p \mid (\beta - \gamma)[a(\beta + \gamma) + b],$$
$$p \mid (\gamma - \alpha)[a(\gamma + \alpha) + b].$$

由于质数 $p \nmid (\alpha - \beta)$，且 $p \nmid (\beta - \gamma), p \nmid (\gamma - \alpha)$，故有

$$p \mid [a(\alpha + \beta) + b],$$
$$p \mid [a(\beta + \gamma) + b],$$
$$p \mid [a(\gamma + \alpha) + b].$$

因此有

$$p \mid a(\alpha - \gamma).$$

又 $p \nmid (\alpha - \gamma)$，所以有 $p \mid a$．同时，由

$$p \mid (a\alpha^2 + b\alpha + c), 且 p \mid a, p \mid b,$$

得到 $p \mid c$．这样对任意的整数 x，都有

$$p \mid (ax^2 + bx + c).$$

说明：本题反复用到了一个整除性质：若 $p \mid ab$，且 $(p, b) = 1$，则 $p \mid a$．

例 5.16 已知 k, m, n 都是正整数，而且 $m + k + 1$ 是大于 $n + 1$ 的质数，并记 $C_s = s(s + 1)$．试证明乘积

$$(C_{m+1} - C_k)(C_{m+2} - C_k) \cdots (C_{m+n} - C_k)$$

能被乘积 $C_1 C_2 \cdots C_n$ 整除．（第 9 届 IMO）

分析：如果乘积 $(C_{m+1} - C_k)(C_{m+2} - C_k) \cdots (C_{m+n} - C_k)$ 有一个因数为零，即有一正整数 i，使得 $m + i = k$，那么这种乘积就是零，结论是显然成立的．此外由于 i 是从 1 到 n 的，所以如果乘积中有正有负，那就必然会出现一个乘积是零，这说明乘积的因子或者都是正的，或者都是负的．这样我们就可以分两种情况讨论，将 $C_s = s(s + 1)$ 代入乘积中，整理并使其与组合数建立起联系．

证明：经过上面的分析可以知道，乘积

$$(C_{m+1} - C_k)(C_{m+2} - C_k) \cdots (C_{m+n} - C_k)$$

只能有两种情况，或者等于零（此时结论正确）；或者所有的因子符号相同.

不妨假设所有的因子都是正的（当所有的因子都是负的时，方法类似）.

由已知：
$$C_m - C_n = m(m+1) - n(n+1)$$
$$= (m-n)(m+n+1),$$

所以
$$(C_{m+1} - C_k)(C_{m+2} - C_k)\cdots(C_{m+n} - C_k)$$
$$= (m+1-k)(m+1+k+1)(m+2-k) \cdot (m+2+k+1)\cdots(m+n-k)(m+n+k+1)$$
$$= \frac{(m+n-k)!}{(m-k)!} \cdot \frac{(m+n+k+1)!}{(m+k+1)!}.$$

又由于 $C_1 \cdot C_2 \cdots C_n = n!(n+1)!$，因此
$$\frac{(C_{m+1} - C_k)(C_{m+2} - C_k)\cdots(C_{m+n} - C_k)}{C_1 C_2 \cdots C_n}$$
$$= \frac{(m+n-k)!}{(m-k)!n!} \cdot \frac{(m+n+k+1)!}{(m+k+1)!(n+1)!}$$
$$= I_1 \cdot I_2.$$

现在来证明 I_1 和 I_2 都是整数.
$$I_1 = \frac{(m+n-k)!}{[(m+n-k)-n]!n!} = C_{m+n-k}^n$$

是二项展开式的系数，故必为整数.
$$I_2 = \frac{(m+n+k+1)!}{(n+1)!(m+k)!} \cdot \frac{1}{m+k+1}$$
$$= C_{m+n+k+1}^{n+1} \cdot \frac{1}{m+k+1},$$

因为 $m+k+1$ 是 $(m+n+k+1)!$ 中的一个因子；

又因 $m+k+1$ 是大于 $n+1$ 的素数，所以 $(m+k)!$、$(n+1)!$ 必不包含 $m+k+1$ 这个因子，故整数 $C_{m+n+k+1}^{n+1}$ 能被 $m+k+1$ 整除.

从而 I_2 也是一个整数，这样 $(C_{m+1} - C_k)(C_{m+2} - C_k)\cdots(C_{m+n} - C_k)$ 能被乘积 $C_1 \cdot C_2 \cdots C_n$ 整除.

例 5.17　设 p 是一个大于 3 的素数，试证：p^2 被 12 除所得的余数比为 1.

证明： 因 p 是大于 3 的素数，故 p 必为 $6k+1$ 的形式，这里 k 为正整数，此时
$$p^2 = (6k \pm 1)^2 = 36k^2 \pm 12k + 1 = 12(3k^2 \pm k) + 1,$$
即 p^2 被 12 除所得的余数必为 1.

例 5.18　求出所有的素数，使它的 4 次方的所有正因数的和为某整数的平方.

解： 设 $\sigma(p^4) = 1 + p + p^2 + p^3 + p^4 = n^2$. 因为
$$(2n)^2 = 4n^2 = 4 + 4p + 4p^2 + 4p^3 + 4p^4 > 4p^4 + 4p^3 + p^2 = (2p^2 + p)^2,$$
$$(2n)^2 = 4n^2 < 4p^4 + 4p^3 + 9p^2 + 4p + 4 = (2p^2 + p + 2)^2,$$

即 $(2p^2+p)^2 < (2n)^2 < (2p^2+p+2)^2$，所以

$$(2n)^2 = (2p^2+p+1)^2 = 4p^4+4p^3+5p^2+2p+1.$$

已知

$$(2p)^2 = 4p^4+4p^3+4p^2+4p+4,$$

把上两式相减，得 $p^2-2p-3=0$，解得 $p=3$（舍去 -1）．

例 5.19 求所有这样的素数，它既是两素数之和，同时又是两素数之差．

分析：考虑素数 2 是素数中唯一的偶数．

解：设所求的素数是 p，因它是两素数之和，故 $p>2$．

从而 p 是奇数，因此和为 p 的两个素数中有一个为 2．

同时差为 p 的两素数中，减数也是 2，即 $p=q+2,p=r-2$，其中 q,r 是素数．

于是 $p-2,p,p+2$ 均为素数．

而 $p-2,p,p+2$ 是三个连续的奇数，必定有一个是 3 的倍数．

因为这个数是素数，故只能是 3．

经检验，只有 $p-2=3,p=5,p+2=7$ 时，才满足条件，故 $p=5$．

例 5.20 证明：对任意六个连续正整数，存在一个质数，使得这个质数能且只能整除六个数之一．（2003 年德国数学竞赛）

证明：记这六个数为 $n,n+1,\cdots,n+5$．

若 n 不能被 5 整除，则 $n+1$ 至 $n+4$ 中却只有一个数能被 5 整除．

若 n 能被 5 整除，则 $n+1$ 至 $n+4$ 中有两个整数不能被 2,5 整除，其中至少有一个不能被 3 整除．

因此，该数至少有一个大于 5 的质因子，且该质数不能整除另外五个整数．

例 5.21 求能使 $p,p+10,p+14$ 都是素数的一切 p．

解：$p=3$ 时，$p+10=13,p+14=17$ 都是素数．所以 $p=3$ 时是一个解．

因为 $p=3k+1$ 时，$p+14=3k+15=3(k+5)$ 是合数．

因为 $p=3k+2$ 时，$p+10=3k+12=3(k+4)$ 是合数．

所以，能使得 $p,p+10,p+14$ 同时都是素数的一切 p 就只有 3．

例 5.22 试证：形如 $4n+3$ 的素数有无穷多．

证明：如果形如 $4n+3$ 的素数有限，则可假定它们的全体是 $p_1,p_2,\cdots,p_k$．

令 $N=4P_1P_2\cdots P_K-1=4(P_1P_2\cdots P_K-1)+3$，则 N 也是形如 $4n+3$ 的数，而且任何 $p_i(1\le i\le k)$ 都除不尽 N．

由于除 2 以外，素数都是奇数，而奇素数用 4 除所得必是 1 或 3，又两个被 4 除余 1 的数 $4l+1$ 与 $4m+1$ 的乘积 $(4l+1)(4m+1)=4(4lm+l+m)+1$ 仍然是一个 $4n+1$ 形式的数，故由 N 是形如 $4n+3$ 的数，推知 N 的素因数不可能都是 $4n+1$ 的形式，即还有 $4n+3$ 形如的素因数，但又不能是 $p_1,p_2,\cdots,p_k$ 中的一个，这就与假设相矛盾，所以形如 $4n+3$ 的素数有无穷多．

例 5.23　试证：小于 n^2 的所有奇素数是不包含在下列等差数列中的所有奇数（1 除外）：

$$r^2, r^2 + 2r, r^2 + 4r, \cdots（直到 n^2），$$

而 $r = 3, 5, 7, \cdots$（直到 $n - 1$）.

证明： 小于 n^2 的不等于 1 的奇数要么是奇合数，要么是奇素数.

设 a 是奇合数，$a < n^2$，那么 a 必有一个最小的素因数 r，且 $3 \leqslant r < n$（如果最小的素因数 $\geqslant n$，则与 $a < n^2$ 矛盾）. 于是存在奇数 $b \geqslant r$，使得 $a = rb$. 因为 b 为奇数，所以 b 总可以写成 r 与一个偶数的和，即 $b = r + 2k$. 因此，$a = r(r + 2k)$，即 a 是等差数列中的某个数.

另外，等差数列中除 1 以外的数都是奇合数，这个结论是显然成立的. 因此，除等差数列外的小于 n^2 的奇数都是奇素数，而小于 n^2 的奇素数不在等差数列中，即小于 n^2 的奇素数恰是不包含在等差数列中的所有奇数（1 除外）.

例 5.24　设 n 是大于 2 的偶数，试证：对于任何充分大的奇素数 p，存在满足 $p_1 < p_2 < \cdots < p_{n-3}$ 的素数 $p_1, p_2, \cdots, p_n$，使得

$$p = \frac{p_1 + p_2 + \cdots + p_n + n - 1}{p_1 + n - 2}$$

证明： 对于正整数 k，设 q_k 是第 k 个素数. 当其是偶数时，记 $p_i = q_i (i = 1, 2, \cdots, n - 3)$，则 $p_1 = 2$ 而 $p_2, \cdots, p_{n-3}$ 均为奇素数. 令 $m = (p - 1)n - (p_2 + \cdots + p_{n-1} + 1)$. 可知当 p 是充分大的奇素数时，m 是充分大的奇数. 因此，存在奇素数 r_1, r_2, r_3，使得

$$m = r_1 + r_2 + r_3.$$

再令 $p_{n-3} = r_1, p_{n-1} = r_2, p_n = r_3$，则有

$$m = p_{n-2} + p_{n-1} + p_n$$

并结合 $p_1 = 2$，可得

$$p = \frac{p_1 + p_2 + \cdots + p_n + n - 1}{p_1 + n - 2}$$

例 5.25　设数列 $\{g(n)\}$ 满足 $g(1) = 1, g(n + 1) = g^2(n) + 4g(n) + 2$，试证：如果 n 是偶数，则 $\{g(n)\}$ 中仅有素数 $g(2) = 7$.

证明： 设 $h(n) = g(n) + 2$，则有 $h(1) = 3, h(n + 1) = h^2(n)$.

由此推出

$$h(n) = 3^{2^{n-1}}$$

于是

$$g(n) = 3^{2^{n-1}} - 2.$$

当 $n = 2$ 时，$g(2) = 7$ 为素数.

当 $n > 2$ 且 n 为偶数时，$3 \mid (2^{n-2} - 1)$.

由于

$$7 \mid (3^6 - 1)$$

所以

$$7 \mid \left[3^{2(2^{n-2})} - 1\right]$$

又

$$g(n) = 3^2\left(3^{2(2^{n-2}-1)} - 1\right) + 7$$

故 $7 \mid g(n)$. 这说明当 $n > 2$ 且 n 为偶数时，$g(n)$ 是合数. 因此，如果 n 为偶数，则 $\{g(n)\}$ 中仅有素数 $g(2) = 7$.

5.5.2 典型练习题

练习 5.1 容易验证 91，92，93，94，95，96，97 是 7 个相邻的合数，试写出 9 个相邻的合数.

练习 5.2 验证 539 是否为合数.

练习 5.3 设 p 是合数 n 的最小素因子，证明：若 $p > n^{\frac{1}{3}}$，则 $\dfrac{n}{p}$ 是素数.

练习 5.4 说明下列质数判别法正确：

若 n 为大于 5 的奇数，且存在互质的整数 a 和 b，使 $a - b = n$ 和 $a + b = p_1 p_2 \cdots p_n$（其中 $p_1 p_2 \cdots p_n$ 是小于 $\sqrt{n}$ 奇质数）则 n 是质数.

练习 5.5 试证：当正整数是 $n > 1$，$n^4 + 4^n$ 为合数.

练习 5.6 （1）德布埃尔（De Bouelles）曾断言：对所有 $n \geq 1$，$6n - 1$ 和 $6n + 1$ 中至少有一个是素数. 试举出反例.

（2）试证：有无穷多个 n，使得 $6n - 1$ 和 $6n + 1$ 同时为合数.

练习 5.7 判断命题是否正确：p 和 q 整除 n，则 $\dfrac{n}{pq}$ 是质数.

练习 5.8 哪些素数 p 可使 $p^2 - 2$，$2p^2 - 1$ 和 $3p^2 + 4$ 都是素数?

练习 5.9 设 $p > 5$，若 p 及 $2p + 1$ 均为素数，试证：$4p + 1$ 必为合数.

练习 5.10 设 N 是任意给定的正整数，$p_1, p_2, \cdots, p_s$ 是所有不超过 N 的素数，试证：

$$\prod_{i=1}^{s}\left(1 - \frac{1}{p_i}\right)^{-1} > \sum_{n=1}^{N} \frac{1}{n}$$ 并由此推出素数有无穷多个.

练习 5.11 对于任意的整数 $n > 1$，证明：总可以找到 n 个连续的合数.

练习 5.12 试证：形若 $6n + 5$ 的素数有无穷多.

练习 5.13 证明在 $n > 2$ 时，n 与 $n!$ 之间一定有一个质数.

练习 5.14 若素数 $p \geq 7$，试证：p^2 除以 30 所得的余数必为 1 或 19.

练习 5.15 求能使 p 和 $8p^2 + 1$ 都是素数的一切 p.

练习 5.16 设奇数 $n \geq 3$，试证：n 是素数的充要条件为：n 不能表为三个或三个以上的相邻正整数之和.

练习 5.17 设 m 为大于的正整数，试证：当且仅当 m 为大于 5 的合数时，$m \mid (m - 1)!$.

练习 5.18 试找出所有奇数 n，使得 $n^2 \mid (n - 1)!$.

练习 5.19 设 $n > 2$，试证：$n - 1$ 个连续正整数 $n! + 2, n! + 3, \cdots, n! + n$ 中，每个数都有一个素因数，并且该因数不能整除其他 $n - 2$ 个数的任何一个.

练习 5.20 全体素数按大小顺序排成的序列为 $p_1 = 2, p_2 = 3, p_3, p_4, \cdots$，试证：$p_n \leq 2^{2^{n-1}}$.

练习 5.21　试找出 6 个小于 160 而成等差数列的素数，并证不可能有 7 个皆小于 200 的素数成等差数列.

练习 5.22　求出满足等式 $x^y + 1 = z$ 的所有的质数 x, y, z.

第 6 章　最大公因数

凡事豫则立，不豫则废．言前定则不跲，事前定则不困，
行前定则不疚，道前定则不穷．

——《礼记·中庸》

正如要认识一盒粉笔，我们在有一个大概外观的认识后，就进入到细致的研究，第一步往往是从中找出一支粉笔进行单独研究，第二步是拿出几支，比如将这盒粉笔中与这支粉笔相同颜色的都拿出来进行研究，然后再与其他颜色的进行对比，最后就认识了整盒粉笔．研究数学也不外如是．

就数学研究而言，第一步所研究的是只有一个整数的时候，主要看它们的因数和倍数所具有的一些性质．第二步自然是研究有若干个整数的时候，它们具有些什么性质呢？接下来就要来进行讨论．本章主要讨论多个整数的最大公因数，下一章讨论相应的最小公倍数．

6.1　公　因　数

用整除性对因数进行定义的话，因数指的是能够整除已知数的所有整数．这里涉及的已知数是一个整数，对于若干个整数的情况，需要考虑的就是他们所具有的共同的性质，比如他们所同时拥有的因数，就被称作公因数，也即公因数指的就是公共的因数．

比如整数 9 的因数有 $\{-1,-3,-9,1,3,9\}$，这是因为
$$-1\mid 9,\ -3\mid 9,\ -9\mid 9,\ 1\mid 9,\ 3\mid 9,\ 9\mid 9,$$
也可以表示为：
$$9=1\times 9, 9=3\times 3$$
$$9=(-1)\times(-9), 9=(-3)\times(-3)$$

整数 12 的因数有 $\{-1,-2,-3,-4,-6,-12,1,2,3,4,6,12\}$，这是因为：
$$-1\mid 12,\ -2\mid 12,\ -3\mid 12,\ -4\mid 12,\ -6\mid 12,\ -12\mid 12$$
$$1\mid 12,\ 2\mid 12,\ 3\mid 12,\ 4\mid 12,\ 6\mid 12,\ 12\mid 12$$
也可以表示为：
$$12=1\times 12, 12=2\times 6, 12=3\times 4$$
$$12=(-1)\times(-12), 12=(-2)\times(-6), 12=(-3)\times(-4)$$
它们当中所共有的数是 $-1,-3,1,3$，这样就说 $-1,-3,1$ 和 3 是 9 和 12 的公因数，这是因为 $-1,-3,1$ 和 3 既能被 9 整除，又能被 12 整除．

将上述讨论数学化，可以得到：

某数 d 称为另外一组［两个及以上 (a,b)］数的公因数，需要满足同时是这一组数中所有数的因数，即要 $d\mid a$，且 $d\mid b$．

任意给定两个或两个以上的整数，必然存在公因数，这是因为 -1 和 1 是任意整数的因数，所以 -1 和 1 必然都可以作为任意两个或两个以上的整数的公因数．更多数的情况之下，我们对负的因数的需要和讨论并不是太多，所以后面再涉及的时候我们主要以非负因数为主．

那么在所有非负因数中，除了 1 之外还有没有其他的公因数呢？

6.2 最大公因数

6.2.1 最大公因数

顾名思义，最大公因数指的是若干个不全为零的整数的公因数中数值最大的一个，后面若不做特别说明，本书所指的一些整数都是不全为零的．以两个整数 a,b 为例，它们的最大公因数指的就是同时整除 a,b 的最大的数，一般记作 $\gcd(a,b)$．

比如 1 和 3 是 9 和 12 的公因数，其中能够同时整除 9 和 12 的最大的数是 3，所以 $\gcd(9,12) = 3$．

在这一概念中，需要特别留意不全为零的前提，因为如果 $a = b = 0$，那么它们有无限多个因数，且可以无限大，这是无法求出其最大公因数的．

不全为零也就是说至少有一个不为零，这个时候，比如 $a = 0, b > 0$，这时候

$$\gcd(a,b) = b$$

而当 $a = 0, b < 0$ 时

$$\gcd(a,b) = -b$$

综合起来，$a = 0, b \neq 0$ 时

$$\gcd(a,b) = |b|$$

那么要怎样才可以求出若干个数的最大公因数呢？对于个数较少且数值不大的整数来说，要求出它们的最大公因数的一个最基本的方法就是按照定义要求，将它们所有的公因数都找出来，这样就可以很容易地看出最大的那个数，那个数就是所求的最大公因数．

比如要求出 225 和 120 的最大公因数，我们需要分别求出 225 和 120 的所有因数，这需要我们一步一步将这两个数进行分解，也就是转换为乘法去思考：

$$225 = 3 \times 3 \times 5 \times 5$$
$$120 = 2 \times 2 \times 2 \times 3 \times 5$$

也就是说 225 的因数有 $\{3, 5, 9, 15, 25, 45, 75, 225\}$，120 的因数有 $\{2, 3, 4, 6, 8, 10, 12, 15, 20, 30, 40, 60, 120\}$，于是他们的公因数有 $\{3, 5, 15\}$，其中最大的是 15.

我们也可以从末尾的 5 和 0 看出，这两个数都被 5 整除，又由 $2 + 2 + 5 = 9$ 和 $1 + 2 + 0 = 3$ 知道，这两个数都能被 3 整除．

于是可以想办法计算这两个数分别除以 5 和 3 后的数可以得到 225 和 120 的公因数有 1，3，5，15，自然的，选择最大的数 15 作为其最大公因数即可，也即：

$$\gcd(225,120) = 15$$

这样的计算从算法原理上来说较为简单，但却费时耗力，尤其对于当整数个数较多数值较大的情况下是很麻烦的．比如要计算

$$\gcd(1160718174,316258250)$$

这就需要将 1160718174 的所有因数找出来，然后再将 316258250 的所有因数找出来，

再从其中找到公共的因数，最后将公因数中的最大数找出来.

将上述讨论数学化，可以得到：

某数 d 称为另外一组［两个及以上 (a,b)］数的最大公因数，需要满足同时是这一组数中所有数的因数，并且任意除自己以外的公因数都比 d 要小.即同时满足下列条件：

(i) $d \mid a$，且 $d \mid b$.

(ii) 若 $c \mid a$，且 $c \mid b$，则 $c \leqslant d$.

条件（i）说明 d 是 a,b 的公因数，条件（ii）说明，d 是 a,b 的公因数中最大的一个.

注意，若 a 和 b 不同时为零，那么 a 和 b 的公因子集合是以 a，b，$-a$ 和 $-b$ 中最大者为其上界的整数集.因此，根据整数的良好原理，a 和 b 的最大公因子存在而且是唯一的.另外，$(0,0)$ 没有意义.而如 (a,b) 有意义，则它是整数.事实上，必成立 $(a,b) \geqslant 1$，因为对任何 a 和 b，$1 \mid a$，$1 \mid b$.

6.2.2　互　素

在最大公因数的定义中，有一个很特别的情形：如果两个数 a,b 只有 1 这个公因数，那么 $\gcd(a,b) = 1$，此时我们称这两个数互素.比如 $\gcd(3,10) = 1$.

例 6.1　试证：若 $(a,b) = d$，则 $\gcd\left(\dfrac{a}{d}, \dfrac{b}{d}\right) = 1$.

证明： 设 $c = \left(\dfrac{a}{d}, \dfrac{b}{d}\right)$.我们需证 $c = 1$.为此，我们证明，$c \leqslant 1$，$c \geqslant 1$.由于 c 是两个整数的最大公因子，我们已注意到，每个最大公因子都大于等于 1，故得后一不等式.为了说明 $c \leqslant 1$，我们利用 $c \left| \dfrac{a}{d} \right.$ 和 $c \left| \dfrac{b}{d} \right.$，即知存在 q 和 r，使 $cq = \dfrac{a}{d}$，$cr = \dfrac{b}{d}$，或 $(cd)q = a$，$(cd)r = b$.这两个式子表明，cd 是 a 和 b 的一个公因子，因此它不大于 a 和 b 的最大公因子 d，故有 $cd \leqslant d$.又因 d 是整数，可得 $c \leqslant 1$.因此，$c = 1$.

6.3　欧几里得算法

从数学方法上来将最大公因数的算法进行优化，其中优化得最好的一个就是欧几里得算法.

6.3.1　欧几里得算法

要计算两个整数 a,b 的最大公因数 $\gcd(a,b)$，$0 < a < b$，根据带余除法，我们可以得到以下的 $k+1$ 个算式：

$$b = a q_1 + r_1, \quad 0 \leqslant r < a,$$
$$a = r_1 q_2 + r_2, \quad 0 \leqslant r_2 < r_1,$$
$$r_1 = r_2 q_3 + r_3, \quad 0 \leqslant r_3 < r_2,$$
$$r_2 = r_3 q_4 + r_4, \quad 0 \leqslant r_4 < r_3,$$

$$\cdots\cdots$$
$$r_{k-2} = r_{k-1}q_k + r_k, 0 \leqslant r_k < r_{k-1},$$
$$r_{k-1} = r_k q_{k+1} + r_{k+1}, r_{k+1} = 0.$$

因为每进行一次带余除法，其余数至少会减少 1，所以经过有限多次的带余除法后，总会得到一个余数是零的等式，也即必定存在一个大于零的整数 k，可以使得

$$r_{k+1} = 0.$$

此时，所求的

$$\gcd(a,b) = r_k.$$

证明：对于两个不全为零的整数 $a,b,a < b$，且 $b = aq + r$，那么：

$$\gcd(a,b) = \gcd(a,r)$$

所以对于算法中的 $k + 1$ 个算式来说必然有：

$$\gcd(a,b) = \gcd(a,r_1)$$
$$\gcd(a,r_1) = \gcd(r_1,r_2)$$
$$\gcd(r_1,r_2) = \gcd(r_2,r_3)$$
$$\gcd(r_2,r_3) = \gcd(r_3,r_4)$$
$$\cdots\cdots$$
$$\gcd(r_{k-2},r_{k-1}) = \gcd(r_{k-1},r_k)$$
$$\gcd(r_{k-1},r_k) = r_k$$

显然，上面的式子都是相等的，联立起来即可得到：

$$\gcd(a,b) = r_k.$$

欧几里得算法又被称作辗转相除法，事实上，将这一算法推广到高等代数的学习中，常常可以用来求出两个多项式的最大公因式．欧几里得算法可以在多项式时间内完成，具体来说，要计算 $\gcd(a,b)$，最多需要进行 $O(\log b)$ 步的除法，因此，欧几里得算法有多项式时间复杂度．

例 6.2 计算 $\gcd(963,657)$.

解：由带余除法得：

$$963 = 657 \times 1 + 306.$$
$$657 = 306 \times 2 + 45.$$
$$306 = 45 \times 6 + 36.$$
$$45 = 36 \times 1 + 9.$$
$$36 = 9 \times 4.$$

所以：

$$\gcd(963,657) = 9.$$

例 6.3 用辗转相除法求 $a = 288, b = 158$ 的最大公因数和 m,n，使 $ma + nb = (a,b)$.

解：由于

$$288 = 158 \times 1 + 130$$
$$158 = 130 \times 1 + 28$$

$$130 = 28 \times 4 + 18$$
$$28 = 18 \times 1 + 10$$
$$18 = 10 \times 1 + 8$$
$$10 = 8 \times 1 + 2$$
$$8 = 2 \times 4$$

因此, $(288,158) = 2$.

再由

$$
\begin{aligned}
2 &= 10 - 8 \times 1 \\
&= 10 - (18 - 10) \\
&= 10 \times 2 - 18 \\
&= (28 - 10 \times 1) \times 2 - 18 \\
&= 28 \times 2 - 18 \times 3 \\
&= 28 \times 2 - (130 - 28 \times 4) \times 3 \\
&= (-130) \times 3 + 28 \times 14 \\
&= (-130) \times 3 + (158 - 130 \times 1) \times 14 \\
&= 158 \times 14 - 130 \times 17 \\
&= 158 \times 14 - (288 - 158 \times 1) \times 17 \\
&= 158 \times 31 - 288 \times 17
\end{aligned}
$$

故 $m = -17, n = 31$.

6.3.2　欧几里得算法的方法论意义

欧几里得算法的一个基本思想是：我们的目标是计算若干个（假设从最简单的两个开始）整数的最大公因数. 问题是这两个整数的数值较大，导致难以通过最大公因数的定义去寻找. 基本思路是将数值较大这个麻烦解决掉，问题将迎刃而解.

要将两个整数变小，但是它们的最大公因数又不能改变，或者说我们要能够通过某种逆向运算将其反推回去而得出原先要计算的最大公因数.

于是有人做了这样的思考：面对两个整数，我们能做的最简单的是加减乘除运算，显然，通过加减运算与求因数没有太大关系，而乘法运算只会将数越算越大，要想减小数值，那么我们可能要从除法入手了.

当面对要计算

$$\gcd(a,b)$$

的时候，整数 a,b，不妨假设 $a < b$，可能存在的所有的除法运算最简单的是：

$$b \div a = q$$

此时，我们可以得到 $a \mid b$ 的结论，于是

$$\gcd(a,b) = a$$

否则，如果 $a \nmid b$，那么我们由带余除法知道：

$$b \div a = q \cdots r (r < a)$$

将其转换成乘法表达式是：

$$b = aq + r$$

这里出现了四个整数：a, b, q, r. 这四个数之间会不会蕴含某种特殊的关系呢？考察几个简单的计算：

$$20 = 12 \times 1 + 8$$
$$30 = 18 \times 1 + 12$$

我们在已知 $\gcd(20, 12) = 4$ 的前提下，发现 $4 \mid 8$. 在已知 $\gcd(30, 18) = 6$ 的前提下，发现 $6 \mid 12$. 这样就让我们想到，是不是有这么一种可能：$\gcd(a, b) \mid r$.

进一步继续刚刚的猜想：

$$12 = 8 \times 1 + 4$$
$$18 = 12 \times 1 + 6$$

可以看出，猜想的结论成立，再进一步：

$$8 = 4 \times 2 + 0$$
$$12 = 6 \times 2 + 0$$

结论依然成立，并且此时出现的这两个数 4 和 6 恰好就是我们所要求的最大公因数. 于是将猜想整理出来就是：

对于两个不全为零的整数 $a, b, a < b$，且 $b = aq + r$，那么：

$$\gcd(a, b) = \gcd(a, r)$$

证明： 假设 d 是 a, b 的公因数，那么 d 必然整除 a, b，这样 d 也就整除 aq，而 $r = b - aq$，所以 d 整除 r，也就是说 a, b 的公因数也是 a, r 的公因数. 同理可得 a, r 的公因数也是 a, b 的公因数. 换句话说，a, b 的公因数集与 a, r 的公因数集是同一个集合，那么其中最大的那个数自然也是同一个数了.

这样的事可以一直继续做下去，直到出现了某一运算中的 $r = 0$，此时也就可以得出所要求的最大公因数了. 将这套方法系统化，那就是欧几里得算法（Euclidean Algorithm），古希腊数学家欧几里得在其《几何原本》中就记载了这个方法.

6.4 裴蜀定理

6.4.1 裴蜀定理

逆向观察欧几里得算法的进程，我们可以看出：

$$r_{k-1} = r_k q_{k+1}$$
$$r_{k-2} = r_{k-1} q_k + r_k = r_k q_{k+1} q_k + r_k$$
$$\cdots\cdots$$

以此类推，式中每一个 r_i 都必然可以被 r_k 线性表示.

反过来，可以得到

$$r_k = r_{k-2} - r_{k-1} q_k$$

也就是说，可以用 r_{k-1} 和 r_{k-2} 表示 r_k，一直继续这个可能下去，那么必然可以得到一个结局，就是可以用 a, b 来表示 r_k，也即：

$$r_k = sa + tb$$

其中的 s 和 t 是由 r_i 和 q_i 线性表示的，而 r_i 和 q_i 都是整数，所以 s 和 t 都是整数. 这就是著名的裴蜀定理.

裴蜀定理： 设整数 a,b 为不全为零的整数，则存在 $s,t \in Z$，使得
$$\gcd(a,b) = sa + tb.$$

为了更好地说明上述定理，我们用以下的例子来进行解释：

例 6.4 求一组整数 s,t，使得
$$\gcd(963,657) = s963 + t657.$$

解： 由用带余除法求 $\gcd(963,657)$ 的过程可知：
$$\begin{aligned}
\gcd(963,657) &= 9 = 45 - 36 \times 1\\
&= 45 - (306 - 45 \times 6) \times 1 = 45 \times 7 - 306 \times 1\\
&= (657 - 306 \times 2) \times 7 - 306 = 306 \times (-15) + 657 \times 7\\
&= (963 - 657 \times 1) \times (-15) + 657 \times 7\\
&= (-15) \times 963 + 22 \times 657
\end{aligned}$$

因此，只需要取 $s = -15, t = 22$，则 $\gcd(963,657) = s963 + t657$.

特别的，当 a,b 互素的时候，这个定理将成为：

设整数 a,b 为不全为零的整数，则 $\gcd(a,b) = 1$ 当且仅当存在 $s,t \in Z$，使得
$$sa + tb = 1.$$

现在将裴蜀定理中的量稍作变换，比如将 s,t 表示成平时常用的 x,y，令 $\gcd(a,b) = d$，则
$$\gcd(a,b) = sa + tb.$$
将变成
$$ax + by = d.$$

这提示着我们还可以通过求两个数 a,b 的最大公因数，对方程 $ax + by = d$ 解的情况进行判别，并可据此求出相关解.

6.4.2　相关推论

利用欧几里得算法，我们还可以得到许多类似的推论，比如以下这些：

推论 1 如果 $d \mid a, d \mid b$，那么 $d \mid \gcd(a,b)$.

证明： 由裴蜀定理知，存在 $s,t \in Z$，使得
$$\gcd(a,b) = sa + tb.$$
用 d 去除上式两端，因为 $d \mid a, d \mid b$，故 $d \mid sa + tb$，因此
$$d \mid \gcd(a,b).$$

推论 2 设 $a,b,c \in Z$，那么 $\gcd(ac,bc) = \gcd(a,b) \times |c|$.

推论 3 设 $d \in Z^+$，那么 $\gcd(a,b) = d$ 的充要条件是
$$\gcd\left(\frac{a}{d}, \frac{b}{d}\right) = 1.$$

推论 4 设 $d \in Z^+$，a,b 是不全为零的整数，那么 $\gcd(a,b) = d$ 的充要条件是
(1) $d \mid a, d \mid b$；

（2）如果 $c \in Z$ 满足 $c \mid a, c \mid b$ 那么 $c \mid d$.

推论 5　如果 $\gcd(a,b) = 1$，那么 $\gcd(a,bc) = \gcd(a,c)$.

推论 6　如果 $\gcd(a,b) = 1$，且 $a \mid bc$，那么 $a \mid c$.

证明：因为 $\gcd(a,b) = 1$，由裴蜀定理知，存在 $s,t \in Z$，使得

$$\gcd(a,b) = sa + tb = 1.$$

在式子的两端同时乘以 c，将得到：

$$sac + tbc = c$$

显然，$a \mid sac$，题设中有 $a \mid bc$，故 $a \mid tbc$，也即 $a \mid sac + tbc$，因此 $a \mid c$.

注意：题设中如果 $\gcd(a,b) = 1$ 不可省. 事实上，若 a, b 不互素，那么结论未必成立. 例如 6 整除 8×9，但是 6 既不整除 8，又不整除 9.

推论 7　如果 $\gcd(a,b) = 1$，且 $a \mid c, b \mid c$，那么 $ab \mid c$.

推论 8　如果 $\gcd(a,b) = \gcd(a,c) = 1$，那么 $\gcd(a,bc) = 1$.

推论 9　设 $a_1, a_2, \cdots, a_n$ 是 n 个不全为零的整数，那么

$$\gcd(a_1, a_2, \cdots, a_n) = \gcd((a_1, a_2), a_3, \cdots, a_n)$$

推论 10　设 $a_1, a_2, \cdots, a_n$ 是 n 个不全为零的整数，存在 $s_1, s_2, \cdots, s_n \in Z$，使得

$$\gcd(a_1, a_2, \cdots, a_n) = a_1 s_1 + a_2 s_2 + \cdots + a_n s_n.$$

6.5　典型问题

6.5.1　典型例题

例 6.5　求 36 和 24 的最大公因数.

解：把这两个数分别分解素因素：

$$36 = 2 \times 2 \times 3 \times 3, \ 24 = 2 \times 2 \times 2 \times 3.$$

把这两个数的素因数比较一下，可以看出素因数 2，2，3 是这两个数所公有的，它们的乘积就是这两个数的最大公因数：

$$2 \times 2 \times 3 = 12.$$

求几个正整数的最大公因数，先把这些正整数分别分解素因素，然后取出它们所公有的素因数（相同的素因数按照公有的个数取）相乘.

例 6.6　求 48，60 和 72 的最大公因数.

解：把这三个数分别分解素因数：

$$48 = 2 \times 2 \times 2 \times 2 \times 3 = 2^4 \times 3,$$
$$60 = 2 \times 2 \times 3 \times 5 = 2^2 \times 3 \times 5,$$
$$72 = 2 \times 2 \times 2 \times 3 \times 3 = 2^3 \times 3^2.$$

把上面三个数的素因数比较一下，可以看出素因数 2，2，3（或 2^2，3）是三个数所共有的，它们的乘积就是这三个数的最大公因数

$$2 \times 2 \times 3 = 12 \text{ 或 } 2^2 \times 3 = 12.$$

为通俗起见，先讨论最大公因数和最小公倍数，再讨论算术基本定理．求几个数的最大公因数，先把这些数分别分解素因数，并且写成乘方形式；然后在各个公有的素因数里，求出指数最小的乘方相乘．

例 6.7　求 1008，1260，882 和 1134 的最大公因数．

解：把这四个数分别分解素因数：

$$1008 = 2 \times 2 \times 2 \times 2 \times 3 \times 3 \times 7 = 2^4 \times 3^2 \times 7,$$
$$1260 = 2 \times 2 \times 3 \times 3 \times 5 \times 7 = 2^2 \times 3^2 \times 5 \times 7,$$
$$882 = 2 \times 3^2 \times 7^2,$$
$$1134 = 2 \times 3^4 \times 7.$$

所以 $(1008，1260，882，1134) = 2 \times 3^2 \times 7 = 126$.

例 6.8　计算 gcd $(5767，4453)$.

解：由于

$$5767 = 4453 \times 1 + 1314,$$
$$4453 = 1314 \times 3 + 511,$$
$$1314 = 511 \times 2 + 292,$$
$$511 = 292 \times 1 + 219,$$
$$292 = 219 \times 1 + 73,$$
$$219 = 73 \times 3.$$

所以

$$(5767,4453) = (4453,1314)$$
$$= (1314,511)$$
$$= (1314,511)$$
$$= (511,292)$$
$$= (292,219)$$
$$= (219,73)$$
$$= 73$$

也即：$(5767,4453) = 73$.

例 6.9　令 $a = -1859, b = 1573$，求 gcd (a,b).

解：因为

$$1859 = 1 \times 1573 + 286$$
$$1573 = 5 \times 286 + 143$$
$$286 = 2 \times 143$$

所以 $(a,b) = (-1859,1573) = 143$.

例 6.10　求 6731 和 2809 的最大公因式．

解：因为

$$6731 = 2809 \times 2 + 1113$$
$$2809 = 1113 \times 2 + 583$$
$$1113 = 583 \times 2 + 530$$
$$583 = 530 \times 1 + 53$$
$$530 = 53 \times 10$$

所以 $(6731,2809) = 53$.

例 6.11 求 525 与 231 的最大公因数.

解： 因为

$$525 = 2 \times 231 + 63$$
$$231 = 3 \times 63 + 42$$
$$42 = 2 \times 21$$

所以 $(525,231) = 21$.

例 6.12 求 $\gcd(136,221,391)$.

解： $(136,221,391) = (136,(221,391)) = (136,17) = 17$.

例 6.13 求 $\gcd(963,657)$.

解： 由带余除法得

$$963 = 657 \times 1 + 360,$$
$$657 = 306 \times 2 + 45,$$
$$306 = 45 \times 6 + 36,$$
$$45 = 36 \times 1 + 9,$$
$$36 = 9 \times 4,$$

故 $(963,657) = 9$.

例 6.14 求一组整数 s,t，使得 $963s + 657t = (963,657)$.

解： 由 $(963,657)$ 的求解得

$$
\begin{aligned}
(963,657) = 9 &= 45 - 36 \times 1 \\
&= 45 - (306 - 45 \times 6) \times 1 \\
&= (657 - 306 \times 2) \times 7 - 306 = 306 \times (-15) + 657 \times 7 \\
&= (963 - 657 \times 1) \times (-15) + 657 \times 7 \\
&= (-15) \times 963 + 22 \times 657.
\end{aligned}
$$

因此取 $s = -15, t = 22$，则可满足 $963s + 657t = (963,657)$.

例 6.15 对 525 与 231 求满足裴蜀定理条件的 x,y.

解： 因为

$$21 = 63 - 1 \times 42 = 63 - 1 \times (231 - 3 \times 63)$$
$$= 4 \times 63 - 1 \times 231 = 4 \times (525 - 2 \times 231) - 1 \times 231$$

$$= 4 \times 525 - 9 \times 231,$$

即 $525 \times 4 + 231(-9) = 21$, $x = 4$, $y = -9$.

例 6.16　求 735000, 421160 和 238948 的最大公因数.

解：按辗转相除法可得：

$$735000 = 238948 \times 3 + 18156,$$
$$238948 = 18156 \times 13 + 2920,$$
$$18156 = 2920 \times 6 + 636,$$
$$2920 = 636 \times 4 + 376,$$
$$636 = 376 \times 1 + 260,$$
$$376 = 260 \times 1 + 116,$$
$$260 = 116 \times 2 + 28,$$
$$116 = 28 \times 4 + 4,$$
$$28 = 7 \times 4 + 0,$$

所以 735000 和 238948 的最大公因数是 4, 由于 $4 \mid 421160$, 故得
$(735000, 238948, 421160) = 4$.

例 6.17　用辗转相除法求 gcd (198, 252).

解：因为

$$252 = 198 \times 1 + 54$$
$$198 = 54 \times 3 + 36,$$
$$54 = 36 \times 1 + 18,$$
$$36 = 18 \times 2,$$

所以

$$(252,198) = (198,54)$$
$$= (54,36)$$
$$= (36,18)$$
$$= 18.$$

为方便起见, 上面一系列计算也可以简写成如下形式：

252	1	198
198	3	162
54	1	36
36	2	36
18		0

例 6.18　用辗转相除法求
gcd (8127, 11352, 21672, 27090).

解：

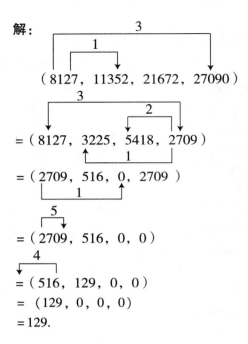

$= (8127, 3225, 5418, 2709)$

$= (2709, 516, 0, 2709)$

$= (2709, 516, 0, 0)$

$= (516, 129, 0, 0)$

$= (129, 0, 0, 0)$

$= 129.$

例 6.19 对于任意正整数 n，求证：$\dfrac{12n+7}{14n+8}$ 是既约分数.

证明：

$\because 14n + 8 = (12n + 7) \times 1 + 2n + 1,$

　$12n + 7 = (2n + 1) \times 6 + 1.$

$\therefore (14n + 8, 12n + 7)$

　$= (12n + 7, 2n + 1)$

　$= (2n + 1, 1) = 1.$

故当 $n \in N_+$ 时，$\dfrac{12n+7}{14n+8}$ 是既约分数.

例 6.20 用辗转相除法求最大公因数：

(i) 51425, 13310.

(ii) 353430, 530145, 165186.

(iii) 81719, 52003, 33649, 310107.

(i) **解：** 由于

	351425	13310	3
1	39930	11495	
	11495	1815	6
3	10890	1815	
	605	0	,

所以（51425，13310）＝605.

（iii）解：由于

	353430	530145	2
1	353430	353430	
	0	176715	，

得到（353430，530145）＝176715.

又由于

	176715	165186	1
14	165186	161406	
	11529	3780	3
20	11340	3780	
	189	0	，

得到（176715，165186）＝189.

所以（353430，530145，165186）＝189.

（iii）解：由于

	81719	52003	1
1	52003	29716	
	29716	22287	1
3	22287	22287	
	7423	0	

得到（81719，52003）＝7429.

由于

	33649	7429	4
1	29716	3933	
	3933	3496	1
8	3496	3496	
	437	0	，

得到（33649，7429）=437.

由于

	30107	437	68
1	29716	391	
	391	46	8
2	368	46	
	23	0	

，

得到（310107，437）= 23.

所以（81719，52003，33649，310107）= 23.

例 6.21 求下面的最大公因数：

(1) $(n! + 1, (n + 1)! + 1)$；   (2) $(nk, k(n + 2))$（k 为正整数）.

解： (1) 由于 $(n + 1, (n + 1)! + 1) = 1$,

所以 $(n! + 1, (n + 1)! + 1) = ((n! + 1)(n + 1), (n + 1)! + 1)$

$$= ((n! + 1)(n + 1) - (n + 1)! - 1, (n + 1)! + 1)$$

$$= (n, (n + 1)! + 1).$$

设 d 为 $n! + 1$ 与 $(n + 1)! + 1$ 的最大公因数，由上式，知 $d \mid n$，即 $d \mid n!$.

又因为 $d \mid (n! + 1)$，所以 $d = 1$.

即 $(n! + 1, (n + 1)! + 1) = 1$.

(2) 当 $n = 2a$ 时，

原式 $= k(n, n + 2) = k(2a, 2a + 2) = k(2a, 2) = 2k(a, 1) = 2k$

当 $n = 2a + 1$ 时，

原式 $= k(n, n + 2) = k(2a + 1, 2a + 3) = k(2a + 1, 2) = k$.

例 6.22 设 $(a, 4) = (b, 4) = 2$ 试求 $(a + b, 4)$.

解： 由题意，$a = 4k_1 + 2, b = 4k_2 + 2$

则

$$(a + b, 4) = (4k_1 + 4k_2 + 4, 4) = 4(k_1 + k_2 + 1, 1) = 4.$$

例 6.23 设 d 是一正整数，那么 $d = (a, b)$ 的充要条件是 $(a/b, b/d) = 1$

证明： 由题知可得

$$\left(\frac{a}{b}, \frac{b}{d}\right)d = \left(\frac{a}{d} \cdot d, \frac{b}{d} \cdot d\right) = (a, b),$$

因此，如果 $d = (a, b)$，那么 $(a/b, b/d) = 1$.

反过来，如果 $(a/b, b/d) = 1$，那么 $d = (a, b)$.

例 6. 24   试证: $\left(\dfrac{a}{(a,c)},\dfrac{b}{(b,a)},\dfrac{c}{(c,b)}\right) = 1$

证明: 因为

$$
\begin{aligned}
(a,b)(b,c)(c,a) &= (ab,ac,b^2,bc)(c,a)\\
&= (abc,a^2b,ac^2,a^2c,b^2c,ab^2,bc^2,abc)\\
&= (a^2b,a^2c,ab^2,b^2c,ac^2,bc^2,abc)
\end{aligned}
$$

$$
\begin{aligned}
&(a(a,b)(b,c),b(b,c),(c,a),c(c,a),(a,b))\\
&= (a(ab,ac,b^2,bc),b(bc,ab,c^2,ab),c(ac,bc,a^2,ab))\\
&= (a^2b,a^2c,ab^2,abc,b^2c,ab^2,bc^2,ab^2,ac^2,bc^2,a^2c,abc)\\
&= (a^2b,a^2c,ab^2,b^2c,ac^2,bc^2,abc)
\end{aligned}
$$

所以

$$(a(a,b)(b,c),b(b,c)(c,a),c(c,a)(a,b)) = (a,b)(b,c)(c,a),$$

即

$$\left(\frac{a}{(a,c)},\frac{b}{(b,a)},\frac{c}{(c,b)}\right) = 1.$$

例 6. 25   设 $(a,b) = 1$, 证明 $(a^2 + b^2,ab) = 1$.

证明: 由 $(a,b) = 1$, 得 $(a^2,b) = 1$, 由此得 $(a^2 + b^2,b) = (a^2,b) = 1$;
同理, 因此 $(a^2 + b^2,ab) = 1$.

例 6. 26   设 m,n 都是正整数, m 为奇数, 证明: $(2^m - 1,2^n + 1) = 1$.

证明: 设 $(2^m - 1,2^n + 1) = d$ 则存在 $s,t \in Z$.
使得 $2^m = sd + 1,2^n = td - 1$. 由此得 $2^{mn} = (sd + 1)^n = (td - 1)^m$.
由二项式展开式得
$2^{mn} = pd + 1 = qd - 1(p,q \in Z)$.
所以 $(q - p) \cdot d = 2$, 故 $d \mid 2$,
因此, $d = 1$ 或 2.
又因为 $2^m - 1,2^n + 1$ 都是奇数, 所以 $d = 1$.

例 6. 27   若 n 是自然数, 证明 $(2n + 1,4n + 1) = 1$.

分析: 要证 $2n + 1$ 与 $4n + 1$ 互质, 将 $2n + 1$ 看成 a, $4n + 1$ 看作 b, 欲使
$$s(2n + 1) + t(4n + 1) = 1,$$
那么 s,t 应取什么值呢?
假设 $s(2n + 1) + t(4n + 1) = 1$,
则 $2ns + s + 4nt + t = 1$,

$$
\begin{cases}
2ns + 4nt = 0,\\
s + t = 1,
\end{cases}
$$

即

$$\begin{cases} s + 2t = 0, \\ s + t = 1, \end{cases}$$

解得 $t = -1, s = 2$.

故可得如下证明.

证明： 因为 $2(2n+1) + (-1)(4n+1) = 1$，根据上述分析，
$$(2n+1, 4n+1) = 1.$$

例 6.28 设 n 为正整数，证明：$(12n+5, 9n+4) = 1$.

证明： 由于 $(-3) \cdot (12n+5) + 4 \cdot (9n+4) = 1$，故 $(12n+5, 9n+4) = 1$.

例 6.29 设 n 是正整数，证明：$(21n+4, 14n+3) = 1$.

证明： 由欧几里得算法，得
$$\begin{aligned} (21n+4, 14n+3) &= (21n+4-14n-3, 14n+3) \\ &= (7n+1, 14n+3) = (7n+1, 14n+3-2(7n+1)) \\ &= (7n+1, 1) = 1. \end{aligned}$$

例 6.30 若 $(a, b_1) = (a, b_2) = \cdots = (a, b_n) = 1$. 求证 $(a, b_1 b_2 \cdots b_n) = 1$.

证明： 反复利用性质即可得：

$\because (a, b_1) = 1, \therefore (a, b_1 b_2 \cdots b_n) = (a, b_2 b_3 \cdots b_n)$，

$\because (a, b_2) = 1, \therefore (a, b_2 b_3 \cdots b_n) = (a, b_3 b_4 \cdots b_n)$，

...

$\because (a, b_{n-2}) = 1, \therefore (a, b_{n-2} b_{n-1} b_n) = (a, b_{n-1} b_n)$，

$\because (a, b_{n-1}) = 1, \therefore (a, b_{n-1} b_n) = (a, b_n)$，

$\because (a, b_n) = 1, \therefore (a, b_n) = 1$.

例 6.31 若 $a_1, a_2, \cdots, a_m$ 中的任一数和 $b_1, b_2, \cdots, b_n$ 中的任一数都互质. 求证：
$$(a_1 a_2 \cdots a_m, b_1 b_2 \cdots b_n) = 1.$$

证明： 因为 $(a_1, b_1) = (a_1, b_2) = \cdots = (a_1, b_n) = 1$，所以
$$(a_1, b_1 b_2 \cdots b_n) = 1.$$

由
$$(a_2, b_1) = (a_2, b_2) = \cdots = (a_2, b_n) = 1,$$
知
$$(a_2, b_1 b_2 \cdots b_n) = 1,$$

由
$$(a_m, b_1) = (a_m, b_2) = \cdots = (a_m, b_n) = 1,$$
知
$$(a_m, b_1 b_2 \cdots b_n) = 1.$$

同理，设 $b_1 b_2 \cdots b_n = b$，因为

$$(a_1,b) = (a_2,b) = \cdots = (a_m,b) = 1,$$

所以

$$(a_1a_2\cdots a_m,b) = 1,$$

即 $(a_1a_2\cdots a_m,b_1b_2\cdots b_n) = 1.$

例 6.32　若 $(a,b) = 1$，求证 $(a-b,a+b) = 1$ 或 2.

证明： 设 $(a-b,a+b) = d,$

则 $d\mid a-b$ 且 $d\mid a+b$，于是 $d\mid 2a$ 且 $d\mid 2b.$

所以 $d\mid (2a,2b).$

但 $(2a,2b) = 2(a,b) = 2,$

所以 $d\mid 2,$

因此 $d = 1$ 或 2.

例 6.33　设 a,m,n 是正整数，$m\neq n$，试证：$(2^{2^m}+1,a^{2^n}+1) = 1$ 或 2.

证明： 由 $m\neq n$，不妨设 $m > n$，且 $m = n+r(r\geqslant 1)$，

则 $a^{2^m}-1 = a^{2^{n+r}}-1 = (a^{2^n})^{2^r}-1 = (a^{2^n}+1)M$（$M$ 是正整数），

即

$$a^{2^m}+1 = (a^{2^n}+1)M+2.$$

所以：$(2^{2^m}+1,a^{2^n}+1) = (2^{2^n}+1,2)$

即

$$(2^{2^m}+1,a^{2^n}+1) = 1$$

例 6.34　正整数 a 和 b 互素，证明：$a+b$ 与 a^2+b^2 的最大公因数等于 1 或 2.

分析： 设 d 为 $a+b$ 和 a^2+b^2 的最大公因数，由于 $(a+b)^2 - (a^2+b^2) = 2ab$ 及 $2a(a+b) - 2ab = 2a^2$，可知 $d\mid 2a^2$，及 $d\mid 2b^2$，再由 $(a,b) = 1$，可证得本题.

证明： 设 d 是 a^2+b^2 及 $a+b$ 的最大公因数，

则有 $d\mid (a^2+b^2),d\mid (a+b),$

于是由 $(a+b)^2 - (a^2+b^2) = 2ab$，得 $d\mid 2ab.$

又因为 $2a(a+b) - 2ab = 2a^2,2b(a+b) - 2ab = 2b^2.$

所以 $d\mid 2a^2,d\mid 2b^2,$

因此，d 是 $2a^2$ 和 $2b^2$ 的公约数.

由题意设，$(a,b) = 1$，则有 $(a^2,b^2) = 1,$

所以 $2a^2$ 和 $2b^2$ 不可能被大于 2 的数整除.

因而 $d\leqslant 2.$

即 $a+b$ 与 a^2+b^2 的最大公因数是 1 或 2.

例 6.35　设 m,n 是正整数，试证：$(2^m-1,2^n-1) = 2^{(m,n)}-1.$

证明： 当 $m = n$ 时，结论显然成立.

现不妨设 $m > n$，则由欧几里得算法可得

$$m = nq_1 + r_1, n = r_1q_2 + r_2, \cdots, r_{n-2} = r_{n-1}q_n + r_n, r_{n-1} = r_nq_{n-1},$$
$$(m,n) = (n,r_1) = (r_1,r_2) = \cdots = (r_{n-1},r_n) = (r_n,0) = r_n$$

由于

$$2^m - 1 = 2^{nq_1+r_1} - 1 = 2^{r_1}(2^{nq_1} - 1) + (2^{r_1} - 1)$$
$$= (2^n - 1)N + (2^{r_1} - 1)$$

这里, N 是正整数,

故 $(2^m - 1, 2^n - 1) = (2^n - 1, 2^{r_1} - 1)$.

以此类推,

可得

$$(2^m - 1, 2^n - 1) = (2^n - 1, 2^{r_1} - 1)$$
$$= (2^{r_1} - 1, 2^{r_2} - 1)$$
$$= (2^{r_{n-1}} - 1, 2^{r_n} - 1)$$
$$= 2^{r_n} - 1$$
$$= 2^{(m,n)} - 1.$$

例 6.36 如果 $(a,b) = 1$, 那么 $(a,bc) = (a,c)$.

证明: 因为 $(a,bc) \mid ac, (a,bc)$, 所以 $(a,bc) \mid (ac,bc)$. 于是

$$(ac, bc) = (a, b) |c| = |c|,$$

因此 $(a,bc) \mid c$.

又因为 $(a,bc) \mid a$, 所以 $(a,bc) \mid (a,c)$.

反过来显然有 $(a,c) \mid a, (a,c) \mid bc$,

因此

$$(a,c) \mid (a,bc)$$

于是 $(a,bc) = (a,c)$.

例如, $(20, 973 \times 15) = (20,15) = 5$.

例 6.37 设 a,b,c,k 都是正整数, 且 $ab = c^k, (a,b) = 1$. 试证:
$$a = (a,c)^k, b = (b,c)^k.$$

证明: 由 $\left(\dfrac{a}{(a,c)}, \dfrac{c}{(a,c)}\right) = 1$ 知

$$\left(\left(\dfrac{a}{(a,c)}\right)^k, \left(\dfrac{c}{(a,c)}\right)^k\right) = 1$$

所以

$$(a^k, c^k) = (a,c)^k$$

又由 $(a,b) = 1$ 知 $(a^{k-1}, b) = 1$

所以

$$a = a(a^{k-1}, b) = (a^k, ab) = (a^k, c^k) = (a,c)^k.$$

类似可证, $b = (b,c)^k$.

例 6.38 设 a,b,c 是不全为零的整数, 若存在 $q \in Z$, 使得 $a = bq + c$, 则 $(a,b) \leqslant$

(b,c).

证明：由 $(b,c)\mid b,(b,c)\mid c$ 及 $a=bq+c$，知
$$(b,c)\mid a$$
因此 (b,c) 是 a 和 b 的公因数，但 (a,b) 是 a 和 b 的最大公因数，所以
$$(b,c)\leqslant(a,b)$$
类似地，可以得到 $(a,b)\leqslant(b,c)$，于是 $(a,b)=(b,c)$.

例 6.39　设 a,b 是整数，若 $9\mid(a^2+ab+b^2)$，证明：$3\mid(a,b)$.
证明：由 $9\mid(a^2+ab+b^2)$.
即 $9\mid(a-b)^2+3ab\Rightarrow3\mid(a-b)^2+3ab\Rightarrow3\mid(a-b)^2\Rightarrow3\mid a-b\Rightarrow9\mid(a-b)^2$
因而 $9\mid3ab\Rightarrow3\mid ab$，由此得 $3\mid a$ 或 $3\mid b$.
若 $3\mid a$，由 $3\mid(a-b)$ 得 $3\mid b$；若 $3\mid b$，由 $3\mid(a-b)$ 得 $3\mid a$.

例 6.40　设 $x,y\in Z,17\mid 2x+3y$，证明：$17\mid(9x+5y)$.
证明：因为 $4\cdot(2x+3y)+(9x+5y)=17(x+y)$，故 $17\mid(9x+5y)$.

例 6.41　n 为整数，求证：$30\mid n^5-n$.
分析：$30=6\times5$，又 6 与 5 互质. 可以分别证明 $6\mid n^5-n$ 以及 $5\mid n^5-n$ 即可.
证明：
$$\begin{aligned}n^5-n&=n(n^4-1)\\&=n(n^2-1)(n^2+1)\\&=(n-1)n(n+1)(n^2+1).\end{aligned}$$
因为 $(n-1)n(n+1)$ 是三个连续整数的乘积，可以被 $3!=6$ 整除.
所以 $6\mid n^5-n$. 又
$$\begin{aligned}n^5-n&=(n-1)n(n+1)(n^2+1)\\&=(n-1)n(n+1)[(n^2-4)+5]\\&=(n-2)(n-1)n(n+1)(n+2)+5(n-1)n(n+1).\end{aligned}$$
由于连续五个整数中必有一个为 5 的倍数，所以 $5\mid n^5-n$.
因为 $(5,6)=1$，所以 $30\mid n^5-n$.

例 6.42　x,y,z 均为整数. 若 $11\mid(7x+2y+5z)$，求证：$11\mid(3x-2y+12z)$.
证明：因为
$$4(3x-7y+12z)+3(7x+2y-5z)$$
$$=(12x-28y+48z)+(21x+6y-15z)$$
$$=11(3x-2y+3z),$$
而 $11\mid11(3x-2y+3z)$.
且 $11\mid 7x+2y-5z$，
因此 $11\mid4(3x-7y+12z)$.
又 $(11,4)=1$，

所以 $11 \mid (3x - 2y + 12z)$.

例 6.43 当 n 为正奇数时，求证：$60 \mid f(n) = 6^n - 3^n - 2^n - 1$.

证明： $f(n) = 6^n - 3^n - 2^n - 1$
$$= (6^n - 3^n) - (2^n + 1).$$

因为 n 为正奇数，所以 $3 \mid (6^n - 3^n)$，$3 \mid (2^n + 1)$，从而 $3 \mid f(n)$.

同理 $4 \mid f(n)$，$5 \mid f(n)$.

因为 3，4，5 两两互质，所以 $60 \mid f(n) = 6^n - 3^n - 2^n - 1$.

例 6.44 若 $(a, a_1) = (a, a_2) = \cdots = (a, a_{n-1}) = 1$，并且 $a \mid a_1 a_2 \cdots a_n (n \geq 2)$，求证：$a \mid a_n$.

证明： 用数学归纳法.

当 $n = 2$ 时，因为 $(a, a_1) = 1$ 且 $a \mid a_1 a_2$，由定理 3 知 $a \mid a_2$，命题成立.

假设当 $n = k$ 时命题成立，即
$$(a, a_1) = (a, a_2) = \cdots = (a, a_{k-1}) = 1,$$

且 $a \mid a_1 a_2 \cdots a_k$ 时，则有 $a \mid a_k$.

我们证明当 $n = k$ 时，即
$$(a, a_1) = (a, a_2) = \cdots = (a, a_{k-1}) = (a, a_k) = 1.$$

且 $a \mid a_1 a_2 \cdots a_{k-1} a_k a_{k+1}$，则有 $a \mid a_{k+1}$ 成立.

事实上由 $a \mid a_1 a_2 \cdots a_{k-1}(a_k a_{k+1})$，以及归纳假设知 $a \mid (a_k a_{k+1})$. 又 $a \mid (a, a_k) = 1$，于是可得 $a \mid a_{k+1}$ 成立.

根据数学归纳原理得证：对任意自然数 $n > 2$ 都有 $a \mid a_n$.

例 6.45 设 m, n 是正整数，且 $(m, n) = 1$，试证：$m! n! \mid (m + n - 1)!$.

证明： 注意到
$$\frac{(m + n - 1)!}{(m - 1)! n!} = c_{m+n-1}^n, \quad \frac{(m + n - 1)!}{m(n - 1)!} = c_{m+n-1}^m$$

都是整数，即存在整数 Q, Q'，使得
$$(m + n - 1)! = (m - 1)! n! Q = m(n - 1)! Q',$$

所以 $nQ = mQ'$，

由于 $(m, n) = 1$，

故 $n \mid Q'$.

即存在整数 Q''，

使得 $Q' = nQ''$，代入得 $(m + n - 1)! = m! n! Q''$.

因此
$$m! n! \mid (m + n - 1)!.$$

例 6.46 设 $m, n > 0, mn \mid (m^2 + n^2)$，证明：$m = n$.

证明： 设 $(m, n) = d$，则 $m = m_1 d, n = n_1 d$，其中 $(m_1, n_1) = 1$.

于是，已知条件化为 $m_1 n_1 \mid (m_1^2 + n_1^2)$，

由此得 $m_1 \mid (m_1^2 + n_1^2)$.

故 $m_1 \mid n_1^2$ 但是 $(m_1, n_1) = 1$，结合 $m_1 \mid n_1^2$，可知必须 $m_1 = 1$.

同理 $n^1 = 1$ 因此 $m = n$. 证毕.

例 6.47　设 k 为正奇数，证明：$1 + 2 + \cdots + n$ 整除 $1^k + 2^k + \cdots + n^k$.

证明： 因为 $1 + 2 + \cdots + n = \dfrac{n(n+1)}{2}$，且 $(n, n+1) = 1$.

所以结论等价于证明

$$n \mid 2(1^k + 2^k + \cdots + n^k), \quad (n+1) \mid 2(1^k + 2^k + \cdots + n^k)$$

事实上，由于 k 是奇数，利用配对法可得

$$2(1^k + 2^k + \cdots + n^k) = [1^k + (n-1)^k] + [2^k + (n-2)^k] + \cdots + [(n-1)^k + 1^k] + 2n^k,$$

上式的每个加项显然都是 n 的倍数，故其和也是 n 的倍数. 同理得

$$2(1^k + 2^k + \cdots + n^k) = (1^k + n^k) + [2^k + (n-1)^k] + \cdots + (n^k + 1^k)$$

上式 $n+1$ 的倍数，故 $n(n+1) \mid 2(1^k + 2^k + \cdots + n^k)$. 证毕.

例 6.48　某公司销售某种货物，去总收入 36963 为元，今年每件货物的售价（单价）不变，总收入 59570 元. 如果单价（以元为单位）是大于 1 的整数，问今年与去年各销售这种货物多少件？

解： 单价是 36963 与 59570 的公约数，由辗转相除法得出 $(36963, 59570) = 37$.

36963	1	59570
22607	1	36963
14356	1	22607
8251	1	14356
6105	1	8251
4292	2	6150
1813	1	2146
1665	5	1813
148	2	333
148	4	296
—		37

因为 37 的约数只有 1 与其本身，所以 36963，59570 的大于 1 的公约数只有 37，即单价为 37 元.

于是今年售出 $59570 \div 37 = 1610$ 件，去年售出 $36963 \div 37 = 999$ 件.

例 6.49　两个容器，一个容量为 27L，另一个为 15L，如何利用它们一桶油中倒出 6L 油来？

解： 因为 $(27, 15) = 3$

$$27 = 1 \times 15 + 12,$$
$$15 = 1 \times 12 + 3,$$
从而 $3 = 15 - 1 \times 12 = 15 - 1 \times (27 - 1 \times 15)$,
即 $3 = 2 \times 15 - 27$,
于是 $6 = 4 \times 15 - 5 \times 27$.

这表明需要往小容器里倒 4 次油，每次倒满就向大容器里倒，大容器满了就往桶里倒，这样在大容器第二次倒满时小容器里剩下的就是 6L.

例 6.50　设有两个容器 A 和 B，A 的容量为 27 升，B 的为 15 升. 如何利用它们从一桶油中倒出来 9 升油来？

解：易知 $(27,15) = 3$，且 $27 = 1 \times 15 + 12, 15 = 1 \times 12 + 3$
即

$$3 = 15 - 1 \times 12 = 15 - 1(27 - 1 \times 15) = 2 \times 15 - 27$$

故

$$9 = 6 \times 15 - 3 \times 27.$$

这表明向容器 B 内倒 6 次油，每次倒满后就向容器 A 内倒，容器 A 满了就向桶内倒，这样在容器 A 第三次倒满时，容器 B 内剩下的就是 9 升.

例 6.51　一块钢板，长 1 丈 3 尺 5 寸，宽 1 丈零 5 寸. 现在把它截成同样大小的正方形，正方形要最大的，并且不准剩下钢板. 求正方形的边长.

解：因为正方形要最大的，所以就要求正方形最大的边长是多少. 要求正方形最大的边长，就要求 135 寸和 105 寸的最大公因数. 由于

$$135 = 3^3 \times 5, 105 = 3 \times 5 \times 7,$$

所以 $(135,105) = 15$.

答：正方形的边长是 1 尺 5 寸.

例 6.52　有钢铁 3 根，一长 13 尺 5 寸，一长 24 尺 3 寸，一长 55 尺 8 寸. 现在要把它们截成相等的小段，每根都不许剩下，截成的小段要最长，求每小段长几寸？一共可以截成多少段？

解：由于
$$135 = 3^3 \times 5, 243 = 3^5, 558 = 2 \times 3^2 \times 31,$$
所以 $(135,243,558) = 9$，又有
$$\frac{135}{9} + \frac{243}{9} + \frac{558}{9} = 15 + 27 + 62 = 104.$$

答：截成小段的每段长 9 寸，一共可以截成 104 段.

例 6.53　已知一个角的度数为 $\dfrac{180°}{n}$，其中 n 不是被 3 整除的正整数，证明这个角可以用圆规与直尺三等分.

解：因为 n 与 3 互质，故有

$$1 = un - 3v$$

其中 u,v 均为正整数，从而 $\dfrac{60^\circ}{n} = u \cdot 60^\circ - \dfrac{180^\circ}{n} \cdot v$

即产生 $\dfrac{60^\circ}{n}$ 为 $\dfrac{180^\circ}{n}$ 的 $\dfrac{1}{3}$.

例 6.54　证明存在一个有理数 $\dfrac{a}{d}$，其中 $d < 100$，能使 $\left[k\dfrac{c}{d}\right] = \left[\dfrac{73}{100}k\right]$，对于 $k = 1$，2，3，$\cdots$，100 成立，这时 $[x]$ 表示实数 x 的整除部分，即不超过 x 的最大整数.

解：因为 73 与 100 互质，因此有才 c,v 存在使

$$73d - 100c = 1$$

事实上，设 $36x + 28y = 1$，由于 $k < 100$

$$\frac{73}{100}k - \frac{kc}{d} = \frac{k(733d - 100c)}{100d} = \frac{k}{100d}$$

所以 $0 < \dfrac{73k}{100} - \dfrac{kc}{d} < \dfrac{1}{d}$

注意由 $\left[\dfrac{kc}{d}\right] = n$，得 $\dfrac{kc}{d} < n + 1 = \dfrac{(n+1)}{d}d$，所以

$$\frac{73k}{100} < \frac{kc+1}{d} \leqslant \frac{(n+1)d}{d} = n + 1$$

从而 $\left[\dfrac{73\text{k}}{100}\right] = n = \left[\dfrac{kc}{d}\right]$

可以用连续求两个数的最大公约数的方法去完成，即先求出 $d_1 = (a_1, a_2)$，再求 $d_2 = (d_1, a_3) = (a_1, a_2, a_3)$，这样继续下去，最后得出

$$d_{n-1} = (d_{n-2}, a_n) = \cdots = (a_1, a_2, \cdots, a_n).$$

例 6.55　数列 1001，1004，1009 的通项是 $a_n = n^2 + 1000$，其中 $n \in N_+$. 对于每一个 n，用 d_n 表示 a_n 与 a_{n+1} 的最大公因数，求 d_n 的最大值，其中 n 取一切正整数.

解：设 $d_n = (1000 + n^2, 1000 + (n+1)^2)$，则 $d_n \mid (1000 + n^2)$，$d_n \mid (1000 + (n+1)^2)$. 所以 $d_n \mid (1000 + (n+1)^2 - 1000 - n^2)$，

即 $d_n \mid (2n + 1)$.

又因为 $2(1000 + n^2) = n(2n + 1) + 2000 - n$，

所以 $d_n \mid (2000 - n)$，$d_n \mid ((2n+1) + 2(2000 - n))$，即 $d_n \mid 4001$.

所以 $1 \leqslant d_n \leqslant 4001$.

又当 $n = 2000$ 时，有 $1000 + n^2 = 1000 + (2000)^2 = 1000 \cdot 4001$，$1000 + (n+1)^2 = 1000 + (2000 + 1)^2 = 1001 \cdot 4001$，所以 d_n 能得倒 4001，即 $(d_n)_{\max} = 4001$.

例 6.56　正整数 $a_1, a_2, \cdots a_{49}$ 的和为 999，令 d 为 $a_1, a_2, \cdots a_{49}$ 的最大公因数，问 d 的最大值为多少？

分析： 由 $a_i \geq d,(i = 1,2,\cdots,49)$ 以及 d 是 999 的约数，可以确定 d 的取值范围，逐一检验排除后即可得到符合题意的 d.

解： 由于 $d = (a_1,a_2,\cdots,a_{49})$，则 $d \mid (a_1 + a_2 + \cdots a_{49})$，即 $d \mid 999$，

故 d 是 $999 = 3^3 \cdot 37$ 的约数.

又因为 $d \mid a_k, k = 1,2,\cdots,49$，则 $a_k \geq d$，

于是必有 $999 = a_1 + a_2 + \cdots + a_{49} \geq 49d$，得到 $d \leq \dfrac{999}{49} < 21$.

又由 d 是 999 的约数可知 d 只能是 $1,3,9$.

又因为 999 能写成 49 个数的和：$\underbrace{9 + 9 + 9 + \cdots + 9}_{48个9} + 567 = 999$.

其中每一个数都能被 9 整除，所以 d 的最大值为 9.

例 6.57 设 $S_n = \displaystyle\sum_{k=1}^{n} (k^5 + k^7)$，求 S_n 与 S_{3n} 的最大公因数.

解： 由于 $S_1 = 1^5 + 1^7 = 2$，

$$S_2 = (1^5 + 1^7) + (2^5 + 2^7) = 2 \cdot 81 = 2 \cdot (1 + 2)^4,$$

$$S_3 = 2 \cdot 3^4 + 3^5 + 3^7 = 2 \cdot 6^4 = 2 \cdot (1 + 2 + 3)^4,$$

$$s_4 = 2 \cdot 6 + 4^5 + 4^7 = 2^5 \cdot 5^4 = 2 \cdot (1 + 2 + 3 + 4)^4.$$

由此猜想 $S_n = 2 (1 + 2 + \cdots + n)^4$.

下面用数学归纳法证明.

$n = 1$ 时，显然成立.

假设 $n = k$ 时，$S_n = 2 (1 + 2 + \cdots + n)^4$ 成立.

对于 $n = k + 1$ 时，$S_{k+1} = 2 (1 + 2 + \cdots + n)^4 + (k + 1)^5 + (k + 1)^7$

$$= \frac{1}{8} k^4 (k + 1)^4 + (k + 1)^5 + (k + 1)^7$$

$$= \frac{1}{8} (k + 1)^4 [k^4 + 8(k + 1) + 8 (k + 1)^3]$$

$$= \frac{1}{8} (k + 1)^4 (k + 2)^4$$

$$= 2 [1 + 2 + \cdots + k + (k + 1)]^4.$$

所以 $n = k + 1$ 时，$S_n = 2 (1 + 2 + \cdots + n)^4$ 成立.

于是对所有自然数 n，$S_n = 2 (1 + 2 + \cdots + n)^4$ 成立.

因此，$S_n = 2 \left[\dfrac{n(n + 1)}{2} \right]^4$，$S_{3n} = 2 \left[\dfrac{3n(3n + 1)}{2} \right]^4$.

(1) 当 $n = 2k$ 时，

$d = (S_n,S_{3n})$

$= \left(2 \left[\dfrac{2k(2k + 1)}{2} \right]^4, 2 \left[\dfrac{6k(6k + 1)}{2} \right]^4 \right)$

$= (2k^4 (2k + 1)^4, 2 \cdot 81k^4 (6k + 1)^4)$.

因为 $(2k + 1,6k + 1) = 1$，所以 $d = 2k^4 ((2k + 1)^4,81)$，

当 $k = 3t + 1$ 时, $(2k + 1)^4 = (6t + 3)^4 = 81 (2t + 1)^4$,

所以 $d = 2 \cdot 81k^4 = 2 \cdot 81 \cdot \dfrac{n^4}{2^4} = \dfrac{81}{8}n^4$.

当 $k \neq 3t + 1$ 时, $d = 2k^4 = \dfrac{n^4}{8}$.

(2) 当 $n = 2k + 1$ 时,

$S_n = 2 \left[(2k + 1)(k + 1) \right]^4, S_{3n} = 2 \left[3(2k + 1)(3k + 2) \right]^4$.

因为 $(3k + 2, 2k + 1) = 1, (3k + 2, k + 1) = 1$, 所以 $d = 2 (2k + 1)^4 (3^4, (k + 1)^4)$,

当 $k = 3t + 2$ 时, $k + 1 = 3(t + 1)$, 所以 $d = 2n^4 \cdot 3^4 = 162n^4$.

当 $k \neq 3t + 2$ 时, $d = 2n^4$.

例 6.58　平面上整点（纵、横坐标均是整数的点）到直线 $y = \dfrac{5}{3}x + \dfrac{4}{5}$ 的距离中的最

小值是 (　　)

　A. $\dfrac{\sqrt{34}}{170}$　　　　　B. $\dfrac{\sqrt{34}}{85}$　　　　　C. $\dfrac{1}{20}$　　　　　D. $\dfrac{1}{30}$

解： 先由点到直线的距离公式得到距离再作分析.

将已知直线化为 $25x - 15y + 12 = 0$, 设平面上整点 (x_0, y_0) 到直线的距离为 $d = \dfrac{|25x_0 - 15y_0 + 12|}{5\sqrt{34}}$. 而 $(25, 15) = 5$, 由裴蜀定理知 $25x_0 - 15y_0$ 表示 5 的所有倍数. 当

$25x_0 - 15y_0 = -10$ 时, d 取最小值 $\dfrac{2}{5\sqrt{34}} = \dfrac{\sqrt{34}}{85}$. 故选 B

例 6.59　设 $f(x) = 3x + 2$. 证明: 存在正整数 m, 使得 $f^{(100)}(m)$ 能被 1988 整除, 其中 $f^{(k)}(x)$ 表示 $\underbrace{f(f(\cdots f(x)\cdots))}_{k \text{个} f}$.

证明： 先由不动点知识求出 $f^{(100)}(m)$ 的表达式, 再由题设数字特征分析出 $(3^n, 1988) = 1(n$ 为自然数). 由题设知迭代不动点为 $x = -1$, 再由数学归纳法知 $f^{(100)}(n) = 3^{100}(n + 1) - 1$, 则 $f^{(100)}(m) = 3^{100}(m + 1) - 1 = 3^{100}m + 3^{100} - 1$.

因为 $(3^{100}, 1988) = 1$, 所以, 由裴蜀定理推论 2, 知存在自然数 u, v, 使得 $1988u - 3^{100}v = 1$

于是, $1988 | (1 + 3^{100}v)$. 从而, $1988 | \left[(3^{100} - 1)(1 + 3^{100}v) \right]$.

令 $m = (3^{100} - 1)v$, 则

$1988 | f^{(100)}(m)$.

例 6.60　设 k, n 为正整数, $(k, n) = 1$, 且 $0 < k < n$. 再 $M = \{1, 2, \cdots, n - 1\}$. 现对集合 M 中的每一个 i 染上蓝色或白色, 且满足:

(a) i 和 $n - i$ 同色;

(b) 当 $i \neq k$ 时, i 和 $|k - i|$ 同色.

证明所有的数均同色.

分析： 由 $(k,n)=1$ 借助裴蜀定理分类可证所有的 $i\in M$ 与 k 同色. 讨论过程要注意由 (b) 可得两个结论：一是 M 中 k 的整倍数与同色；二是 $i>k$ 时，i 与 $i-qk$ 同色（$i-qk>0$）.

证明： 据题设，可知存在整数 x,y 使得 $i=xk+yn$.

由 $1\leqslant i\leqslant n-1$，知 x,y 的取值无外乎以下三种情形：

$(1)x>0,y=0$ $\qquad\qquad$ $(2)x>0,y<0$ $\qquad\qquad$ $(3)x<0,y>0$

(1) 时结论成立，而 (3) 可化为 (2) 讨论.

因为由 (a) 可知 i 和 $n-i=(-x)k+(-y+1)n$ 同色. 若 $-y+1=0$，则化为 (1).

若 $-y+1<0$，则化为 (2).

综上，只需讨论 (2)，对于 (2)（必有 $x>-y$），此时又分为三类：(i) $k=i$；(ii) $k>i$；(iii) $k<i$.

对于 (i) 显然成立.

当 (iii) 出现时，由带余除法知 $i=qk+i'(0\leqslant i'<k,q,i\in Z)$. 若 $i'=0$ 则由 (1) 结论成立.

若 $1\leqslant i'<k$，由条件 (b) 知 i 和 $i'=i-qk=(x-q)k+yn$ 同色，用 i' 代替 i 讨论就化为 (ii).

当 (ii) 出现时，由条件 (b) 可知 i 和 $k-i=(-x+1)k-yn$ 同色；

再由条件 (a) 可知 i 和 $i''=n-(k-i)=(x-1)k+(y+1)n$ 同色.

若 $y+1=0$，则结论成立.

若 $y+1<0$，又化为情形 (2).

继续对 i'' 分 (i)、(ii)、(iii) 三种情形考虑，这样经过有限次讨论后（因为 $y<0$，且每讨论一次 y 增加 1，这样若干次后变为 0），总可以得到 i 和 $x'k$ 同色，即 i 和 k 同色.

例 6.61 集合 $M=\{u\mid u=12m+8n+4l,m,n,l\in Z\}$ 与 $N=\{u\mid u=20p+16q+12r,p,q,r\in Z\}$ 的关系为（　　）

A. $M=N$ $\qquad\qquad\qquad\qquad\qquad$ B. $M\subseteq N,N\subseteq M$

C. $M\subset N$ $\qquad\qquad\qquad\qquad\qquad$ D. $M\supset N$

解： 从两个集合中元素的表达式入手.

因为 $12m+8n+4l=4(3m+2n+l)$，$20p+16q+12r=4(5p+4q+3r)$，及 $(3,2,1)=1$，$(5,4,3)=1$，所以，由裴蜀定理知 $3m+2n+l$ 与 $5p+4q+3r$ 均可表示所有整数. 于是，$M=N=\{k\mid k=4l,l\in Z\}$. 故选 A

例 6.62 设 $n\geqslant m\geqslant 1$，n,m 为整数. 证明：$\dfrac{(m,n)}{n}C_n^m$ 为整数.

证明： 本题的关键是如何表示 (m,n).

由裴蜀定理，知存在 $r,t\in Z$，使得 $mr+nt=(m,n)$. 故

$$\frac{(m,n)}{n}C_n^m=\frac{mr+nt}{n}C_n^m=r\frac{m}{n}C_n^m+tC_n^m=rC_{n-1}^{m-1}+tC_n^m\in Z.$$

例 6.63　m 个盒子中各放若干个球，每次在其中 $n(n < m)$ 个盒中各加一球．证明：当 $(m, n) = 1$ 时，无论开始的分布情形如何，总可按上述方法进行有限次加球后，使得各盒子中的球数相等．

证明： 从极端情形出发，考虑球数最多和最少的盒子．

因为 $(m, n) = 1$，所以，由裴蜀定理，知存在 $u, v \in Z_+$ 使得

$$un = um + 1 = v(m - 1) + v + 1.$$

上式表明，对盒子连续加球 u 次，可使 $m - 1$ 个盒子各增加 v 个球，一个盒子增加了 $v + 1$ 个球．

这样可将多增加了一个球的盒子选择为原来球数最少的那一个，经过 u 次加球之后，原球数最多的盒子中的球数与球数最少的盒子中的球数之差减少 1．

因此，经过有限次加球后，各盒子中的球数相等．

例 6.64　已知 p, q 是正整数，且 $(p, q) = 1$．若一个整数 n 可以写成 $n = px + qy$ 的形式，其中 x, y 为非负整数，则称此数为"好数"；反之，则称为"坏数"．

（1）证明：一定存在正整数 c，使得两个整数 n 和 $c - n$ 中恰有一个是好数，而另一个是坏数．

（2）求非负整数中坏数的个数．

分析： 由题设特征 $(p, q) = 1$ 想到应用裴蜀定理．

（1）证明： 由定理知存在整数 u, v，使得 $n = pu + qv$．于是，对任意的整数 t 有 $n = p(u - tp) + q(v + tp)$．

从而可将 n 表示成

$$n = px + qy \, (x, y \in Z, 0 \leqslant x \leqslant q - 1). \qquad ①$$

易知当且仅当式①中的整数 $y \geqslant 0$ 时，n 是好数．

将整数 z 表示成 $z = px + qy \, (x, y \in Z, 0 \leqslant x \leqslant q - 1)$，则整数

$z' = (pq - p - q) - z = p(q - 1 - x) + q(-1 - y) \, (0 \leqslant q - 1 - x \leqslant q - 1)$，

当且仅当 y 为非负整数时，$-1 - y$ 为负整数．从而，z 与 z' 中恰有一个好数一个坏数．于是，取 $c = pq - p - q$，满足（1）的结论．

（2）解： 由（1）及所有负整数均是坏数，知所有大于 $pq - p - q$ 的整数均为好数，而在 $0, 1, \cdots, pq - p - q$ 中有一半为坏数，一半为好数，亦即共有

$$\frac{pq - p - q + 1}{2} = \frac{(p - 1)(q - 1)}{2}$$

个非负整数为坏数．

例 6.65　确定所有的正整数对 (n, p)，满足 p 是一个素数，$n \leqslant 2p$，且 $(p - 1)^n + 1$ 能被 n^{p-1} 整除．

解： 先从 n 寻找解题突破口，再利用费马小定理求解．

显然，$(1, p)$ 和 $(2, 2)$ 满足题意．

下面考虑 $n \geqslant 2, p \geqslant 3$ 的情形．

因为 $(p - 1)^n + 1$ 是奇数，所以 n 也是奇数．从而，$n < 2p$．记 q 为 n 的任一素因子，

则 $q \mid [(p-1)^n + 1]$. 于是,$(p-1)^n \equiv -1 (\bmod q)$,且 $(q, p-1) = 1$.

由 q 的选取知
$$(n, p-1) = 1.$$

由裴蜀定理知存在整数 u, v 使得 $un + v(q-1) = 1$.

据费马小定理,知
$$p - 1 \equiv (p-1)^{un} (p-1)^{v(q-1)} \equiv (-1)^u \times 1^v (\bmod q).$$

因为 u 必为奇数,所以,$p - 1 \equiv -1 (\bmod q)$.

这表明 $q \mid p$,进而有 $q = p$,故证得 $n = p$.

于是,
$$p^{p-1} \mid [(p-1)^p + 1] = p^2(p^{p-2} - C_p^1 p^{p-3} + \cdots + C_p^{p-3} p - C_p^{p-2} + 1).$$

在上式的小括号中,除最后一项外均可被 p 整除($p \mid C_p^k, 1 \leqslant k \leqslant p-1$). 这表明 $p - 1 \leqslant 2$,即得 $p = 3$.

综上,所有的解为 $(2,2)$,$(3,3)$ 和 $(1,p)$,其中 p 为任意素数.

例 6.66 若 a, b, n 是正整数,证明:

(i) $(a^n, b^n) = (a, b)^n$.

(ii) $(na, nb) = n(a, b)$.

(i) 证:假设 $(a, b) = d$,则必然有 $\left(\dfrac{a}{d}, \dfrac{b}{d}\right) = 1$,因此 $\left[\left(\dfrac{a}{d}\right)^n, \left(\dfrac{b}{d}\right)^n\right] = 1$,即
$$\left(\frac{a^n}{d^n}, \frac{b^n}{d^n}\right) = 1$$

所以 $(a^n, b^n) = d^n$.

由于假设 $(a, b) = d$,因此得到
$(a^n, b^n) = (a, b)^n$.

(ii) 证:假设 $(a, b) = d$,那么 a 和 b 可以写成 $a = a_1 d$,$b = b_1 d$,并且 $(a_1, b_1) = 1$,因此 $(na, nb) = (na_1 d, na_2 d) = nd$.

所以由假设得到 $(na, nb) = n(a, b)$.

例 6.67 利用上一题关于最大公因数的性质求下列最大公因数:

(i) 216,64,1000.

(ii) 24000,36000,144000.

解:(i) 上一题关于最大公因数的性质显然可以推广到多于两个数的情形,所以
$$(216, 64, 1000) = (6^3, 4^3, 10^3) = (6, 4, 10)^3 = 2^3 = 8.$$

(ii)
$$\begin{aligned}
(24000, 36000, 144000) &= 1000 \times (24, 36, 144) \\
&= 1000 \times 12 \times (2, 3, 12) \\
&= 1000 \times 12 \\
&= 12000
\end{aligned}$$

例 6.68　证明：假设 a 和 b 是任意两个正整数，并且

$$a = p_1^{\alpha_1} p_2^{\alpha_2} \cdots p_k^{\alpha_k}, \alpha_i \geqslant 0, i = 1, 2, \cdots, k,$$

$$b = p_1^{\beta_1} p_2^{\beta_2} \cdots p_k^{\beta_k}, \beta_i \geqslant 0, i = 1, 2, \cdots, k,$$

这里 $p_1, \cdots p_k$ 为不同的素因数. 又假设 r_i 是 α_i 和 β_i 中较小的数，δ_i 是 α_i 和 β_i 中较大的数，则

$$(a, b) = p_1^{r_1} p_2^{r_2} \cdots p_k^{r_k},$$

$$(a, b) = p_1^{\delta_1} p_2^{\delta_2} \cdots p_k^{\delta_k}.$$

证：因为 r_i 是 α_i 和 β_i 两个数中小的那一个，所以 $\dfrac{p_i^{\alpha_i}}{p_i^{r_i}}$ 和 $\dfrac{p_i^{\beta_i}}{p_i^{r_i}}$ 中必有一个数是 1，而另一个数是整数，所以 $\left(\dfrac{p_i^{\alpha_i}}{p_i^{r_i}}, \dfrac{p_i^{\beta_i}}{p_i^{r_i}} \right) = 1$. 令 $i = 1, 2, \cdots, k$，就得到下面这个 k 个等式：

$$\left(\frac{p_1^{\alpha_1}}{p_1^{r_1}}, \frac{p_1^{\beta_1}}{p_1^{r_1}} \right) = 1,$$

$$\left(\frac{p_2^{\alpha_2}}{p_2^{r_2}}, \frac{p_2^{\beta_2}}{p_2^{r_2}} \right) = 1,$$

$$\cdots\cdots$$

$$\left(\frac{p_k^{\alpha_k}}{p_k^{r_k}}, \frac{p_k^{\beta_k}}{p_k^{r_k}} \right) = 1.$$

由于 $p_1, p_2, \cdots p_k$ 为不同的素数，所以由上面的等式得到

$$\left(\frac{p_1^{\alpha_1}}{p_1^{r_1}} \frac{p_2^{\alpha_2}}{p_2^{r_2}} \cdots \frac{p_k^{\alpha_k}}{p_k^{r_k}}, \frac{p_1^{\beta_1}}{p_1^{r_1}} \frac{p_2^{\beta_2}}{p_2^{r_2}} \cdots \frac{p_k^{\beta_k}}{p_k^{r_k}} \right) = 1,$$

即 $\left(\dfrac{a}{p_1^{r_1} p_2^{r_2} \cdots p_k^{r_k}}, \dfrac{b}{p_1^{r_1} p_1^{r_1} \cdots p_k^{r_k}} \right) = 1$，所以 $(a, b) = p_1^{r_1} p_2^{r_2} \cdots p_k^{r_k}$.

又由 r_i 和 δ_i 的定义可知 $r_i + \delta_i = \alpha_i + \beta_i$，即

$$\delta_i = (\alpha_i + \beta_i) - r_i, 1 \leqslant r \leqslant k.$$

于是可以得到

$$\{a, b\} = \frac{a \cdot b}{(a, b)} = \frac{p_1^{\alpha_1} p_2^{\alpha_2} \cdots p_k^{\alpha_k} \cdot p_1^{\beta_1} p_2^{\beta_2} \cdots p_k^{\beta_k}}{p_1^{r_1} p_2^{r_2} \cdots p_k^{r_k}} = p_1^{\alpha_1 + \beta_1 - r_1} p_2^{\alpha_2 + \beta_2 - r_2} \cdots p_k^{\alpha_k + \beta_k - r_k} = p_1^{\delta_1} p_2^{\delta_2} \cdots p_k^{\delta_k}$$

此题说明，两个正整数的公共素因数的较小次幂的乘积就是这两个正整数的最大公因数. 两个正整数的所有素因数的较大次幂的乘积就是这两个正整数的最小公倍数，对于多于这两个数的情形，这个结果同样适用.

例 6.69　有一间长方形的屋子长 5.25 米，宽 3.25 米. 现用方砖铺地，要恰好铺满整个屋子，问所用方砖最大边长是多少？

解：要用方砖恰好铺满整个屋子，所有方砖的最大边长就应该是屋子的长与宽的最大公因数. 为了去掉小数，我们用厘米作为长度单位，屋子的长为 525 厘米，宽 325 厘米.

由于

$$(525, 325) = 5 \times (105, 65) = 5 \times 5 \times (21, 13) = 5 \times 5 = 25,$$

所以 $(525,325) = 25$.

答：所用方砖的最大边长是 25 厘米.

例 6.70 一块长方形的木料长 3 尺 5 寸 7 分，宽 1 尺另 5 分，厚 8 寸 4 分，要把它锯成同样大小的方木块，木块的体积要最大，问木块的边长是多少？

解： 木块的边长显然是木料的长、宽、厚的最大公因数. 我们用分作为长度单位，由于

$$357 = 3 \times 7 \times 17,$$
$$105 = 3 \times 5 \times 7,$$
$$84 = 2^2 \times 3 \times 7,$$

所以 $(357,105,84) = 3 \times 7 = 21$.

答：木块的边长是 2 寸 1 分.

例 6.71 甲、乙、丙三个班的人数分别是 54 人，48 人，72 人，现要在各班内分别组织体育锻炼小组，但各小组的人数要相同，问锻炼小组的人数最多是多少？这时甲、乙、丙三班各有多少个小组？

解： 由于各班学生都要组织在锻炼小组内，而且各小组的人数相同，所以每组的人数必须是三个班学生人数的公因数，求小组的最多人数就是求三个班学生人数的最大为公因数，由于

$$54 = 2 \times 3 \times 3 \times 3 = 2 \times 3^3,$$
$$48 = 2 \times 2 \times 2 \times 2 \times 3 = 2^4 \times 3,$$
$$72 = 2 \times 2 \times 2 \times 3 \times 3 = 2^3 \times 3^2,$$

所以 $(54,48,72) = 2 \times 3 = 6$. 又 $\frac{54}{6} = 9, \frac{48}{6} = 8, \frac{72}{6} = 12$.

答：锻炼小组人数最多为 6 人. 这时甲班有 9 个组，乙班有 8 个组，丙班有 12 个组.

例 6.72 一箱手榴弹，设每颗手榴弹的重量都是超过一斤的整数斤，去掉箱子重量后净重 201 斤，然后拿出若干颗手榴弹后，净重 183 斤，求证每颗手榴弹的重量为 3 斤.

证： 由于 201 斤和 183 斤都是整数颗手榴弹的重量，所以每颗手榴弹的重量必定是它们的公因数. 它们的最大公因数是

$$(201,18) = 3 \times (67,61) = 3,$$

由于 3 是素数，而 3 的因数只有 1 和 3，因此每颗手榴弹的重量必定是 3 斤.

6.5.2 典型练习题

练习 6.1 $(4, 14)$，$(314, 159)$，$(3141, 1592)$，$(10001, 100083)$ 各是多少？

练习 6.2 设 n 为任意正整数，$(n,1)$ 是什么？$(n,0)$ 是什么？

练习 6.3 若 d 为正整数，(d,nd) 是什么？

练习 6.4 求出 $299x + 274y = 13$ 的一组解.

练习 6.5 证明：若 $a \mid b$，$a > 0$，则 $(a,b) = a$.

练习 6.6　证明：$((a,b),b) = (a,b)$.

练习 6.7　（a）证明：对所有 $n > 0$，有 $(n,n+1) = 1$；

（b）当 $n > 0$ 时，$(n,n+2)$ 可取说明值？

（c）当 $n > 0$ 时，$(n,n+k)$ 可取说明值？

提示：若 $d \mid n$，$d \mid (n+2)$，则 $d \mid 2$.

练习 6.8　若 $N = n_1 n_2 \cdots n_k + 1$，证明：对于 $i = 1,2,\cdots,k$，有 $(n_i,N) = 1$.

提示：若 $d \mid n_i$，$d \mid N$，则 $d \mid (N - n_1 n_2 \cdots n_k)$.

练习 6.9　证明：若 $(a,b) = 1$，$c \mid a$，则 $(c,b) = 1$.

练习 6.10　求 x 和 y，使

（a）$341x + 159y = 1$；　　　　　　（b）$3141x + 1592y = 1$；

（c）$4144x + 7696y = 592$；　　　　（d）$10001x + 100083y = 73$.

提示：（a）$x = -40, y = 79$；（b）$x = 37, y = -73$；（c）$x = 2, y = -1$；（d）$x = -10, y = 1$.

练习 6.11　（a）证明：当且仅当 $(k,n) = 1$，成立 $(k,n+rk) = 1$.

（b）当且仅当 $(k,n) = d$ 时，$(k,n+k) = d$ 成立，这一说法对不对？

（c）当且仅当 $(k,n) = d$ 时，对所有整数 r 有 $(k,n+rk) = d$ 这一说法对不对？

练习 6.12　（a）证明：$(299,249) = 13$；

（b）求出 $299x + 247y = 13$ 的两组解；

（c）求出 $299x + 247y = 52$ 的两组解.

提示：（a）若 $229r + 247s = 13$，则 $299(r+247) + 247(s-229) = 13$.（b）$x = 5 + 19t, y = -6 - 23t$，其中 t 为整数，这给出了方程的所有解；（c）$x = 1 + 19t, y = -1 - 23t$，其中 t 为整数，这给出了方程的所有解.

练习 6.13　解下列一次不定方程：

（1）$19x + 20y = 1909$；

（2）$1485x + 1745y = 15$；

（3）$1492x + 1066y = -4$；

（4）$8x - 18y + 10z = 16$.

练习 6.14　（a）若 $x^2 + ax + b = 0$ 有一整数根，证明此根整除 b；

（b）若 $x^2 + ax + b = 0$ 有一有理数根，证明此根实际上是一整数.

提示：若 r/s 为一根，则 $(r/s)^2 + a(r/s) + b = 0$，或 $r^2 + ars + bs^2 = 0$. 利用定理 4 证明：若 $s \mid r^2$，$(r,s) = 1$；则 $s \mid 1$.

练习 6.15　证明：若 $c \mid ab$，$(c,a) = d$，则 $c \mid db$.

提示：$(c/d,a/d) = 1$，$(c/d) \mid (a/d)b$，再用推论.

练习 6.16　证明：若 d 为奇数，$d \mid (a+b)$，$d \mid (a-b)$，则 $d \mid (a,b)$.

提示：先证 $d \mid 2a$ 和 $d \mid 2b$，再用推论.

练习 6.17　证明：“若 $a \mid b$，则 $(a,b) = 1$”未必成立.

练习 6.18　已知 $m > 0, n > 0$，且 m 是奇数，试证：$(2^m - 1, 2^n + 1) = 1$.

练习 6.19　设 $ax_o + by_o$ 是形如 $ax + by$（x,y 是任意整数，a,b 是两个不全为零的整数）的数中最小的正整数，试证：$(a,b) = ax_o + by_o$.

练习6.20 试证：$(2^p - 1, 2^q - 1) = 1$ 的充要条件是 $(p, q) = 1$.

练习6.21 设 $n \geq 2$，试证：存在 n 个合数可组成一个等差数列，而且其中任意两个数互素.

练习6.22 求下面的最大公因数：

(1) $(2^t + 1, 2^t - 1)(t > 0)$；　　　(2) $(n - 1, n^2 + n + 1)$.

练习6.23 试给出四个正整数，它们的最大公因数是1，但任何三个数都不互素.

练习6.24 求 $(353430, 530145, 165186)$.

练习6.25 试证：$(a, b, c)(ab, bc, ca) = (a, b)(b, c)(c, a)$.

练习6.26 证明：$6 \mid n(n + 1)(2n + 1)$，其中 n 是任何整数.

练习6.27 证明：任意 n 个连续整数中（$n \geq 1$），有一个且只有一个数被 n 除尽.

练习6.28 证明：若 $m - p \mid (mn + qp)$，则 $m - p \mid (mq + np)$.

练习6.29 证明：若 $p \mid (10a - b)$ 和 $p \mid (10c - d)$，则 $p \mid (ad - bc)$.

练习6.30 证明：若 $(a, b) = 1$，则 $(a + b, a - b) = 1$ 或 2.

练习6.31 证明：若 $(a, b) = 1$，则 $(a + b, a^2 - ab + b^2) = 1$ 或 3.

练习6.32 证明：若方程 $x^n + a_1 x^{n-1} + \cdots + a_n = 0$（$n > 0, a_i$ 是整数，$i = 1, \cdots, n$）有有理数解，则此解必为整数.

练习6.33 一个有理数 $\dfrac{a}{b}$，当 $(a, b) = 1$ 时叫作既约分数. 证明：若两个既约分数 $\dfrac{a}{b}, \dfrac{c}{d}$ 的和是一个整数，则 $|b| = |d|$.

练习6.34 如果一个整数不能被任一个素数的平方所整除则称为无平方因子. 证明：对每一个整数 $n \geq 1$，能唯一决定 $a > 0, b > 0$ 使得 $n = a^2 b$，这里 b 无平方因子.

练习6.35 证明：若 b^2 是 n 的最大平方因子，则由 $a^2 \mid n$，可推出 $a \mid b$.

练习6.36 给定 x 和 y，若 $m = ax + by, n = cx + dy$，这里 $ad - bc = \pm 1$，证明 $(m, n) = (x, y)$.

练习6.37 证明：若 $n > 0, a^n \mid b^n$，则 $a \mid b$.

练习6.38 证明：若 $(a, b) = 1$，且 $ab = c^n$，则 $a = x^n, b = y^n, c = xy$.

练习6.39 证明：对于同样的整数 x 和 y，$17 \mid 2x + 3y$ 的充分必要条件是 $17 \mid 9x + 5y$.

练习6.40 设 $5 \mid df(x) = ax^3 + bx^2 + cx + d, g(x) = dx^3 + cx^2 + bx + a$. 证明：若存在 m，使 $5 \mid f(m)$，则存在 n，使 $5 \mid g(n)$.

练习6.41 证明：如果 a 和 b 是正整数，那么等差数列 $a, 2a, 3a, \cdots, ba$ 中能被 b 整除的项的个数等于数 a 和 b 的最大公约数.

练习6.42 假设 a, b, c, d 是整数，证明：若数 $ac, bc + ad, bd$ 都能被某整数 u 整除，则 bc 和 ad 也都能被 u 整除.

练习6.43 证明：若 a, b 是任意两个不全为零的整数，m 为任一正整数，则 $(am, bm) = (a, b)m$.

练习6.44 证明 $(a_1, a_2, \cdots, a_n) = ((a_1, \cdots, a_s), (a_{s+1}, \cdots, a_n))$.

练习6.45 证明 $[b_1, \cdots, b_n] = [[b_1, \cdots, b_s], [b_{s+1}, \cdots, b_n]]$.

练习6.46 证明：若 $a > 0, b > 0, a' > 0, b' > 0, (a, b) = d, (a', b') = d'$，则 $(aa', ab',$

$ba', bb') = dd.$

练习 6.47　证明：若 $n > 0, d \mid 2n^2$，则 $n^2 + d$ 不是完全平方.

练习 6.48　证明：对于给定的 $n > 0$，数对 $\{u, v\}$ 适合 $[u, v] = n$ 的对数为 n^2 的因数的个数.

练习 6.49　证明：对于任何自然数 $n, \dfrac{21n + 4}{14n + 3}$ 是既约分数.

练习 6.50　证明：若 $m > 0, n > 0, (m, n) = 1$，方程 $x^m = y^n$ 的全部整数解可以由 $x = t^n, y = t^m$ 给出，其中 t 取任意整数.

练习 6.51　证明：对于平面上任给的五个整点（即点的坐标都是整数的点）$A_i(x_i, y_i)(i = 1, 2, \cdots, 5)$，必有其中两点的连线的中点也是整点.

练习 6.52　证明：若方程组

$$a_{11}x_1 + a_{12}x_2 + \cdots + a_{1q}x_q = 0,$$
$$a_{21}x_1 + a_{22}x_2 + \cdots + a_{1q}x_q = 0,$$
$$\vdots \qquad \qquad \vdots$$
$$a_{p1}x_1 + a_{p2}x_2 + \cdots + a_{pq}x_q = 0$$

中，未知数的个数 q 与方程的个数 p 间满足 $q = 2p$，而且系数 a_{ij} 仅取 $-1, 0$ 或 $+1$. 则这个方程组必有满足下列条件的解 $(x_1, \cdots, x_q)$：

①所有的 x_j 都是整数；

②对于某些 $j(1 \leq j \leq q), x_j \neq 0$；

③对所有 j $(1 \leq j \leq q)$，$|x_j| \leq q$.

练习 6.53　设 $n > 2, V_n$ 是一个形如 $1 + kn$ 的数集（其中 $k = 1, 2, \cdots$）. 一个数 $m \in V_n$，如果不存在 $p, q \in V_n$，使得 $pq = m$，则称 m 为 V_n 中的不可约数. 证明：存在着一个数 $r \in V_n$，这个数可以用不止一种方式分解成为数集 V_n 中若干不可约数的乘积.

练习 6.54　证明：$\dfrac{[a, b, c]^2}{[a, b][b, c][c, a]} = \dfrac{(a, b, c)^2}{(a, b)(b, c)(c, a)}$.

练习 6.55　证明：一正整数为其诸因数（除本身为）之积的充分必要条件是此数唯一素数的立方，或为两素数之积.

练习 6.56　证明：6 是仅有的无平方因子数.

练习 6.57　证明：$2^{2^n} - 1$ 至少有 n 个不同的素因数.

练习 6.58　证明：$\dfrac{1}{x} + \dfrac{1}{x + 1} + \cdots + \dfrac{1}{x + n}(x > 0)$ 不是整数.

练习 6.59　证明：若 p_n 表第 n 个素数，则 $p_n < 2^{2^n}$.

练习 6.60　证明：若 $m > 0, n > 0, m$ 是奇数，则 $(2^m - 1, 2^n + 1) = 1$.

练习 6.61　证明：若 $(a, b) = 1, m > 0$，则数列 $\{a + bk\}, k = 0, 1, \cdots$ 中存在无限多个数与 m 互素.

练习 6.62　若正整数 a，质数 p 满足 $a < p$. 请你证明 $(a, p) = 1$.

练习 6.63　求证，对任意自然数 $a, a, a + 1, 2a + 1$ 两两互质.

练习 6.64　若 n 为自然数，求证 $(n - 1, n + 1)$ 等于 1 或 2.

练习 6.65　若 $(a, b) = 1. c = a + b$. 求证 $(c, a) = (c, b) = 1$.

练习 6.66 若 $(a,b) = 1$，且 $c \mid a, d \mid b$．求证 $(c,d) = 1$．

练习 6.67 已知 a, b, c, d 都是整数，且 $(b,n) = 1$．若 $n \mid (ad - bc)$，又 $n \mid a - b$．求证 $n \mid c - d$．

练习 6.68 求所有的正整数对 (x,n)，使得 $x^n + 2^n + 1 \mid x^{n+1} + 2^{n+1} + 1$．

练习 6.69 求所有大于 2 的正整数对 (m,n)，使得：存在无穷多个正整数 a，满足 $a^n + a^2 - 1 \mid a^m + a - 1$．

练习 6.70 求所有正整数三元组 (a,b,c) 使得 $\left(b - \dfrac{1}{a}\right)\left(c - \dfrac{1}{b}\right)\left(a - \dfrac{1}{c}\right)$ 是整数．

练习 6.71 设 $m,n \in N^*$，且 $m \equiv 1 (\bmod 2)$，求证：$(2^m - 1, 2^n + 1) = 1$．

练习 6.72 求集合 $\{16^n + 10n - 1 \mid n \in N^*\}$ 中所有元素的最大公因数．

练习 6.73 正整数 $a_i (i = 1, 2, \cdots, 49)$ 满足 $a_1 + a_2 + \cdots + a_{49} = 999$，求 $d = (a_1, a_2, \cdots, a_{49})$ 的最大值．

练习 6.74 C 为复数集，设集合

$A = \{z \mid z^{18} = 1, z \in C\}, B = \{w \mid w^{48} = 1, w \in C\}, D = \{zw \mid z \in A, w \in B\}$．

求 D 的元素个数．

［提示：$z = e^{\frac{2k\pi}{18}i}, w = r^{\frac{2l\pi}{48}i} (k,l \in Z)$．则 $zw = e^{\frac{2(8k+3l\pi)}{144}i}$．因为 $(8,3) = 1$，所以，$8k + 3l$ 可表示所有的整数，故 D 中有 144 个元素］

练习 6.75 设 $m > n > 1$．已知 $2^m - 1$ 个盒子中装有一些球，若给其中的 $2^n - 1$ 个盒子中添加一个球，则称为一次操作．证明：当 $(m,n) = 1$ 时，无论 $2^m - 1$ 个盒子里已有多少个球，经有限次操作后，总能使所有盒子里的球一样多．

［提示：$(m,n) = 1$，存在正整数 a,b，使得 $na - bm = 1$，令 $x = 2^m - 1, y = 2^n - 1$，则 $2^{bm} = (x + 1)^b, 2^{an} = (y + 1)^a$．从而，$(y + 1)^a = 2(x + 1)^b$．由二项式定理展开，知存在正整数 M, N，使得 $yM = xN + 1$，即 $(2^n - 1)M = (2^m - 1)N + 1$］

练习 6.76 若 p,q 互素，且 $p,q \in N$，则存在最小的 $m = pq - p - q + 1$，使得对所有 $n \geq m, n$ 都可写成 $n = px + qy$（x,y 是非负整数）．

［提示：只需证 $pq - p - q$ 不可表示为 $px + qy$．反证法．若 $pq - p - q = ap + bp$（a,b 非负），则 $pq = (1 + a)p + (1 + b)q$．因此，$p \mid (1 + b), q \mid (1 + a)$．令 $1 + a = a'q, 1 + b = b'p (a', b' \geq 1)$，得 $pq = (a' + b')pq$．但 $a' + b' = 1$，矛盾］

练习 6.77 数列 $f_{n+1}(x) = f_1(f_n(x))(f_1(x) = 2x + 1, n \in N)$．试证：对任意的 $n \in \{11, 12, \cdots\}$，必存在一个由 n 唯一确定的 $m_0 \in \{0, 1 \cdots 1993\}$，使得 $1995 \mid f_n(m_0)$．

［提示：$f_n(m) = 2^n(m + 1) - 1$．由 $(2^n, 1995) = 1$，知存在 $1 \leq u_0 \leq 1994, u_0 \in Z$，使得 $2^n u_0 + 1995 k_1 = 1, k \in Z$．从而，$1995 \mid (2^n u_0 - 1)$．取 $m_0 = u_0 - 1$，则 $1995 \mid f_n(m_0)$（$0 \leq m_0 \leq 1993$）．唯一性可由带余除法证明］

练习 6.78 给定正整数 a, b, c，定义函数 $f(x,y,z) = ax + by + cz (x,y,z \in Z)$．试求 $f(x,y,z)$ 的最小正整数值．

［提示：记 $f_{\min} = ax_0 + by_0 + cz_0 = d_0$．设 $(a,b,c) = d$，显然，有 $d \mid d_0$．从而，$d \leq d_0$．另一方面，由裴蜀定理，知存在整数 x, y, z，使得 $ax + by + cz = d$．因为 d_0 最小，所以，$d_0 \leq d$．因此，$d_0 = d$，即 $f_{\min} = (a,b,c)$］

第 7 章　最小公倍数

信念是鸟，它在黎明仍然黑暗之际，感到了光明，唱出了歌.

——泰戈尔

现实中，也经常会遇到这样的问题：军训阅兵时，需要将队伍排成 8 行、10 行、12 行、15 行时，队形都是矩形，最少需要多少人参加这次阅兵？要解决这一问题，就需要寻找这些数的共有的倍数中的最小一个，也就是本章的内容了．

和上一章最大公因数所讨论的一样，要认识事物的另一个方面是先从所有可能的选项中找出一个具体选项，随后将与此选项相关、类似的都找出来进行研究．第一步所研究的是只有一个整数的时候，一个整数的倍数所具有的一些性质．第二步自然是有若干个整数的时候，研究它们具有些什么性质。接下来就要来进行讨论．本章主要讨论多个整数的最小公倍数，许多讨论可以和上一章的最大公因数相对应．

7.1 公倍数

用整除性对倍数进行定义的话，倍数指的是能够被已知数整除的所有整数．如 6 是 3 的倍数，$2a$ 是 a 的倍数等．这里涉及的已知数是一个整数，对于若干个整数的情况，需要考虑的就是它们所具有的共同的性质，比如它们所同时拥有的倍数，就被称作公倍数，也即公倍数指的就是公共的倍数．

比如整数 2 的倍数有 $\{2, 4, 6, 8, 10, 12, \cdots\}$，这是因为
$$2 \mid 2, \ 2 \mid 4, \ 2 \mid 8, \ 2 \mid 10, \ 2 \mid 12 \cdots$$

也可以表示为
$$2 = 1 \times 2, 4 = 2 \times 2, 6 = 3 \times 2$$
$$8 = 4 \times 2, 10 = 5 \times 2, 12 = 6 \times 2$$

整数 3 的倍数有 $\{3, 6, 9, 12, 15, \cdots\}$，这是因为
$$3 \mid 3, 3 \mid 6, 3 \mid 9, 3 \mid 12, 3 \mid 15, \cdots$$

也可以表示为
$$3 = 1 \times 3, 6 = 2 \times 3, 9 = 3 \times 3, 12 = 4 \times 3, 15 = 5 \times 3$$

它们当中所共有的数是 6，12…这样就说 6，12 是 2 和 3 的公倍数，这是因为 6，12……既能被 2 整除，又能被 3 整除．

将上述讨论数学化，可以得到：

某数 d 称为另外一组（两个及以上 $[a, b]$）数的公倍数，需要满足同时是这一组数中所有数的因数，即 $a \mid d$，且 $b \mid d$．

7.2 最小公倍数

7.2.1 最小公倍数

顾名思义，最小公倍数指的是若干个不为零的整数的公倍数中数值最小的一个正整数，后面若不做特别说明，本书所指的一些整数都是不为零的．以两个整数 a, b 为例，它们的最小公倍数指的就是同时被 a, b 整除的最小的数，一般在计算操作中记作 $\operatorname{lcm}(a, b)$，

而在数学运算时经常使用 $[a,b]$.

为了让读者对两种记法都有一定的印象，在上一章介绍和这一章的例题习题中，我们大多数使用括号的记法，而在这一章的介绍中，将加大使用 lcm 这种运算符号的比例.

比如 6 和 12 是 2 和 3 的公倍数，其中能够同时整除 2 和 3 的最小的数是 6，所以 $\mathrm{lcm}(2,3) = 6$.

由上面的描述我们很容易得到：
$$\mathrm{lcm}(a,b) = \mathrm{lcm}(|a|,|b|).$$

所以以后讨论的时候可以将所有的负整数和零都排除在外，只需要讨论正整数的最小公倍数即可.

任意给定两个或两个以上的整数 a,b，必然存在公倍数，这是因为给定两个数一定可以进行乘法计算，而其乘积 $a \times b$ 必然就是这两个数的一个公倍数了. 那么在所有非负倍数中，有没有比 $a \times b$ 小的公倍数呢？

若干个正整数的公倍数有无穷多个，而最小公倍数只有一个.

如若不然，假设
$$\mathrm{lcm}(a,b) = m$$
$$\mathrm{lcm}(a,b) = n$$

其中 m,n 是两个整数.

若 $m < n$，则与 $\mathrm{lcm}(a,b) = n$ 矛盾；

若 $m > n$，则与 $\mathrm{lcm}(a,b) = m$ 矛盾；

所以必然有 $m = n$，也即最小公倍数唯一存在.

那么要怎样才可以求出若干个数的最小公倍数呢？对于个数较少且数值不大的整数来说，要求出它们的最小公倍数的一个最基本的方法就是按照定义要求，将它们所有的公倍数都找出来，这样就可以很容易地看出最大的那个数，那个数就是所求的最小公倍数.

比如要求出 4 和 6 的最小公倍数，我们需要分别求出 4 和 6 的较小的倍数，这需要我们先求出 4 和 6 的乘积：
$$4 \times 6 = 24$$

然后再看是否存在比 24 小的数也刚好是 4 和 6 的公倍数，也即只需要找到不超过 24 的公倍数即可. 在这里，很容易就能看得出来，12 就是这样的数，所以 12 是 4 和 6 的最小公倍数.

对于若干个已知正整数，用定义求其最小公倍数也只需要找到不超过所有数的乘积的公倍数即可.

7.2.2 最小公倍数的几个性质

性质1 若干个数的任意一个公倍数一定是其最小公倍数的倍数.

该性质的数学描述是：

设 m 是 $a_1,a_2,\cdots a_n$ 的一个公倍数，q 是 $a_1,a_2,\cdots a_n$ 的任意一个公倍数，则 $\mathrm{lcm}(a_1,a_2,\cdots a_n) = m$ 的充分必要条件是 $m \mid q$.

证明：

（1）必要性：

用反证法证明.

假设 $m \nmid q$，因为
$$\mathrm{lcm}\,(a_1,a_2,\cdots a_n) = m$$

所以 $m < q$.

设 $q = mx + r(0 \leqslant r < m)$，则由 $a_k \mid m, a_k \mid q$ 可得
$$a_k \mid q - mx = r(k = 1,2,\cdots n)$$

所以 r 也是 $a_1,a_2,\cdots a_n$ 的公倍数，而 $r < m$，这与
$$\mathrm{lcm}\,(a_1,a_2,\cdots a_n) = m \text{ 矛盾.}$$

故 $m \mid q$.

（2）充分性：

假设 $\mathrm{lcm}\,(a_1,a_2,\cdots a_n) = A \neq m$

那么 $m \mid A$，于是 $m < A$. 而这与
$$\mathrm{lcm}\,(a_1,a_2,\cdots a_n) = A$$

矛盾，故 $m = A$，也即
$$\mathrm{lcm}\,(a_1,a_2,\cdots a_n) = m$$

性质2　公倍数是最小公倍数的充要条件是以这一公倍数除以各已知数后所得的数互素.

该性质的数学描述是：

若 m 是 $a_1,a_2,\cdots a_n$ 的公倍数，也即 $a_k \mid m(k = 1,2,\cdots n)$，则
$$\mathrm{lcm}(a_1,a_2,\cdots a_n) = m$$

的充要条件是
$$\gcd\left(\frac{m}{a_1},\frac{m}{a_2},\cdots,\frac{m}{a_n}\right) = 1$$

证明：

（1）必要性：

用反证法证明.

设
$$\gcd\left(\frac{m}{a_1},\frac{m}{a_2},\cdots,\frac{m}{a_n}\right) = q > 1$$

则 $q \mid \dfrac{m}{a_k}(k = 1,2,\cdots n)$，于是可得 $q\,a_k \mid m$，那么
$$a_k \mid \frac{m}{q}(k = 1,2,\cdots n)$$

而 $\dfrac{m}{q} < m$，这与
$$\mathrm{lcm}(a_1,a_2,\cdots a_n) = m$$

矛盾，故

$$\gcd\left(\frac{m}{a_1}, \frac{m}{a_2}, \cdots, \frac{m}{a_n}\right) = 1$$

（2）充分性：

假设

$$\operatorname{lcm}(a_1, a_2, \cdots a_n) = k < m$$

则由性质 1 可知：$k \mid m$.

不妨假设 $m = kq(q > 1)$，因为 $a_k \mid k(k = 1, 2, \cdots n)$，所以 $a_k \mid \frac{m}{q}$，故

$$q \mid \frac{m}{a_k}(k = 1, 2, \cdots n)$$

而 $q > 1$，这与

$$\gcd\left(\frac{m}{a_1}, \frac{m}{a_2}, \cdots, \frac{m}{a_n}\right) = 1$$

矛盾，故

$$\operatorname{lcm}(a_1, a_2, \cdots a_n) = m$$

推论 1 当提取出一组数的某一公因数（如 d）之后所得新数组的最小公倍数的 d 倍就是原数组的最小公倍数.

该推论的数学描述是：

$$\operatorname{lcm}(k a_1, k a_2, \cdots k a_n) = k \operatorname{lcm}(a_1, a_2, \cdots a_n)$$

例 7.1 对于勾股数组 3，4，5 来说，有如下关于最小公倍数的结论成立：
$$\operatorname{lcm}(6, 8, 10) = 2\operatorname{lcm}(3, 4, 5)$$
$$\operatorname{lcm}(9, 12, 15) = 3\operatorname{lcm}(3, 4, 5)$$
$$\operatorname{lcm}(15, 20, 25) = 5\operatorname{lcm}(3, 4, 5)$$

关于该推论的证明留给读者自己进行. 值得一提的是该推论的作用就是如果能找出一组数的最大公因数，则可将其提取出去，这样对于求解最小公倍数是有很大的帮助的，事实上，最小公倍数之所以难以计算的原因之一就是数值过大，先行提取了最大公因数后自然降低了其数值.

在这一想法的基础上，再进行最小公倍数计算的时候就会想到将其和最大公因数相联系，如后面的性质就是这样.

性质 3 两个数的最大公因数和最小公倍数的乘积刚好是这两个数的乘积.

该性质的数学描述是：

$$\operatorname{lcm}(a_1, a_2)\gcd(a_1, a_2) = a_1 \times a_2$$

也即

$$(a_1, a_2)[a_1, a_2] = a_1 \times a_2$$

证明：假设 $\operatorname{lcm}(a_1, a_2) = m$，则由性质 2 可得

$$\gcd\left(\frac{m}{a_1}, \frac{m}{a_2}\right) = 1$$

所以

$$\gcd\left(\frac{m\,a_2}{a_1\,a_2},\frac{m\,a_1}{a_1\,a_2}\right)=1$$

由最大公因数的学习可知

$$\mathrm{lcm}(a_1,a_2)\gcd(a_1,a_2)=a_1\times a_2$$

例 7.2　因为 $\gcd(5,15)=5$，所以

$$\mathrm{lcm}(5,15)\gcd(5,15)=5\times 15$$

而 $5\times 15=75$，故据此可得：

$$\mathrm{lcm}(5,15)=75\div 5=15$$

因为 $\gcd(25,15)=5$，所以

$$\mathrm{lcm}(25,15)\gcd(25,15)=25\times 15$$

而 $25\times 15=375$，故据此可得：

$$\mathrm{lcm}(25,15)=375\div 5=75$$

推论 2　两个互素的整数的最小公倍数是其乘积.

该推论的数学描述是：

若 $\gcd(a_1,a_2)=1$，则 $\mathrm{lcm}(a_1,a_2)=a_1\times a_2$.

例 7.3　因为 $\gcd(3,4)=1$，所以

$$\mathrm{lcm}(3,4)\gcd(3,4)=3\times 4=\mathrm{lcm}(3,4)$$

推论 3　若两个数互素，则两数中一个的若干倍（如 k 倍）与另一个的最小公倍数是原两数最小公倍数的 k 倍.

该推论的数学描述是：

若 $\gcd(a_1,a_2)=1$，则 $\mathrm{lcm}(a_1,a_2a_3)=a_3\mathrm{lcm}(a_1,a_2)$.

例 7.4　因为 $\gcd(3,4)=1$，所以

$$\mathrm{lcm}(6,4)=\mathrm{lcm}(2\times 3,4)=3\mathrm{lcm}(2,4)=3\times 4=12$$

需要特别注意的是，这里提取出的数是新乘进去的倍数，而不是原先互素的两个数.

推论 4　两个数的若干次幂的最小公倍数是这两个数的最小公倍数的若干次幂.

该推论的数学描述是：

$$\mathrm{lcm}(a_1{}^n,a_2{}^n)=\mathrm{lcm}(a_1,a_2)^n$$

事实上，

$$\mathrm{lcm}(a_1{}^n,a_2{}^n)=\frac{a_1{}^n\times a_2{}^n}{\gcd(a_1{}^n,a_2{}^n)}=\frac{a_1{}^n\times a_2{}^n}{\gcd(a_1,a_2)^n}$$

$$=\left(\frac{a_1\times a_2}{\gcd(a_1,a_2)}\right)^n=\mathrm{lcm}(a_1,a_2)^n$$

例 7.5　因为 $\mathrm{lcm}(3,4)=12$，所以

$$\mathrm{lcm}(9,16)=144$$

$$\mathrm{lcm}(27,64)=1728$$

$$\text{lcm}(81,256) = 20736$$
$$\text{lcm}(243,1024) = 248832$$
$$\text{lcm}(3^n,4^n) = \text{lcm}(3,4)^n = 12^n$$

性质 4　若干个数的最小公倍数可以转换为这些数中一部分的最小公倍数与剩余部分的最小公倍数.

该性质的数学描述是：

若 $\text{lcm}(a_1,a_2,\cdots,a_k) = m_k$，则
$$\text{lcm}(a_1,a_2,\cdots,a_n) = \text{lcm}(m_k,a_{k+1},\cdots,a_n)(1 \leqslant k < n)$$

例 7.6　求 $\text{lcm}(3,4,5)$.

因为 $\text{lcm}(3,4) = 12$，所以
$$\text{lcm}(3,4,5) = \text{lcm}(12,5)$$

而 $\gcd(12,5) = 1$，故
$$\text{lcm}(12,5) = 12 \times 5 = 60$$

在该性质的保证之下，可以将计算较多个数的最小公倍数的问题转化为较少个数的最小公倍数，用部分代替整体来进行计算，最后再统一起来考虑得到一开始需要计算的结果即可.

推论 5　将若干个数分为两部分，分别计算出其最小公倍数，再计算出这两个数的最小公倍数即为原若干个数的最小公倍数.

该推论的数学描述是：

若 $\text{lcm}(a_1,a_2,\cdots,a_k) = m_k$，$\text{lcm}(a_{k+1},\cdots,a_n) = q_k$，则
$$\text{lcm}(a_1,a_2,\cdots,a_n) = \text{lcm}(m_k,q_k)$$

例 7.7　求 $\text{lcm}(3,4,5,6,7,8)$.

因为 $\text{lcm}(3,4,5) = 60$，$\text{lcm}(6,7,8) = 168$，所以
$$\text{lcm}(3,4,5,6,7,8) = \text{lcm}(60,168)$$
$$= \text{lcm}(12 \times 5,12 \times 14) = 12\text{lcm}(5,14)$$

而 $\gcd(14,5) = 1$，故 $\text{lcm}(5,14) = 5 \times 14$，所以
$$\text{lcm}(3,4,5,6,7,8) = 12 \times 5 \times 14 = 840$$

性质 5　求若干个数的最小公倍数时，可将其中一部分的公因数提取出来.

该性质的数学描述是：

若 $\gcd(x,a_m) = 1(m = k+1,k+2,\cdots n)$，则
$$\text{lcm}(xa_1,xa_2,\cdots,xa_k,a_{k+1},\cdots,a_n) = x\text{lcm}(a_1,a_2,\cdots,a_k,a_{k+1},\cdots,a_n)$$

例 7.8　求 $\text{lcm}(2,4,9,12,17,18)$.

解由性质 5 可得：
$$\text{lcm}(2,4,9,12,17,18) = 2 \times \text{lcm}(1,2,9,6,17,9)$$
$$= 2 \times 2 \times \text{lcm}(1,1,9,3,17,9)$$
$$= 2 \times 2 \times 3 \times \text{lcm}(1,1,3,1,17,3)$$

$$= 2 \times 2 \times 3 \times 3 \times \text{lcm}(1,1,1,1,17,1)$$
$$= 2 \times 2 \times 3 \times 3 \times 17$$
$$= 612$$

性质6　三个数的最小公倍数与任取其中两个相乘所得的三个积的最大公因数的乘积与这三个数的乘积相等.

该性质的数学描述是：

$$\text{lcm}(a_1,a_2,a_3)\gcd(a_1 a_2,a_2 a_3,a_1 a_3) = a_1 a_2 a_3$$

证明：由前述几个性质可得：

$$\text{lcm}(a_1,a_2,a_3) = \text{lcm}(a_1,\text{lcm}(a_2,a_3))$$

$$= \frac{a_1 \times \text{lcm}(a_2,a_3)}{\gcd(a_1,\text{lcm}(a_2,a_3))}$$

$$= \frac{\dfrac{a_1 \times a_2 \times a_3}{\gcd(a_2,a_3)}}{\gcd\left(a_1,\dfrac{a_2 \times a_3}{\gcd(a_2,a_3)}\right)}$$

$$= \frac{a_1 \times a_2 \times a_3}{\gcd(a_2,a_3)} \times \frac{\gcd(a_2,a_3)}{\gcd(a_1\gcd(a_2,a_3),a_2 \times a_3)}$$

$$= \frac{a_1 \times a_2 \times a_3}{\gcd(a_1 a_2,a_2 a_3,a_1 a_3)}$$

推论6　若三个数互素，则这三个数的最小公倍数就是其乘积.

该推论的数学描述是：

若 $\gcd(a_1,a_2) = \gcd(a_1,a_3) = \gcd(a_2,a_3) = 1$，则

$$\text{lcm}(a_1,a_2,a_3) = a_1 a_2 a_3$$

例7.9　求证 $\text{lcm}(a_1,a_2,a_3) = \dfrac{a_1 a_2 a_3 \gcd(a_1 a_2 a_3)}{\gcd(a,b)\gcd(a,c)\gcd(b,c)}$.

证明：由性质4可知：

$$\text{lcm}(a_1,a_2,a_3) = \frac{a_1 \times a_2 \times a_3}{\gcd(a_1 a_2,a_2 a_3,a_1 a_3)} = \frac{a_1 \times a_2 \times a_3\gcd(a_1,a_2,a_3)}{\gcd(a_1 a_2,a_2 a_3,a_1 a_3)\gcd(a_1,a_2,a_3)}$$

而 $\gcd(a_1 a_2,a_2 a_3,a_1 a_3)\gcd(a_1,a_2,a_3)$

$= \gcd(a_1 a_2\gcd(a_1,a_2,a_3),a_2 a_3\gcd(a_1,a_2,a_3),a_1 a_3\gcd(a_1,a_2,a_3))$

$= \gcd(a_1^2 a_2,a_2^2 a_1,a_1 a_2 a_3,a_1 a_2 a_3,a_2^2 a_3,a_3^2 a_2,a_1^2 a_3,a_1 a_2 a_3,a_3^2 a_1)$

$= \gcd(a_1^2 a_2,a_2^2 a_1,a_1 a_2 a_3,a_1 a_2 a_3,a_2^2 a_3,a_3^2 a_2,a_1^2 a_3,a_1 a_2 a_3,a_3^2 a_1)$

$= \gcd(\gcd(a_1^2 a_2,a_1 a_2 a_3),\gcd(a_2^2 a_1,a_2^2 a_3),\gcd(a_2^2 a_3,a_1 a_2 a_3),\gcd(a_1^2 a_3,a_3^2 a_1))$

$= \gcd(a_1 a_2\gcd(a_1,a_3),a_2^2\gcd(a_1,a_3),a_2 a_3\gcd(a_1,a_3),a_1 a_3\gcd(a_1,a_3))$

$= \gcd(a_1,a_3)\gcd(a_1 a_2,a_2^2,a_2 a_3,a_1 a_3)$

$= \gcd(a_1,a_3)\gcd(\gcd(a_1 a_2,a_2^2),\gcd(a_2 a_3,a_1 a_3))$

$= \gcd(a_1,a_3)\gcd(a_2\gcd(a_1,a_2),a_3\gcd(a_2,a_1))$

$= \gcd(a_1,a_3)\gcd(a_2,a_3)\gcd(a_1,a_2)$

由此，命题得证.

7.3 最小公倍数的主要求法

根据最小公倍数的定义、性质等，对照最大公因数的求法，就可以获得最小公倍数的几种主要求法.

7.3.1 分解素因数法

由最小公倍数的概念可以知道，作为几个数的最小公倍数首先要是这几个数的一个公倍数，其次还要是这几个数的任意公倍数的因数. 因此，要求最小公倍数，就可以考虑将这几个数进行因数分解后得出，具体步骤如下：

首先，将已知各数进行因数分解，写出其标准分解式.

然后，将标准分解式中所有素因数及其最高次幂找出，并将得到的幂连乘即可.

例 7.10 求 $\mathrm{lcm}(2940,756,168)$.

解：按分解素因数法进行计算：

$$2940 = 2^2 \times 3 \times 5 \times 7^2$$
$$756 = 2^2 \times 3^3 \times 7$$
$$168 = 2^3 \times 3 \times 7$$

所以，$\mathrm{lcm}(2940,756,168) = 2^3 \times 3^3 \times 5 \times 7^2 = 52920$.

7.3.2 提取公因数法

根据推论 1 和性质 5 可知，要计算几个数的最小公倍数可以通过提取公因数的方法进行. 具体步骤如下：

首先，先提取出这几个数的最大公因数，此时所得的商互素，但是不一定两两互素.

然后，在不互素的商中再次提取公因数，其他的商正常留下，直到剩下的所有的商两两互素.

最后把提取出来的各个数及各个商数连乘即可.

7.3.3 先求最大公因数法

根据性质 3，可以将最大公因数和最小公倍数通过两数之乘积进行转化，也即要求最小公倍数，转化成求最大公因数，随后再用其乘积除以最大公因数来获得. 具体步骤如下：

首先，求出两个数的最大公因数，再求出这两个数的乘积.

然后，用这个乘积除以这两个数的最大公因数即可.

特别的，这一方法可以继续往后进行，比如涉及三个数的时候，可以用性质 6、推论 6 以及例 7.9 的结论计算.

7.4　典型问题

7.4.1　典型例题

例 7.11　求 108，28 和 42 的最小公倍数.

解： 由于

$$108 = 2^2 \times 3^3,$$
$$28 = 2^2 \times 7,$$
$$42 = 2 \times 3 \times 7,$$

故由分解素因数法可得

$$[108,28,42] = 2^2 \times 3^3 \times 7 = 756.$$

例 7.12　求 198，240 和 360 的最小公倍数.

解： 由于

$$198 = 2 \times 3^2 \times 11,$$
$$240 = 2^4 \times 3 \times 5,$$
$$360 = 2^3 \times 3^2 \times 5,$$

故由分解素因数法可得

$$[198,240,360] = 2^4 \times 3^2 \times 5 \times 11 = 7920.$$

注意：（1）几个正整数里，如果最大的一个正整数是其他各个正整数的倍数，则最大的那一个正整数就是这几个正整数的最小公倍数. 例如 15，30 和 60 的最小公倍数是 60.

（2）几个正整数里，如果任意二个数都是互素的，则这几个正整数的最小公倍数就是它们的相乘积. 例如 15，32 和 49 的最小公倍数就是

$$15 \times 32 \times 49 = 23520.$$

例 7.13　求 24871 和 3468 的最小公倍数.

解： 由先求最小公倍数法可得：

$$[24871,3468] = \frac{24871 \times 3468}{17} = 5073684.$$

例 7.14　求下列各数的最小公倍数：

（i）391，493.

（ii）209，665，4025.

（iii）1965，1834，30261，55020.

解：（i）由于

	391	493	
	306	391	3
1			
	85	102	5
3	85	85	
	0	17	

得到：

$$(391, 493) = 17$$

所以所求为：

$$[391, 493] = \frac{391 \times 493}{17} = 11339.$$

（ii）由于

	209	665	
3	190	627	5
	19	38	
2		38	
	19	0	

故可得到

$$(209, 665) = 19,$$

所以所求为：

$$[209, 665] = \frac{209 \times 665}{19} = 7135.$$

又由

	7135	4025	
	4025	3290	1
1	3290	735	4
	2940	400	
2	350	35	10
	350		
	0	35	

得到 (7315,4025) = 35, 因此
$$[7315,4025] = \frac{7315 \times 4025}{35} = 841225.$$

(iii) 由

	1965	1834	
14	1834	1834	1
	131	0	

得到 (1965,1834) = 131, 因此
$$[1965,1834] = \frac{1965 \times 1834}{131} = 27510.$$

由

	27510	30261	
1	27510	27510	10
	0	2751	

得到 (27510,30261) = 2751, 因此
$$[27510,30261] = \frac{27510 \times 30261}{2751} = 302610.$$

又由

	302610	55020	
2	275100	55020	5
	27510	0	

得到 (302610,55020) = 27510, 因此
$$[302610,55020] = \frac{302610 \times 55020}{27510} = 605220.$$

综合以上结果可得:
$$[1965,1834,30261,55020] = 60520.$$

例 7.15　求 $[120,330,525]$.

解: 因为
$$120 = 2^3 \times 3 \times 5,$$
$$330 = 2 \times 3 \times 5 \times 11,$$
$$525 = 3 \times 5^2 \times 7,$$

所以由分解素因数法可得
$$[120,330,525] = 2^3 \times 3 \times 5^2 \times 7 \times 11 = 462000.$$

例 7.16 求 $[180,240,525]$.

解：由分解素因数法可得
$$[180,240,525] = 15 \times 2^2 \times 3 \times 4 \times 35 = 25200.$$

例 7.17 求 $[24871,3468]$.

解：用辗转相除法求 $(24871,3468)$ 可得
$$(24871,3468) = 17.$$

根据先求最大公因数法 $[a,b] = \dfrac{ab}{(a,b)}$，可得
$$[24871,3468] = \frac{24871 \times 3468}{17} = 5073684.$$

例 7.18 求 $[300,160,720,540]$.

解：由提取公因数法可得
$$
\begin{aligned}
[300,160] &= \frac{300 \times 160}{(300,160)}\\
&= \frac{300 \times 160}{20}\\
&= 2400.\\
[2400,720] &= \frac{2400 \times 720}{(2400,720)}\\
&= \frac{2400 \times 720}{240}\\
&= 7200,\\
[7200,540] &= \frac{7200 \times 540}{(7200,540)}\\
&= \frac{7200 \times 540}{180}\\
&= 21600.
\end{aligned}
$$

所以 $[300,160,720,540] = 21600$.

例 7.19 求 $[2,4,12,9,17,18]$.

解：由提取公因数法可得
$$
\begin{aligned}
[2,4,12,9,17,18] &= 2[1,2,6,9,17,9]\\
&= 2 \times 2[1,1,3,9,17,9]\\
&= 4 \times 3[1,1,1,3,17,3]\\
&= 12 \times 3[1,1,1,1,17,1]
\end{aligned}
$$

$$= 36 \times 17 = 612$$

例 7.20　求 $[2940,756,168]$.

解：因为

$$2940 = 2^2 \times 3 \times 5 \times 7^2,$$
$$756 = 2^2 \times 3^3 \times 7,$$
$$168 = 2^3 \times 3 \times 7,$$

所以由分解素因数法可得：

$$[2940,756,168] = 2^2 \times 3^3 \times 5 \times 7^2 = 52920.$$

例 7.21　求 $[162,216,378,108]$.

解：由提取公因数法可得

$$\begin{aligned}
[162,216,378,108] &= 2[81,108,189,54] \\
&= 2 \times 9[9,12,21,6] \\
&= 18 \times 3[3,4,7,2] \\
&= 54 \times 2[3,2,7,1] \\
&= 108 \times 3 \times 2 \times 7 \times 1 \\
&= 4536
\end{aligned}$$

这一过程通常简写成下面的形式，叫作短除式．

$$
\begin{array}{c|cccc}
2 & 162 & 216 & 378 & 108 \\
\hline
9 & 81 & 108 & 189 & 54 \\
\hline
3 & 9 & 12 & 21 & 6 \\
\hline
2 & 3 & 4 & 7 & 2 \\
\hline
& 3 & 2 & 7 & 1
\end{array}
$$

因为 $3,2,7,1$ 两两互质，所以

$$[162,216,378,108] = 2 \times 9 \times 3 \times 2 \times 3 \times 2 \times 7 \times 1 = 4536.$$

例 7.22　求 $[24\,871,3\,468]$.

解：由辗转相除法，可得

$$(24\,871,3\,468) = 17$$

从而由先求最大公因数法可得：

$$[24\,871,3\,468] = 24\,871 \times 3\,468 \div 17 = 5\,073\,684.$$

例 7.23　证明：正整数 $a_1,a_2,\cdots,a_n$ 两两互素的充分必要条件是

$$[a_1,a_2,\cdots,a_n] = a_1 a_2 \cdots a_n.$$

证明：

（1）必要性：

由于 $(a_1,a_2) = 1$，故

$$[a_1,a_2] = \frac{a_1 a_2}{(a_1,a_2)} = a_1 a_2.$$

再由 $(a_1, a_3) = (a_2, a_3) = 1$ 知

$$(a_1, a_2, a_3) = 1.$$

故可得

$$[a_1, a_2, a_3] = [[a_1, a_2], a_3] = [a_1 a_2, a_3] = a_1 a_2 a_3.$$

如此继续下去即得结论.

（2）充分性：

设 $[a_1, a_2, \cdots, a_n] = a_1 a_2 \cdots a_n$，

当 $n = 2$ 时，由 $[a_1, a_2] = a_1 a_2$ 易得 $(a_1, a_2) = 1$.

假设 $n = k$ 时充分性成立．即

$$[a_1, \cdots, a_k, a_{k+1}] = [m_k, a_{k+1}].$$

其中 $m_k = [a_1, \cdots, a_k]$.

若 $[a_1, a_2 \cdots, a_k, a_{k+1}] = a_1 a_2 \cdots a_k a_{k+1}$. 由于

$$[a_1, \cdots, a_k, a_{k+1}] = [m_k, a_{k+1}] = \frac{m_k a_{k+1}}{(m_k, a_{k+1})}.$$

因此

$$\frac{m_k}{(m_k, a_{k+1})} = a_1 a_2 \cdots a_k$$

显然

$$m_k \leqslant a_1 a_2 \cdots a_k, (m_k, a_{k+1}) \geqslant 1.$$

若上式成立，必有

$$(m_k, a_{k+1}) = 1$$

且 $m_k = a_1 a_2 \cdots a_k$，于是 $(a_i, a_{k+1}) = 1 (1 \leqslant i \leqslant k)$.

由此及归纳假设得 $(a_i, a_j) = 1 (1 \leqslant i, j \leqslant k, i \neq j)$.

因此，当 $n = k + 1$ 时，结论成立.

由归纳原理得充分性得证.

例 7.24 两数之和为 667，它们的最小公倍数除以最大公因数所得的商等于 120，求这两数.

分析： 遇到这类最大公因数与最小公倍数有等式关系的问题时，一般设题目中两数为 $x, y; d$ 为 x, y 的最大公因数，$x = dx_1, y = dy_1$，最小公倍数为 $dx_1 y_1$，则可建立关于 d, x_1, y_1 的方程，从而解出 x, y.

解： 设所求的数为 x 和 y，且 $d = (x, y)$，则

$$x = dx_1, y = dy_1, (x_1, y_1) = 1,$$

于是

$$[x, y] = \frac{xy}{d} = dx_1 y_1$$

由题意，有

$$d(x_1 + y_1) = 667 \qquad \qquad ①$$

$$\frac{dx_1 y_1}{d} = 120 \qquad \qquad ②$$

由 $x_1y_1 = 120 = 2^3 \cdot 3 \cdot 5$，又 $667 = 1 \cdot 23 \cdot 29$，

因此由①，②可得 $x_1 = 8, y_1 = 15$，或 $x_2 = 24, y_2 = 5$，

相应地 $d_1 = 29, d_2 = 23$，

因此，所求的数有两组

$x_1 = 232, y_1 = 435$ 或 $x_2 = 552, y_2 = 115$．

7.4.2 典型练习题

练习 7.1 求 $[391, 493]$．

练习 7.2 求 $[6731, 2809]$．

练习 7.3 求 $[513, 135, 3114]$．

练习 7.4 求 $[8127, 11352, 21672, 27090]$．

练习 7.5 证明：若对自然数 a, b 成立 $[a, b] = (a, b)$，则 $a = b$

练习 7.6 证明：若对自然数 a, b 满足 $(a, b) = 1$，则 $[a, b] = ab$．

练习 7.7 若自然数 a, b 满足 $[a^2, b] = [a, b]$．证明：$a^2 \mid b$．

练习 7.8 设 a, b 为正整数，若 $a \mid b$，求证：$(a, b) = a, [a, b] = b$．

练习 7.9 设 a, b 为正整数，若 $(a, b) = [a, b]$，求证：$a = b$．

练习 7.10 设 a, b 为正整数，且 $a < b$，若 $(a, b) = 15, [a, b] = 180$，求 a, b．

练习 7.11 （1）一个数除以 36 和 38 都余 5，求这个数．

（2）一个数除以 7 余 5，除以 13 差 8，求这个数．

练习 7.12 在 1 至 100 的正整数中，

（1）能同时被 13 和 31 整除的有多少个？

（2）不能同时被 13 和 31 整除的有多少个？

（3）既不能被 13 整除，又不能被 31 整除的有多少个？

练习 7.13 若 $[a, b] = m, n \mid a, n \mid b$，求证：$\left[\dfrac{a}{n}, \dfrac{b}{n}\right] = \dfrac{m}{n}$．

练习 7.14 若 a, b, c, d 两两互质，则 $[a, b, c, d] = abcd$．

练习 7.15 分别用分解质因数法、提取质因数法和辗转相除法，求

$$(360, 204), [360, 204].$$

练习 7.16 异分母分数加减法法则可定义为：

$$\frac{a}{b} - \frac{d}{c} = \frac{ad - bc}{bd},$$

或定义为：

$$\frac{a}{b} - \frac{d}{c} = \frac{ab' - cd'}{m} \ (\text{其中 } [b, d] = m, m \div b = b', m \div d = d')．$$

试证明这两个法则是一致的．

练习 7.17 甲、乙、丙、丁四个齿轮相互啮合，齿轮齿数分别为 84，36，60，48，在转动过程中同时啮合的各齿到下次再次同时啮合时，各齿轮分别转过了多少周？

练习 7.18 金星和地球在某一时刻相对于太阳处于某一确定位置．已知金星绕太阳一周为 225 日，问这两个行星至少经过多少日才同时回到原来位置？

练习 7.19 甲、乙、丙三人在环形跑道上跑步，他们同时同地同向出发，经过 8 分钟三人第一次同时回到出发地，已知甲、乙、丙三人每分钟分别跑 120 米、180 米、150 米，求跑道长.

练习 7.20 正整数 $n \geqslant a_1 > a_2 > \cdots > a_k > 0$ 满足 $[a_i, a_j] \leqslant n (1 \leqslant i \leqslant j \leqslant n)$，求证：$i a_i \leqslant n, \forall 1 \leqslant j \leqslant k$.

练习 7.21 正整数 $a_0 < a_1 < a_2 < \cdots < a_n$，求证：

$$\frac{1}{[a_0, a_1]} + \frac{1}{[a_1, a_2]} + \cdots + \frac{1}{[a_{n-1}, a_n]} \leqslant 1 - \frac{1}{2^n}.$$

练习 7.22 任取正整数 $m > n$，求证：

$$[m, n] + [m = 1, n + 1] > \frac{2mn}{\sqrt{m - n}}.$$

练习 7.23 求所有的正整数三元组 (a, b, c)，使得 $[a, b, c] = a + b + c$.

练习 7.24 给定正整数 r, s，求满足 $3^r \cdot 7^s = [a, b, c] = [a, b, d] = [a, c, d] = [b, c, d]$ 的正整数四元组.

练习 7.25 求满足 $[a, b] = 1000, [b, c] = 2000$ 以及 $[c, a] = 2000$ 的正整数三元组 (a, b, c) 的个数.

第 8 章　数的进位制

易有太极，是生两仪，两仪生四象，四象生八卦．

—— 《易传·系辞上传》

数学的最原始基础，是给事物赋予一个确定的量．这个量一旦被赋予完成，它就将独立于原来的事物而超越事物本身成为一个本质属性．因此，这个量也就成为用数学认识和研究世界的基础．这就是计数，有了计数基础后，还需要将这些计数用的"数"进行一定的规范，并约定一些相应的规则以便更好地解决事物和数学两个领域的问题，这就是进位制的产生背景．

本章主要讨论计数原理及一些常用的进位制．

8.1　计数及其原理

自然数是用以计量事物的件数或表示事物次序的数．即用数码 0，1，2，3，4，……所表示的数．自然数由 0 开始（若按数学史来说，也可以由 1 开始），一个接一个，组成一个无穷的集体．

为了将这一概念公理化，意大利数学家 G. 皮亚诺发展了的序数理论[①]，并由此得到自然数的公理化定义：

自然数集 N 是指满足以下条件的集合：

①N 中有一个元素，记作 1.

②N 中每一个元素都能在 N 中找到一个元素作为它的后继者．

③1 是 0 的后继者．

④0 不是任何元素的后继者．

⑤不同元素有不同的后继者．

⑥（归纳公理）N 的任一子集 M，如果 $1 \in M$，并且只要 x 在 M 中就能推出 x 的后继者也在 M 中，那么 $M = N$.

若将这一理论数学化，或用程序化的形式进行表示，可以得到：

$$\exists N \left(\begin{array}{l} N \neq \varnothing \\ \wedge\ \exists f{:}N(\ \forall x \forall y((x \in N \wedge y \in N \wedge x \neq y){\Rightarrow}f(x) \neq f(y))) \\ \wedge\ \exists 0 \left(0 \in N \wedge \left(\begin{array}{l} \wedge\ \exists x(x \in N \wedge f(x) = 0) \\ \wedge\ S((S \subseteq N \wedge 0 \in S \wedge\ \forall x(x \in S{\Rightarrow}f(x) \in S)){\Rightarrow}S = N) \end{array} \right) \right) \right)$$

基数理论则把自然数定义为有限集的基数，这种理论提出，两个可以在元素之间建立一一对应关系的有限集具有共同的数量特征，这一特征叫作基数．这样，所有单元素集 $\{x\}$，$\{y\}$，$\{a\}$，$\{b\}$ 等具有同一基数，记作 1．类似的，凡能与两个元素建立一一对应的集合，它们的基数相同，记作 2，等等．自然数的加法、乘法运算可以在序数或基数理论中给出定义，并且两种理论下的运算是一致的．

---

① 序数理论是数学的一个分支，是使用二元关系（如，大于、优于）来研究顺序的直观概念．它提供了一个用于描述诸如"这小于"或"这在其之前"的语句的正式框架．本书介绍了该研究领域，并提供了基本定义．序数理论术语的列表可以在顺序理论词汇表中找到．

8.2 进位计数法

十进位值制是我们最为熟悉的计数法.

8.2.1 十进位值制

十进位值制是我们最为熟悉的计数法. 所谓十进位值制, 又称为十进制, 指的就是日常所说的"满十进一", 是"每相邻的两个计数单位之间的进率都为十"的一种计数法则.

有证据表明, 中国是最早使用十进位值制表示自然数的国家之一, 有考古学发现的商代陶文和甲骨文中, 就已经有一、二、三、四、五、六、七、八、九、十、百、千、万等数字, 而这些数字已经可以用于表示十万以内的所有自然数了.

在十进位值制下, 一般可以约定任意一个 n 位自然数的表示方法为:

$$m = \overline{a_{n-1} \, a_{n-2} \cdots a_2 \, a_1 \, a_0}$$
$$= a_{n-1} \times 10^{n-1} + a_{n-2} \times 10^{n-2} + \cdots + a_2 \times 10^2 + a_1 \times 10 + a_0$$
$$= \sum_{i=0}^{n-1} a_i \times 10^i \, (a_i \in N, 0 \leqslant a_i \leqslant 9, a_{n-1} \neq 0)$$

其中的 10 称为自然数 m 在十进制表示下的"基"或"底".

按照这一规则, 可以衍生出"科学计数法". 这样就可以将一个数表示为:

$$m = a \times 10^b = aEb \, (1 \leqslant |a| < 10)$$

用科学计数法的最大好处是当需要表示一个较大或者较小且数位比较多的数时, 可以节省时间和空间. 当然, 在古中国, 曾有通过汉字来完成表示这类数的需求, 如表 8.1.

表 8.1

数字	汉字	数字	汉字	数字	汉字	数字	汉字	数字	汉字
10^0	一	10^1	十	10^2	百	10^3	千	10^4	万
10^5	十万	10^6	百万	10^7	千万	10^8	亿	10^9	十亿
10^{10}	百亿	10^{11}	千亿	10^{12}	兆	10^{13}	十兆	10^{14}	百兆
10^{15}	千兆	10^{16}	京	10^{17}	十京	10^{18}	百京	10^{19}	千京
10^{20}	垓	10^{21}	十垓	10^{22}	百垓	10^{23}	千垓	10^{24}	秭
10^{25}	十秭	10^{26}	百秭	10^{27}	千秭	10^{28}	穰	10^{29}	十穰
10^{30}	百穰	10^{31}	千穰	10^{32}	沟	10^{33}	十沟	10^{34}	百沟
10^{35}	千沟	10^{36}	涧	10^{37}	十涧	10^{38}	百涧	10^{39}	千涧
10^{40}	正	10^{41}	十正	10^{42}	百正	10^{43}	千正	10^{44}	载
10^{45}	十载	10^{46}	百载	10^{47}	千载	10^{48}	极	10^{49}	十极
10^{50}	百极	10^{51}	千极	10^{52}	恒河沙	10^{53}	十恒河沙	10^{54}	百恒河沙

续　表

数字	汉字	数字	汉字	数字	汉字	数字	汉字	数字	汉字
10^{55}	千恒河沙	10^{56}	阿僧祇	10^{57}	十阿僧祇	10^{58}	百阿僧祇	10^{59}	千阿僧祇
10^{60}	那由他	10^{61}	十那由他	10^{62}	百那由他	10^{63}	千那由他	10^{64}	不可思议
10^{65}	十不可思议	10^{66}	百不可思议	10^{67}	千不可思议	10^{68}	无量	10^{69}	十无量
10^{70}	百无量	10^{71}	千无量	10^{72}	大数	10^{73}	十大数	10^{74}	百大数
10^{75}	千大数	10^{76}	全仕祥	10^{77}	十全仕祥	10^{78}	百全仕祥	10^{79}	千全仕祥
10^{0}	一	10^{-1}	分	10^{-2}	厘	10^{-3}	毫	10^{-4}	丝
10^{-5}	忽	10^{-6}	微	10^{-7}	纤	10^{-8}	沙	10^{-9}	尘
10^{-10}	埃	10^{-11}	渺	10^{-12}	漠	10^{-13}	模糊	10^{-14}	逡巡
10^{-15}	须臾	10^{-16}	瞬息	10^{-17}	弹指	10^{-18}	刹那	10^{-19}	六德
10^{-20}	空虚	10^{-21}	清静	10^{-22}	阿赖耶	10^{-23}	阿摩罗	10^{-24}	涅盘寂静

在这样的表示下，最需要确定的就是任何一个自然数都有唯一的一种表示方法与之相对应，这一想法的数学表示是：如果 m 是一个自然数，那么将 m 表示成十进制数的形式必须唯一．这里面包含了两层意思：一是任何一个自然数 m 都可以表示成十进制数；二是这种表示方法唯一．下面进行证明．

证明：首先进行存在性．

当 $m < 10$ 时，结论明显成立．

当 $m \geqslant 10$ 时，以 m 为被除数，10 为除数做带余除法可得：

$$m = 10 q_1 + r_1, 0 \leqslant r_1 < 10, q_1 < m$$

如果 $q_1 < 10$，则结论成立．

如果 $q_1 \geqslant 10$，则继续上面的步骤，也即以 q_1 为被除数，10 为除数做带余除法可得：

$$q_1 = 10 q_2 + r_2, 0 \leqslant r_2 < 10, q_2 < m$$

如果 $q_2 < 10$，则结论成立．

如果 $q_2 \geqslant 10$，则继续上面的步骤，也即以 q_2 为被除数，10 为除数做带余除法可得：

$$q_2 = 10 q_3 + r_3, 0 \leqslant r_3 < 10, q_3 < m$$

$$\cdots\cdots$$

于是又一次回到了上面的讨论，并且这一讨论可以一直进行下去，直到得到一个递减的正整数数列：

$$m > q_1 > q_2 > q_3 > \cdots$$

由于 m 是确定的自然数，所以这个正整数数列显然只会有有限项，也就是说最后一定存在某一个正整数 k，可以使得：

$$q_{k-1} = 10 q_k + r_k, 0 \leqslant r_k < 10, q_k < 10$$

于是可得：

$$m = q_k 10^k + r_k 10^{k-1} + \cdots + 10 r_2 + r_1$$

也即任何一个自然数 m 都可以表示成十进制的数．

下面证明唯一性.

假设 m 的表示不唯一, 不妨将设 m 有两种表示方分别是:

$$m = b_k 10^k + b_{k-1} 10^{k-1} + \cdots + 10 b_1 + b_0$$
$$m = a_n 10^n + a_{n-1} 10^{n-1} + \cdots + 10 a_1 + a_0$$

由上面的假设可以得到:

$$b_k 10^k + b_{k-1} 10^{k-1} + \cdots + 10 b_1 + b_0 = a_n 10^n + a_{n-1} 10^{n-1} + \cdots + 10 a_1 + a_0$$

所以

$$b_0 - a_0 = (a_n 10^n + a_{n-1} 10^{n-1} + \cdots + 10 a_1) - (b_k 10^k + b_{k-1} 10^{k-1} + \cdots + 10 b_1)$$

又因为 $|b_0 - a_0| < 10, 10 | (b_0 - a_0)$, 所以 $b_0 - a_0 = 0$, 也即 $b_0 = a_0$.

将 $b_0 = a_0$ 代入可得:

$$b_k 10^k + b_{k-1} 10^{k-1} + \cdots + 10 b_1 = a_n 10^n + a_{n-1} 10^{n-1} + \cdots + 10 a_1$$

将上式两边分别除以 10, 可得:

$$b_k 10^{k-1} + b_{k-1} 10^{k-2} + \cdots + b_1 = a_n 10^{n-1} + a_{n-1} 10^{n-2} + \cdots + a_1$$

于是类似前面的步骤再继续做减法:

$$b_1 - a_1 = (a_n 10^{n-1} + \cdots + 10 a_2) - (b_k 10^{k-1} + \cdots + 10 b_2)$$

同理可得 $b_1 = a_1$.

重复上面的做法, 可以依次得到 $b_2 = a_2, b_3 = a_3, \cdots$ 直到全部消去为止. 至此得证.

在数学中证明唯一的方法一般都是假设其不唯一, 并将其表示出来, 然后通过和题设及一些基本公理定理的推导得出矛盾, 据此用反证法的思想可得.

例 8.1 已知 $a_3 > a_1, b_3 \neq 0$, 且 $\overline{a_3 a_2 a_1} - \overline{a_1 a_2 a_3} = \overline{b_3 b_2 b_1}$. 求证:
$$\overline{b_3 b_2 b_1} + \overline{b_1 b_2 b_3} = 1089$$

证明: 由已知可得

$$b_1 = a_1 + 10 - a_3$$
$$b_2 = a_2 - 1 + 10 - a_2 = 9$$
$$b_3 = a_3 + 1 - a_1$$

所以

$$\begin{aligned}
\overline{b_3 b_2 b_1} + \overline{b_1 b_2 b_3} &= (100 b_3 + 10 b_2 + b_1) + (100 b_1 + 10 b_2 + b_3) \\
&= 101(b_1 + b_3) + 20 b_2 \\
&= 1089
\end{aligned}$$

例 8.2 一个六位数 $\overline{2abcde}$ 与 3 的积等于 $\overline{abcde9}$, 求这个六位数.

分析在两个数 $\overline{abcde9}$ 与 $\overline{2abcde}$ 中, a, b, c, d, e 顺序未变, 故把 $\overline{abcde}$ 作为一个整体来考虑, 会使求解过程简便.

解: 由题意可知 $\overline{2abcde} \times 3 = \overline{abcde9}$, 于是

$$(2 \times 10^5 + \overline{abcde}) \times 3 = \overline{abcde} \times 10 + 9$$

故 $\overline{abcde} = 85713$.

所以这个六位数是 285713.

8.2.2　二进位值制

二进位值制是现代计算机科学及相关领域中最基本的计数法.

二进位值制的起源可以追溯到中国道家文化中的太极八卦图，太极八卦图是一套用四组阴阳组成的形而上的哲学符号. 其深邃的哲理能用以解释自然、社会现象.

图 8.1

《易经》中记载："易有太极，是生两仪，两仪生四象，四象生八卦."

太极指的是天地未开、混沌未分阴阳之前的状态. 阐明了宇宙从无极而到太极，以至万物化生的过程.

两仪指的是天地或阴阳，也有各种以此为基础进行的进一步解释.

四象指的是太阳、太阴、少阴、少阳. 从数学角度论四象为：七、九、八、六. 从方位角度论四象为：东、南、西、北. 从一年季节论四象为：春、夏、秋、冬.

八卦表示事物自身变化的阴阳系统，用"—"代表阳，用"— —"代表阴，这两种代表阴阳的符号分别称为阳爻和阴爻，按照大自然的阴阳变化进行组合，组成八种不同形式，叫作八卦. 其中每一卦形代表一定的事物.

乾是由三阳爻组成，是纯阳之卦象，代表天. 震是二阴爻在上，一阳爻在下，代表雷. 坎是上下为阴爻，中间为阳爻，代表水. 艮是一阳爻在上，二阴爻在下，代表山. 坤是由三阴爻组成，为纯阴之卦象，代表地. 巽是二阳爻在上，一阴爻在下，代表风. 离是上下为阳爻，中间为阴爻，代表火. 兑是一阴爻在上，二阳爻在下，代表泽.

从数学的角度，可以看出，这里数字的增长从 1 到 2，再到 4，再到 8，事实上，所有的整数都可以用这样的方法将其表示出来，具体来说就是将一个整数表示成：

$$\sum_{i=0}^{n-1} a_i \times 2^i (a_i = 0, or\ a_i = 1, a_{n-1} \neq 0)$$

的形式，按照十进位值制的理解可以得到对应的定义，在此不再赘述. 为了在表示上和十进位值制进行一定的区分，将二进位值制下的数用 $(m)_2$ 表示.

二进位值制的本质是"满二进一"，这可以和十进位值制之下的数进行一一对应，也正因为如此再结合十进位值制下的唯一表示的性质可以证明用二进位值制可以表示所有的整数.

例 8.3　用二进位值制分别表示十进位值制下的 $1 \sim 10$ 的各数值.

解： 按照进位制的相关定义可得

$$0 = (0)_2$$
$$1 = (1)_2$$
$$2 = (10)_2$$
$$3 = 2 + 1 = (10)_2 + (1)_2 = (11)_2$$
$$4 = (100)_2$$
$$5 = 4 + 1 = (100)_2 + (1)_2 = (101)_2$$
$$6 = 4 + 2 = (100)_2 + (10)_2 = (110)_2$$
$$7 = 6 + 1 = (110)_2 + (1)_2 = (111)_2$$
$$8 = (1000)_2$$
$$9 = 8 + 1 = (1000)_2 + (1)_2 = (1001)_2$$
$$10 = 9 + 1 = (1001)_2 + (1)_2 = (1010)_2$$

事实上，由于计算机能识别的只有物理元件中的"接通"或"断开"，并由此来和逻辑运算中对命题判断的"真"或"假"，所以在计算机中采用的二进位值制，就是用 0 和 1 这两个数码来表示一切数字并进行相应运算的．一般来说，我们将一个十进位值制下的数输入到计算机中，计算机先将其翻译成二进位值制下的数，然后再通过相关运算，得到二进位值制下的结果，最后将这一结果翻译成十进位值制下的数并输出反馈给我们．

在这一基础上，可以按照 2 的乘次幂来表示一切的数，为了计算上的方便，有以下的表可以用于在十进位值制和二进位值制之下的数值间的转化．

表 8.2

乘次幂	数值	乘次幂	数值
1	2	9	512
2	4	10	1024
3	8	11	2048
4	16	12	4096
5	32	13	8192
6	64	14	16384
7	128	15	32768
8	256	……	……

例 8.4 十进位值制下的 24 转化为二进位值制下是多少？

解： 由上表中的数值那一列可以看出不大于 24 的最大的数是 16，而 $16 = 2^4$

又由于

$$24 - 16 = 8$$

表中不大于 8 的最大的数是 8，而

$$8 = 2^3$$

于是可得

$$24 = 16 + 8 = 1 \times 2^4 + 1 \times 2^3 + 0 \times 2^2 + 0 \times 2^1 + 0$$

所以

$$24 = (11000)_2.$$

例 8.5 十进位值制下的 92 转化为二进位值制下是多少？

解：按例 8.4 的步骤可得

$$92 = 64 + 16 + 8 + 4 = 1 \times 2^6 + 0 \times 2^5 + 1 \times 2^4 + 1 \times 2^3 + 1 \times 2^2 + 0 \times 2^1 + 0$$

所以

$$92 = (1011100)_2.$$

例 8.6 求十进位值制下的 1864 转化为二进制下是多少？

$$1864 = 2 \times 932 + 0$$
$$932 = 2 \times 466 + 0$$
$$466 = 2 \times 233 + 0$$
$$233 = 2 \times 116 + 1$$
$$116 = 2 \times 58 + 0$$
$$58 = 2 \times 29 + 0$$
$$29 = 2 \times 14 + 1$$
$$14 = 2 \times 7 + 1$$
$$7 = 2 \times 3 + 1$$
$$3 = 2 \times 1 + 1$$

所以 $1864 = (11101001000)_2$.

例 8.7 计算下面各题：

$$(1001)_2 + (1101)_2$$
$$(11101)_2 - (1011)_2$$
$$(1110)_2 \times (11)_2$$
$$(101101)_2 \div (11111)_2$$

解一：先将二进位值制数 $(m)_2$ 转化为十进位值制下的数，然后按照十进位值制下的数的加减乘除运算进行计算.

因为

$$(1001)_2 = 9$$
$$(1101)_2 = 13$$
$$(11101)_2 = 29$$
$$(1011)_2 = 11$$
$$(1110)_2 = 14$$
$$(11)_2 = 3$$
$$(101101)_2 = 45$$
$$(1111)_2 = 15$$

所以

$$(1001)_2 + (1101)_2 = (9)_{10} + (13)_{10} = (22)_{10} = (10110)_2.$$

为了书写的方便，以后凡涉及十进位值制的数将不再采用类似 $(9)_{10}$ 的写法，而直接写为 9.

$$(11101)_2 - (1011)_2 = 29 - 11 = 18 = (10010)_2$$
$$(1110)_2 \times (11)_2 = 14 \times 3 = 42 = (101010)_2$$
$$(101101)_2 \div (1111)_2 = 45 \div 15 = 3 = (11)_2$$

解二： 模仿十进位值制的相关计算法则，直接用二进位值制进行相关计算．可以模仿十进位值制下的竖式加法直接用竖式计算：

$$
\begin{array}{r}
(1001)_2 \\
+ (1101)_2 \\
\hline
(10110)_2
\end{array}
$$

需特别注意的是在十进位值制下进行竖式计算，进位法则是满十进一，而在二进位值制下进行竖式计算，进位法则是满二进一．类似的，在二进位值制下，若是在减法计算中，向高一位的数"借"1 只可当 2 用．

模仿十进位值制下的减法竖式运算直接用竖式计算二进位值制下的减法运算：

$$
\begin{array}{r}
(11101)_2 \\
- (1011)_2 \\
\hline
(10010)_2
\end{array}
$$

模仿十进位值制下的乘法竖式运算直接用竖式计算二进位值制下的乘法运算：

$$
\begin{array}{r}
(1110)_2 \\
\times (11)_2 \\
\hline
(1110)_2 \\
\times (11100)_2 \\
\hline
(101010)_2
\end{array}
$$

模仿十进位值制下的除法竖式运算直接用竖式计算二进位值制下的除法运算：

$$
\begin{array}{r}
(11)_2 \\
(1111)_2 \overline{)(101101)_2} \\
- (11110)_2 \\
\hline
(1111)_2 \\
- (1111)_2 \\
\hline
(0)_2
\end{array}
$$

解三： 按照求"补数"的方法计算二进制下的减法和除法[①].

8.2.3 五进位值制

有人认为，人类最先认识事物的时候是根据对自身的了解推广得到的，所以最先的进

---

① 陈省身．初等数论（I）［M］．哈尔滨：哈尔滨工业大学出版社，2012：36 - 40.

制是依据人类有两只手，中国也叫一双手，于是就有了二进位值制．随后是看到有两只手两只脚，也即有四肢，因此有了四进位值制．由于每只手有五个手指，每只脚有五个脚趾，于是就有了五进位值制．

　　根据前面对十进位值制和二进位值制的定义，不难想象，所谓五进位值制，指的是以 5 为底的进位制，逢五进一的一种计数方法．在五进位值制之下，用到的数只可能是 0，1，2，3，4 这五个．

　　中国古代的五行学说也是采用的五进制，0 代表土，1 代表水，2 代表火，3 代表木，4 代表金，以此类推，5 又属土，6 属水，减去 5 即得．

　　在 20 世纪，只有肯尼亚和尼日利亚的约鲁巴人仍在使用这种五进制的系统，不过，十进制在各地区已普遍使用，这些部落原本使用的五进制也已慢慢消逝．

　　类似上一节的基础，可以按照 5 的乘次幂来表示一切的数，为了计算上的方便，有以下的表可以用于在五进位值制与十进位值制、二进位值制之下的数值间的转化．

表 8.3

五进位值制	十进位值制	二进位值制
0	0	0000
1	1	0001
2	2	0010
3	3	0011
4	4	0100
10	5	0101
11	6	0110
12	7	0111
13	8	1000
14	9	1001
20	10	1010
21	11	1011
22	12	1100
23	13	1101
24	14	1110
30	15	1111

　　五进位值制下的计算类似二进位值制，在此不再赘述．

8.2.4　八进位值制

　　在 19 世纪中期，阿尔弗雷德·泰勒认为："如果我们以八基数，那么将超越所有的计数法，因为它是一个最好的算术系统．"为了克服二进位值制在表示数上的多而烦琐的位数，有人引进了八进位值制和十六进位值制．

所有现代计算平台使用 16 ~ 32 位，或者 64 位，如果使用 64 位，将进一步划分为八位字节．这种系统三个八进制数字就能满足每字节需要，与最重要的八进制数字代表两个二进制数字（+1 为下一个字节，如果有的话）．16 位字的八进制表示需要 6 位数，但最重要的八进制数字代表（通过）只有一个（0 或 1）．这表示无法提供容易阅读的字节，因为它是在 4 位八进制数字．

所谓八进位值制，指的是以 8 为底的进位制，逢八进一的一种计数方法．在八进位值制之下，用到的数只可能是 0，1，2，3，4，5，6，7 这八个．

类似前面的基础，可以按照 8 的乘次幂来表示一切的数，为了计算上的方便，有以下的表可以用于在八进位值制与十进位值制、二进位值制之下的数值间的转化．

表 8.4

八进位值制	十进位值制	二进位值制
0	0	0000
1	1	0001
2	2	0010
3	3	0011
4	4	0100
5	5	0101
6	6	0110
7	7	0111
10	8	1000
11	9	1001
12	10	1010
13	11	1011
14	12	1100
15	13	1101
16	14	1110
17	15	1111
……	……	……
100	64	1000000
……	……	……
1000	512	1000000000
……	……	……

续　表

八进位值制	十进位值制	二进位值制
10000	4096	1000000000000
……	……	……
100000	32768	1000000000000000

由这一变化趋势，应该能够展示八进位值制下的数和二进位值制下的数在存储数据和进行计算时的优势了．八进位值制下的计算类似二进位值制，在此不再赘述．

8.2.5　十六进位值制

在中国古代的度量衡中，一直就有着十六进位值制，如汉语中"半斤八两"的说法就来源于此，后来十六进位值制在计算机发展中起到了更大的作用．

所谓十六进位值制，指的是以 16 为底的进位制，逢十六进一的一种计数方法．可问题是如果在数的表示中再次使用 10，11，12，13，14，15 这六个数的话，将会使得数与位上产生歧义，所以后来将十六进位值制的表示改为使用数 0，1，2，3，4，5，6，7，8，9，A，B，C，D，E，F 这十个数值和六个字母，其中的 A，B，C，D，E，F 依次对应于10，11，12，13，14，15.

类似前面的基础，可以按照 16 的乘次幂来表示一切的数，为了计算上的方便，有以下的表可以用于在十六进位值制与十进位值制、八进位值制、二进位值制之下的数值间的转化．

表 8.5

十六进位值制	十进位值制	八进位值制	二进位值制
0	0	0	0000
1	1	1	0001
2	2	2	0010
3	3	3	0011
4	4	4	0100
5	5	5	0101
6	6	6	0110
7	7	7	0111
8	8	10	1000
9	9	11	1001
A	10	12	1010
B	11	13	1011
C	12	14	1100
D	13	15	1101

续 表

十六进位值制	十进位值制	八进位值制	二进位值制
E	14	16	1110
F	15	17	1111
10	16	20	10000
11	17	21	10001
12	18	22	10010
13	19	23	10011
……	……	……	……
1A	26	32	11010
1B	27	33	11011
1C	28	34	11100
1D	29	35	11101
1E	30	36	11110
1F	31	37	11111
……	……	……	……
2A	42	52	101010
……	……	……	……
9A	154	232	10011010
……	……	……	……
AA	170	252	10101010

十六进位值制下的计算类似二进位值制，在此不再赘述.

8.2.6　六十进位值制

六十进制是以 60 为基数的进位制，源于公元前 3 世纪的古闪族，后传至巴比伦，现仍用作记录时间、角度和地理坐标. 其他文明也有使用六十进制，如西新几内亚的 Ekagi 族.

其实际应用的主要有计时中的六十进位值制：

$$1 \text{ 小时} = 60 \text{ 分钟}$$
$$1 \text{ 分钟} = 60 \text{ 秒钟}$$

另一个是在角度分割时，一个圆周被均分成 360 度，每一度有 60 角分，一角分等于 60 角秒.

在中国农历中，有六十甲子的概念，以天干与地支两者经一定的组合方式搭配成六十对，为一个周期.

表 8.6

十进制数	天干地支	十进制数	天干地支	十进制数	天干地支	十进制数	天干地支
1	甲子	16	己卯	31	甲午	46	己酉
2	乙丑	17	庚辰	32	乙未	47	庚戌
3	丙寅	18	辛巳	33	丙申	48	辛亥
4	丁卯	19	壬午	34	丁酉	49	壬子
5	戊辰	20	癸未	35	戊戌	50	癸丑
6	己巳	21	甲申	36	己亥	51	甲寅
7	庚午	22	乙酉	37	庚子	52	乙卯
8	辛未	23	丙戌	38	辛丑	53	丙辰
9	壬申	24	丁亥	39	壬寅	54	丁巳
10	癸酉	25	戊子	40	癸卯	55	戊午
11	甲戌	26	己丑	41	甲辰	56	己未
12	乙亥	27	庚寅	42	乙巳	57	庚申
13	丙子	28	辛卯	43	丙午	58	辛酉
14	丁丑	29	壬辰	44	丁未	59	壬戌
15	戊寅	30	癸巳	45	戊申	60	癸亥

以天干和地支按顺序相配，即甲、乙、丙、丁、戊、己、庚、辛、壬、癸与子、丑、寅、卯、辰、巳、午、未、申、酉、戌、亥相组合，从"甲子"起，到"癸亥"止，满六十为一周，称为"六十甲子"．亦称"六十花甲子"．又因起头是"甲"字的有六组，所以也叫"六甲"．

六十甲子衍生出来的《黄帝内经》之五运六气理论及四柱命理学理论之所以能数千年不衰，因为这些理论是探索人体奥秘、预测、诊断、治疗人体疾病的学问．在自然科学领域里，六十甲子的作用也是巨大的．

六十进位值制下的计算类似二进位值制，在此不再赘述．

8.2.7　k 进位值制

事实上，除了上述介绍的十进位值制、二进位值制、五进位值制、八进位值制、十六进位值制和六十进位值制之外，还有很多进位值制在人类认识世界的进程中起到了重要的作用，但限于篇幅，在此不再一一介绍，而代之以 k 进位值制来统一介绍．

对任意一个数码，均可作为某进位值制的底数，也即逢 k 进一，或称为满 k 进一．例如三进位值制、四进位值制等．

在 k 进位值制下，一般可以约定任意一个 n 位自然数的表示方法为：

$$m = \overline{a_{n-1}\, a_{n-2} \cdots a_2\, a_1\, a_0}$$
$$= a_{n-1} \times k^{n-1} + a_{n-2} \times k^{n-2} + \cdots + a_2 \times k^2 + a_1 \times k + a_0$$

$$= \sum_{i=0}^{n-1} a_i \times k^i \, (a_i \in N, 0 \leqslant a_i \leqslant k, a_{n-1} \neq 0)$$

将 k 进位值制下的数用 $(m)_k$ 表示.

有关 k 进位值制下的表示和计算类似于十进位值制与二进位值制的表示和计算, 在此亦不再赘述.

8.3 典型问题

8.3.1 典型例题

例 8.8 把 $(\text{A35B0F})_{16}$ 从十六进制转换为十进制表示:

解: 按照十六进制和十进制的关系可以得到:

$(\text{A35B0F})_{16} = 10 \cdot 16^5 + 3 \cdot 16^4 + 5 \cdot 16^3 + 11 \cdot 16^2 + 0 \cdot 16 + 15 = (10705679)_{10}$.

注: 从十六进制到二进制的转换的一个例子是 $(2\text{FB}3)_{16} = (10111110110011)_2$. 每个十六进制数字转换为一个四位二进制数字块 [与数字 $(2)_{16}$ 相关的初始块 $(0010)_2$ 的起始零被省略了] .

例 8.9 计算 $(1101)_2$ 和 $(1110)_2$ 的乘积.

解: 模仿十进制数的乘法竖式计算, 有:

```
        1101
     × 1110
     ————————
        0000
        1101
       1101
      1101
   ————————————
    10110110
```

注: 首先用 $(1)_2$ 的每个数字乘 $(1101)_2$, 每次做适当数目的移位, 然后把适当的整数相加得到积.

例 8.10 已知 $a_3 > a_1, b_3 \neq 0$, 且 $\overline{a_3 a_2 a_1} - \overline{a_1 a_2 a_3} = \overline{b_3 b_2 b_1}$.

求证: $\overline{b_3 b_2 b_1} + \overline{b_1 b_2 b_3} = 1089$.

证明: 由已知可得

$$b_1 = a_1 + 10 - a_3, b_2 = a_{2|} - 1 + 10 - a_2 = 9, b_3 = a_3 - 1 - a_1.$$

所以

$$\overline{b_3 b_2 b_1} + \overline{b_1 b_2 b_3} = (100b_3 + 10b_2 + b_1) + (100b_1 + 10b_2 + b_3)$$
$$= 101(b_1 + b_3) + 20b_2$$
$$= 1089.$$

例 8.11 把 $110111_{(2)}$ 化为十进位制数.

解：二进位制的计数单位从右至左依次是 $2^0, 2^1, 2^2, \cdots$，可以把 $110111_{(2)}$ 先写成它的不同计数单位的数之和的形式，然后按十进位制计算出结果．

$$110111_{(2)} = 1 \times 2^5 + 1 \times 2^4 + 0 \times 2^3 + 1 \times 2^2 + 1 \times 2^1 + 1 \times 2^0$$
$$= 1 \times 32 + 1 \times 16 + 1 \times 4 + 1 \times 2 + 1 = 55.$$

例 8.12　把 49 化为二进位制数．

解：根据二进位制数"满二进一"的规则，可以用 2 连续去除 49 和所得的商，然后取余数．具体计算方法如下：

$49 = 2 \times 24 + 1$，把余数 1 放在右起第一位，24 是所得商；

$24 = 2 \times 12 + 0$，把余数 0 放在右起第二位，12 是所得商；

$12 = 2 \times 6 + 0$，把余数 0 放在右起第三位，6 是所得商；

$6 = 2 \times 3 + 0$，把余数 0 放在右起第四位，3 是所得商；

$3 = 2 \times 1 + 1$，把余数 1 放在右起第五位，3 是所得商；

$1 = 2 \times 0 + 1$，把余数 1 放在右起第六位．

所以 $49 = 110001_{(2)}$．

这种方法叫作 2 除取余法．通常写成下面的竖式：

$$
\begin{array}{r|l}
2 & 49 \\
2 & 24 \\
2 & 12 \\
2 & 6 \\
2 & 3 \\
2 & 1 \\
\end{array}
\quad \text{余数（从右往左）为：} 1, 1, 0, 0, 0, 1
$$

所以按顺序写出得数为 $49 = 110001_{(2)}$．

例 8.13　现有 1 克、2 克、4 克、8 克、16 克的砝码各一个，若只能将砝码放在天平的一端，问能称出多少种不同质量的物品？若称 23 克的物品，应如何选配上述砝码？

解：注意到砝码质量可以表示成二进位制数：

$$1_{(2)}, 10_{(2)}, 100_{(2)}, 1000_{(2)}, 1000_{(2)},$$

则天平可以称量的物品质量为

$$2^4 k_0 + 2^3 k_1 + 2^2 k_2 + 2 k_3 + k_4, k_i = 0 \text{ 或 } 1, \text{ 且 } k_i \text{ 不全为 } 0 \, (i = 0, 1, 2, 3, 4).$$

所以可称 $1_{(2)}$ 克到 $11111_{(2)}$ 克的任何一种质量的物品．称一种质量的物品时，不同质量的砝码只许用一次，故可称 $11111_{(2)} = 31$ 种不同质量的物品．

因为 $23 = 10111_{(2)}$，所以称 23 克的物品，选配 $10000_{(2)} = 16$ 克，$100_{(2)} = 4$ 克，$10_{(2)} = 2$ 克，$1_{(2)} = 1$ 克的砝码各一个．

8.3.2　典型练习题

练习 8.1　填空：

（1）$2011_{(3)} = $ ＿＿＿＿＿$_{(10)}$；（2）$31404_{(5)} = $ ＿＿＿＿＿$_{(10)}$；

（3）$7137_{(10)} = $ ＿＿＿＿＿$_{(2)}$；（4）$21580_{(10)} = $ ＿＿＿＿＿$_{(8)}$；

（5）$1376_{(5)}$ = _____ $_{(8)}$.

练习 8.2 把 $(1999)_{10}$ 从十进制表示转换为七进制表示. 把 $(6105)_7$ 从七进制表示转换为十进制表示.

练习 8.3 把 $(89156)_{10}$ 从十进制表示转换为八进制表示. 把 $(706113)_8$ 从八进制表示转换为十进制表示.

练习 8.4 把 $(1010111)_2$ 从二进制表示转换为十进制表示. 并把 $(999)_{10}$ 从十进制表示转换为二进制表示.

练习 8.5 把 $(101001000)_2$ 从二进制表示转换为十进制表示. 并把 $(1984)_{10}$ 从十进制表示转换为二进制表示.

练习 8.6 把 $(100011110101)_2$ 和 $(11101001110)_2$ 从二进制转换为十六进制.

练习 8.7 把 $(ABCDEF)_{16}$，$(DEFACED)_{16}$ 和 $(9A0B)_{16}$ 从十六进制转换为二进制.

练习 8.8 比较 $1011011_{(2)}$ 与 $1203_{(4)}$ 的大小.

练习 8.9 求 $(334)_5$ 加上 $(1100111011)_2$ 的和.

练习 8.10 求 $(10001000111101)_2$ 加上 $(11111101011111)_2$ 的和.

练习 8.11 求 $(1111000011)_2$ 减去 $(11010111)_2$ 的差.

练习 8.12 求 $(1101101100)_2$ 减去 $(101110101)_2$ 的差.

练习 8.13 求 $(11101)_2$ 乘以 $(110001)_2$ 的积.

练习 8.14 求 $(1110111)_2$ 乘以 $(10011011)_2$ 的积.

练习 8.15 求 $(110011111)_2$ 除以 $(1101)_2$ 的商和余数.

练习 8.16 求 $(110100111)_2$ 除以 $(11101)_2$ 的商和余数.

练习 8.17 求 $(1234321)_5$ 加上 $(2030104)_5$ 的和.

练习 8.18 求 $(4434201)_5$ 减去 $(434421)_5$ 的差.

练习 8.19 求 $(1234)_5$ 乘以 $(3002)_5$ 的积.

练习 8.20 求 $(14321)_5$ 除以 $(334)_5$ 的商和余数.

练习 8.21 求 $(FEED)_{16}$ 减去 $(CAFE)_{16}$ 的差.

练习 8.22 求 $(FACE)_{16}$ 乘以 $(BAD)_{16}$ 的积.

练习 8.23 求 $(BEADED)_{16}$ 除以 $(ABBA)_{16}$ 的商和余数.

练习 8.24 解释为何在实际中当我们把大的十进制整数分成三位的块并用空格隔开时是在使用基为 1000 的表示.

练习 8.25 证明，如果 b 是小于 -1 的负数，则每个非零整数 n 可以被唯一地写成如下形式：

$$n = a_k b^k + a_{k-1} b^{k-1} + \cdots + a_1 b + a_0.$$

其中 $a_k \neq 0$ 且 $0 \leq a_j \leq |b|$，$j = 0,1,2,\cdots,k$，我们把它写成 $n = (a_k a_{k-1} \cdots a_1 a_0)_b$ 就像基为正数那样.

练习 8.26 求 $(101001)_{-2}$ 和 $(12012)_{-3}$ 的十进制表示.

练习 8.27 求十进制数 -7，-17 和 61 的基于 -2 的表示.

练习 8.28 证明当所有的砝码都放在一个盘子中时，不超过 $2^k - 1$ 的重量可以使用重为 $1,2,2^2 \cdots 2^{k-1}$ 的砝码来测量.

练习 8.29　解释如何从三进制表示转换为九进制表示.

练习 8.30　解释如何从基于 r 的表示转换为基于 r^n 的表示，以及如何从基于 r^n 的表示转换为基于 r 的表示，其中 $r > 1$ 且 n 为正整数.

练习 8.31　证明：如果 $n = (a_k a_{k-1} \cdots a_1 a_0)_b$，则 n 被 b^j 除得的商和余数分别是 $q = (a_k a_{k-1} \cdots a_j)_b, r = (a_{j-1} \cdots, a_1 a_0)_b$.

练习 8.32　一打（dozen）等于 12，一罗（gross）等于 12^2. 用 12 为基或十二进制（duodecimal）算法回答下列问题.

$a)$ 如果从 11 罗 3 打鸡蛋中取出 3 罗 7 打零 4 个鸡蛋，还剩下多少鸡蛋？

$b)$ 如果每卡车有 2 罗 3 打零 7 个鸡蛋，共往超市运送 5 卡车，那么一共运了多少鸡蛋？

$c)$ 如果 11 罗 10 打零 6 个鸡蛋被分成等数量的 3 堆，那么每堆有多少鸡蛋？

练习 8.33　对于十进制展开为 $(a_n a_{n-1} \cdots a_1 a_0)$ 且末尾数字 $a_0 = 5$ 的整数，求其平方的一个众所周知的规律求乘积 $(a_n a_{n-1} \cdots a_1)_{10} [(a_n a_{n-1} \cdots a_1)_{10} + 1]$ 并在最后添上数字 $(25)_{10}$. 例如，$(165)^2$ 的十进制展开由 $16 \cdot 17 = 272$ 开始，所以 $(165)^2 = 27225$. 证明这个规则是有效的.

练习 8.34　一个行为乖张的富豪要把一百万元赠人，获赠者所得钱的元数只能是 1 或 7 的乘幂，他不喜欢有 6 人以上获得同额赠钱，但却不计较领钱人多少. 问如何分配才能使他如愿以偿？

练习 8.35　夏天的一天，4 条腿的青蛙说"我今天吃了 1221 只蚊子."8 条腿的蜘蛛说："你吹牛，我数过，你今天吃了 151 只蚊子." 实际它们都对，而且都是按自己习惯的办法数的. 按照十进位制，你说青蛙今天吃了多少只蚊子？

练习 8.36　现有 10 箱小球，其中混进了几箱次品. 每个标准球的质量是 10 克，每个次品球比标准球少 1 克，但外观无区别，能否称一次就查出装次品球的箱子？

练习 8.37　如果 $52_{(k)}$ 是 $25_{(k)}$ 的两倍，那么 $123_{(k)}$ 在十进位制中表示多少？

第 9 章　算术基本定理

在数学中，我们发现真理的主要工具是归纳和模拟.

——拉普拉斯

　　算术是数学中最古老、最基础和最初等的部分，它研究数的性质及其运算．把数和数的性质、数和数之间的四则运算在运用过程中的经验累积起来，并加以整理，就形成了最古老的一门数学——算术．在古代全部数学就叫作算术，现代的代数学、数论等最初就是由算术发展起来的．后来，算学、数学的概念出现了，它代替了算术的含义，包括了全部数学，算术就变成了其中的一个分支．

　　算术中最基本的运算就是加减乘除四则运算，在前面几章的铺垫下，这一章主要针对如何将一个自然数进行分解以及分解后的情形作出讨论．

9.1　因数分解

9.1.1　素数的整除性质

　　一方面，由素数的定义可以知道，一个素数 p 只可能分解为

$$p = 1 \times p$$

唯一的一种情形（在不计算乘法交换律的前提下），于是按照整除的定义及性质可以得到：

$$1 \mid p$$
$$p \mid p$$

除此之外，再没有任何数能够整除素数 p．

　　另一方面，有必要探讨素数能整除那些数？易知，素数能整除的数是该素数的整数倍．更进一步，如果素数 p 整除两个数的乘积，会是怎样的情形？

　　性质： 若素数 p 整除 ab，则 p 至少整除 a,b 之一．也即要么 p 整除 a，要么 p 整除 b，要么 p 同时整除 a,b．

　　证明： 假设 p 整除 a，证明完成．

　　假设 p 不整除 a．由于 $\gcd(p,a) \mid p$，而 p 是素数，故 $\gcd(p,a) = 1$ 或 $\gcd(p,a) = p$．由于 $\gcd(p,a) \mid a$，而假设了 p 不整除 a，所以 $\gcd(p,a)$ 必等于 1．也即 p,a 互素，证明完成．

　　例 9.1　素数 3 整除两个数 2 和 6 的乘积 12，所以有 3 整除至少 2 和 6 之一．但是这个说法只对素数成立，如对于合数 6 整除两个数 14 和 15 的乘积 210，但是 6 既不整除 14，也不整除 15．

　　值得注意的是，上述性质可以继续推广到若干个数的乘积的情形，也即：若素数 p 整除 $a_1 a_2 \cdots a_r$，则 p 至少整除 $a_1 a_2 \cdots a_r$ 之一．

9.1.2　因数分解

　　因数分解，有时候又称为素因数分解．主要是将一个整数分解为几个因数的乘积的形式．一般来说，需要将其分解为最简形式，也就是分解到每个因数都不能再分解的素数乘积的形式．因数分解在代数学、密码学、计算复杂性理论和量子计算机等领域中有重要意义．

　　由上面的描述我们很容易得到：如果一个整数 p 是素数，则其只可能写成

$$p = 1 \times p$$

或者

$$p = p \times 1$$

的形式. 可以看出上述两式的右边除了顺序有差异外, 而对于乘法必然满足交换律, 因此两式并无本质区别, 也即在不计较因数的先后顺序的情况下它们是相同的. 换句话说, 在不计较因数的先后顺序的情况下, 素数 p 的分解式是唯一的.

如果一个整数 p 是合数, 将其写成几个因数的乘积的时候会不会也满足和素数时候一样的情形呢?

9.2 算术基本定理

算术基本定理是整数论中最为重要的理论之一, 它将整数都变成了一些素数乘积的形式, 也即说明了素数是整数的乘法的构成单元.

9.2.1 算术基本定理

每一个大于 1 的整数都可以被的写成素数的乘积的形式, 在不计较因数的顺序的情况下, 这种写法是唯一的.

对于所要讨论的整数, 我们不妨将其按照素数还是合数进行粗略的分类, 这样对于素数的讨论在上一节已经完成, 现在就看一看对于合数的情形是否满足.

例 9.2 合数的因数分解

$$120 = 2^3 \times 3 \times 5$$
$$1001 = 7 \times 11 \times 13$$
$$\cdots\cdots$$

除了上述的分解方式没找到其他的分解表达式, 但这并不能证明这样的分解式唯一, 而需要进行逻辑证明. 而普通的通过较多数学语言的证明在很多初等数论的书籍上都有, 在此无须再次赘述, 我们将以另一种方式, 尝试用归纳法加以证明.

第一步:

对于整数 2 来说, 有 $2 = 1 \times 2$;

对于整数 3 来说, 有 $3 = 1 \times 3$;

对于整数 5 来说, 有 $5 = 1 \times 5$;

对于整数 7 来说, 有 $7 = 1 \times 7$.

事实上, 由前面的知识我们知道, 对于任意素数 p 都有 $p = 1 \times p$.

故对于素数来说, 定理成立.

对于非素数的整数, 如对于 4 来说, 有 $4 = 2 \times 2$;

对于整数 6 来说, 有 $6 = 3 \times 2$;

对于整数 8 来说, 有 $8 = 4 \times 2$;

对于整数 9 来说, 有 $9 = 3 \times 3$.

也即经过验证, 对于 $n = 2,3,4,5,6,7,8,9$ 可以分解为素数乘积的形式.

第二步：

假设对于 n 可以分解为素数乘积的形式．

第三步：

证明对于 $n + 1$ 也可以分解为素数乘积的形式．

此时按照是否是素数的分类 $n + 1$ 一共有两种可能性，即 $n + 1$ 为素数或 $n + 1$ 为合数，下面依次讨论．

若 $n + 1$ 为素数，则它本身已经分解为素数，或者可以分解为：

$$n + 1 = (n + 1) \times 1$$

结论成立．

若 $n + 1$ 为合数，这表明 $n + 1$ 可以分解为

$$n + 1 = n_1 \times n_2$$

其中 n_1, n_2 是大于 2 且小于 n 的整数．

而由第二步的假设可知，对于 n_1, n_2 来说，定理是成立的，也即可以将 n_1, n_2 进行因数分解．

不妨将该分解式假定为：

$$n_1 = p_1 p_2 p_3 \cdots p_k$$
$$n_2 = q_1 q_2 \cdots q_r$$

将这两个式子相乘可得：

$$n_1 \times n_2 = p_1 p_2 p_3 \cdots p_k q_1 q_2 \cdots q_r$$

所以，对于合数 $n + 1$ 来说，可以分解为素数的乘积的形式．

关于唯一性的证明，其方法与存在性的证明类似，这里不再赘述．

9.2.2　标准分解式

由算术基本定理可以保证，任意一个整数都可以写成素数因数连乘积的形式，为了表述上的方便，特做如下定义：

标准分解式：将表达式

$$n = \prod_{i=1}^{k} p_i^{n_i}$$

称为整数 n 的标准分解式，其中 p_i 是整数 n 的素因数．

如例 9.2 中的表达式就是标准分解式．

例 9.3　求 9828 的标准分解式．

解：因为 9828 是偶数，所以可以一步一步进行分解．

$$9828 = 2 \times 4914$$
$$4914 = 2 \times 2457$$

2457 不是偶数，但可以看出能被 3 整除，故：

$$2457 = 3 \times 819$$
$$819 = 3 \times 273$$
$$273 = 3 \times 91$$

91 不是素数, 可以验证有:
$$91 = 7 \times 13$$

至此, 可以看出
$$9828 = 2 \times 2 \times 3 \times 3 \times 3 \times 7 \times 13$$

也可以用次幂进行表示为:
$$9828 = 2^2 \times 3^3 \times 7 \times 13$$

这就是 9828 的标准分解式, 这样的分解方法称为直接分解法.

要计算一个整数的标准分解式, 还可以用小学时候学习过的短除法等进行.

9.3 典型问题

9.3.1 典型例题

例 9.4 证明: 如果质数 p 是 ab 的因数, 那么 p 一定是 a 或 b 的因数.

解: 如果 p 不是 a 的因数, 那么 p 和 a 的公因数只有 1.

所以, 存在整数 u,v 使得 $1 = ua + vp$.

从而 $b = uab + vbp$.

又 $p \mid ab$, 所以上式右边两项均被 p 整除, 因而左边的 b 被 p 整除.

例 9.5 求 27 与 15 的最大公约数与最大公倍数.

解: 因为 $27 = 3^3, 15 = 3 \times 5$, 所以 $(27,15) = 3, [27,15] = 3^3 \times 5 = 135$

一般的, $a = p_1^{\alpha_1} p_2^{\alpha_2} \cdots p_k^{\alpha_k}$ 与 $b = p_1^{\beta_1} p_2^{\beta_2} \cdots p_k^{\beta_k} (\alpha_i, \beta_i \geqslant 0, i = 1,2,\cdots,k)$ 的最大公约数为
$$P_1^{\gamma_1} P_2^{\gamma_2} \cdots P_k^{\gamma_k} = \min(\alpha_i, \beta_i), i = 1,2,\cdots k,$$

而最小公倍数为
$$p_1^{\delta_1} p_2^{\delta_2} \cdots p_k^{\delta_k}, \delta_i = \max(\alpha_i, \beta_i), i = 1,2,\cdots k,$$

由于
$$\gamma_i + \delta_i = \alpha_i + \beta_i, i = 1,2,\cdots k,$$

所以
$$ab = (a,b)[a,b].$$

于是我们得到: 如果 a,b 是正整数, 那么 $ab = (a,b)[a,b]$

例 9.6 求使 $1989m$ 为平方数的最小的 m.

解: 由于 $1989 = 3^2 \times 17 \times 13$, 所以 $m = 17 \times 13 = 221$.

例 9.7 对于任意给定的自然数 n, 证明: 必有无穷多个自然数 a, 使 $n^4 + a$ 为合数.

基本思路: 配方并应用公式 $a^2 - b^2$.

证明: 取 $a = 4m^2$, 则

$$n^4 + 4m^4 = n^4 + 4m^2n^2 + 4m^4 - 4m^2n^2$$
$$= (2m^2 + n^2)^2 - 4m^2n^2$$
$$= (2m^2 + n^2 - 2mn)(2m^2 + n^2 + 2mn)$$

例 9.8　设 n 为正整数. 证明：若 n 的所有正因子之和是 2 的幂，则这些正因子的个数也是 2 的幂.

证明： 设 $n = p_1^{s_1} p_2^{s_2} \cdots p_k^{s_k}$，其中，$p_1, p_2, \cdots, p_k$ 为不同素数，$s_i \in N_+ (i = 1, 2, \cdots, k)$.

则 n 的所有正因子之和可表示为

$$(1 + p_1 + \cdots + p_1^{s_1})(1 + p_2 + \cdots + p_2^{s_2}) \cdots (1 + p_k + \cdots + p_k^{s_k}).$$

若它是 2 的幂，则它的因子 $f_i = 1 + p_i + p_i^2 + \cdots + p_i^{s_i} (i = 1, 2, \cdots, k)$ 也是 2 的幂.

因此，所有的 p_i, s_i 均为奇数，若存在 $s_i > 1$，则

$$f_i = (1 + p_i)(1 + p_i^2 + p_i^4 + \cdots + p_i^{s_i-1}).$$

又由于 f_i 不含大于 1 的奇因子，故偶数 $s_i - 1$ 必为 $4k + 2$ 的形式. 于是，

$$f_i = (1 + p_i)(1 + p_i^2)(1 + p_i^4 + \cdots + p_i^{s_i-3})$$

由于 $1 + p_i$ 和 $1 + p_i^2$ 均为 2 的幂，故

$$(1 + p_i) \mid (1 + p_i^2),$$

这与

$$1 + p_i^2 = (1 + p_i)(p_i - 1) + 2$$

矛盾. 因此，必有

$$s_i = 1 (i = 1, 2, \cdots, k).$$

故 n 的正因子的个数也是 2 的幂.

例 9.9　设一个正整数满足下列性质：其所有模 4 不余 2 的正因数之和等于 1000. 求满足上述性质的所有正整数.

解： 对于正整数 n，设 $S(n)$ 为 n 的所有模 4 不余 2 的正因数的和，假设 n 的素因数分解为

$$2^m p_1^{m_1} p_2^{m_2} \cdots p_k^{m_k} (m, m_i \in N_+, i = 1, 2, \cdots, k).$$

因为一个整数模 4 余 2 等价于其恰被 2 整除，所以，$S(n)$ 是所有形如

$$2^l p_1^{l_1} p_2^{l_2} \cdots p_k^{l_k} (0 \le l \le m, l \ne 1, 0 \le l_1 \le m_1, \cdots, 0 \le l_k \le m_k)$$

的数的和. 故

$$S(n) = \sum_{l=0, l \ne 1}^{m} 2^l \cdot \sum_{l_1=0}^{m_1} p_1^{l_1} \cdots \sum_{l_k=0}^{m_k} p_k^{l_k}.$$

为了简单起见，对于每个非负整数 m，设

$$f(2, m) = \sum_{l=0, l \ne 1}^{m} 2^l, f(p, m) = \sum_{l=0}^{m} p^l,$$

其中，p 是大于 2 的素数. 则

$$S(n) = f(2, m) f(p_1, m_1) \cdots f(p_k, m_k) = 1000$$

先考虑 $f(2, m)$. 若 $m \ge 9$. 则

$$f(2,m) \geqslant f(2,9) = 2^{10} - 1 - 2^1 = 1021.$$

因此, $m \leqslant 8$.

经计算得 $f(2,1) = 1, f(2,2) = 5, f(2,6) = 125$, 满足 $f(2,m) \mid 1000$.

再考虑 $f(p,m)$, 当 $3 \leqslant p \leqslant 31$ 时, 类似地可以验证

$$f(3,1) = 4, f(3,3) = 40, f(7,1) = 8, f(19,1) = 20$$

均整除 1000. 当 $p \geqslant 32$ 时, 若 $m \geqslant 2$, 则

$$f(p,m) \geqslant f(p,2) \geqslant 1 + 32 + 32^2 > 1000.$$

故 $m = 1$.

经验证 $f(199,1) = 200, f(499,1) = 500$, 均整除 1000.

最后考虑 $f(p,m)(p \geqslant 2)$ 的组合, 只有 $2^6 \times 7^1 = 448$ 及 $2^2 \times 199^1 = 796$ 满足题意.

例 9.10 (1) 已知 p 为大于 3 的素数. 证明 p 的平方被 24 除的余数为 1.

(2) 求所有使 $p^2 + 2543$ 具有少于 16 个不同正因子的素数 p.

证明: (1) 因为大于 3 的素数均可以表示成 $6k \pm 1$ 的形式, 所以,

$$p^2 - 1 = (6k \pm 1)^2 - 1 = 12k(3k \pm 1).$$

又 k 与 $3k \pm 1$ 的奇偶性不同, 则它们的积为偶数. 于是, $p^2 - 1$ 能被 24 整除.

解: (2) 由 (1) 知, 当 $p > 3$ 时, $p^2 + 2543 = p^2 - 1 + 106 \times 24$ 是 24 的倍数. 设

$$p^2 + 2543 = 24k(k \in N, k \geqslant 107).$$

设 $k = 2^r \times 3^s \times k'$, 其中, r, s 为非负整数, k' 是使得 $(k', 6) = 1$ 的正整数. 设 $\tau(n)$ 是 n 的所有正因子的个数. 则

$$\tau(p^2 + 2543) = \tau(2^{3+r} \times 3^{1+s} k') = (4 + r)(2 + s)\tau(k').$$

当 $k' > 1$ 时, $\tau(k') \geqslant 2$, 即

$$\tau(p^2 + 2543) \geqslant 16;$$

当 $k' = 1$ 时,

$$\tau(p^2 + 2543) = (4 + r)(2 + s).$$

如果 $\tau(p^2 + 2543) < 16$, 则 $r \leqslant 3, s \leqslant 1$. 故 $k = 2^r \times 3^s \leqslant 24$, 这与 $k \geqslant 107$ 矛盾.

因此, 对所有的素数 $p > 3$, 有 $\tau(p^2 + 2543) \geqslant 16$.

当 $p = 2$ 时,

$$\tau(2^2 + 2543) = \tau(2547) = \tau(3^2 \times 283) = 6.$$

当 $p = 3$ 时,

$$\tau(3^2 + 2543) = \tau(2552) = \tau(2^3 \times 11 \times 29) = 16.$$

综上, 只有 $p = 2$ 满足题意.

例 9.11 (1) 若正整数 $k(k \geqslant 3)$ 满足: 有 k 个正整数, 使得任意两个不互素, 任意三个互素, 求 k 的所有可能值;

(2) 是否存在一个无穷项的正整数集, 满足 (1) 的条件?

解: (1) 由于素数有无穷多个, 设为 $p_1, p_2, \cdots$. 故满足任意两个不互素及任意三个互素, 可构造 $a_1, a_2, \cdots, a_k$ 如下:

$$a_1 = p_1 p_2 \cdots p_{k-1}, a_2 = p_1 p_k p_{k+1} \cdots p_{2k-3}, a_3 = p_2 p_k p_{2k-2} p_{2k-1} \cdots p_{2k-6},$$

$$a_4 = p_3 p_{k+1} p_{2k-2} p_{2k-5} \cdots p_{3k-10}, \cdots, a_k = p_{k-1} p_{2k-3} p_{2k-6} p_{3k-10} \cdots.$$

则 $(a_i, a_j) \neq (a_i, a_k)(j \neq k)$. 因此, $(a_i, a_j, a_k) = 1$.

故对于所有的 $k(k \geq 3)$ 均满足条件.

（2）不存在满足条件的无穷集合.

假设满足条件的集合 $\{a_1, a_2, \cdots\}$.

则显然 $a_1 > 1$. 设 $a_1 = p_1^{a_1} p_2^{a_2} \cdots p_s^{a_s}(p_1, p_2, \cdots p_s$ 是不同的素数）.

考虑其中的 $s + 1$ 个数 $a_2, a_3, \cdots, a_{s+2}$,

因为这 $s + 1$ 个数中的每一个均与 a_1 不互素, 所以, 每一项均能被 $p_1, p_2, \cdots, p_s$ 之一整除.

因此, 存在两个数（不妨设 a_m, a_n）均能被 p_i 整除. 于是 a_1, a_m, a_n 不互素, 矛盾.

例 9.12　设 b, n 是大于 1 的整数. 若对每一个大于 1 的正整数, 均存在一个整数 a_k, 使得 $k \mid (b - a_k^n)$, 证明: 存在正整数 A, 使得 $b = A^n$.

〔提示: 设 b 的素因数分解式为 $b = p_1^{a_1} p_2^{a_2} \cdots p_s^{a_s}$, 其中 $p_1, p_2, \cdots, p_s$ 是互不相同的素数, $a_i \in N_+ (i = 1, 2, \cdots, s)$〕

下面证明所有的指数 a_i 均被 n 整除. 设 $A = p_1^{\frac{a_1}{n}} p_2^{\frac{a_2}{n}} \cdots p_s^{\frac{a_s}{n}}$. 则 $b = A^n$.

证明: 对于 $k = b^2$, 应用条件得 $b^2 \mid (b - a_k^n)$. 于是, 对于每一个 $i(1 \leq i \leq s)$,
$$p_i^{2a_i} \mid (b - a_k^n).$$

又因为 $p_i^{a_i} \mid b$,

所以,
$$a_k^n \equiv b \equiv 0 (\bmod p_i^{a_i}), \text{且} a_k^n \equiv b \neq 0 (\bmod p_i^{a_i+1}).$$

这表明 $p_i^{a_i} \mid a_k^n$.

从而, a_i 是 n 的倍数.

当 $m > 1$ 时
$$2m^2 + n^2 - 2mn = (m - n)^2 + m^2 > 1$$
因此 $2m^2 - 2mn + n^2$ 是 $n^4 + a$ 的真因数, 即 $n^4 + a$ 为合数.

由 m 的任意性可知结论成立.

例 9.13　n 是正整数, 证明: $n^2 + n + 1$ 不是平方数.

证明: 因为 $n^2 < n^2 + n - 1 < (n + 1)^2$

即 $n^2 + n + 1$ 夹在两个连续整数的平方之间, 所以它不是平方数.

例 9.14　证明: 四个连续自然数的乘积加上 1 一定是平方数.

解: 设这 4 个数为 $n, n + 1, n + 2, n + 3$, 则
$$n(n + 1)(n + 2)(n + 3) + 1 = (n^2 + 3n)(n^2 + 3n + 2) + 1 = (n^2 + 3n + 1)^2$$
是平方数.

例 9.15　求 2000! 的标准分解式中素因数 7 的指数.

解：因为

$$\sum_{k=1}^{\infty} \left[\frac{2000}{7^k}\right] = \left[\frac{2000}{7}\right] + \left[\frac{2000}{7^2}\right] + \left[\frac{2000}{7^3}\right]$$

$$= 285 + 40 + 5 = 330,$$

所以 2000! 的标准分解式中素因数 7 的指数为 330.

例 9.16 求 50! 的标准分解式.

解：50! 的标准分解式中素因数 2，3，5，7，11，13，17，19，23 的指数分别为

$$\sum_{k=1}^{\infty} \left[\frac{50}{2^k}\right] = \left[\frac{50}{2}\right] + \left[\frac{50}{2^2}\right] + \left[\frac{50}{2^3}\right] + \left[\frac{50}{2^4}\right] + \left[\frac{50}{2^5}\right]$$

$$= 25 + 12 + 6 + 3 + 1 = 47,$$

$$\sum_{k=1}^{\infty} \left[\frac{50}{3^k}\right] = 16 + 5 + 1 = 22,$$

$$\sum_{k=1}^{\infty} \left[\frac{50}{5^k}\right] = 10 + 2 = 12,$$

$$\sum_{k=1}^{\infty} \left[\frac{50}{7^k}\right] = 7 + 1 = 8,$$

$$\sum_{k=1}^{\infty} \left[\frac{50}{11^k}\right] = 4,$$

$$\sum_{k=1}^{\infty} \left[\frac{50}{13^k}\right] = 3,$$

$$\sum_{k=1}^{\infty} \left[\frac{50}{17^k}\right] = 2,$$

$$\sum_{k=1}^{\infty} \left[\frac{50}{19^k}\right] = 2,$$

$$\sum_{k=1}^{\infty} \left[\frac{50}{23^k}\right] = 2.$$

而 29，31，37，41，43，47 在 50! 的标准分解式中的指数均为 1，所以

$$50! = 2^{47} \cdot 3^{22} \cdot 5^{12} \cdot 7^8 \cdot 11^4 \cdot 13^3 \cdot 17^2 \cdot 19^2 \cdot 23^2 \cdot 29 \cdot 31 \cdot 37 \cdot 41 \cdot 43 \cdot 47.$$

例 9.17 在 1000! 的末尾共有多少个零？

解：要求 1000! 末尾的零的个数，只需要出 1000! 的标准分解式中素因数 2 和 5 的指数，并取最小的即可. 易知，1000! 的标准分解式中素因数 2 的指数大于素因数 5 的指数，而 5 的指数是

$$h = \sum_{k=1}^{\infty} \left[\frac{1000}{5^k}\right] = \left[\frac{1000}{5}\right] + \left[\frac{1000}{5^2}\right] + \left[\frac{1000}{5^3}\right] + \left[\frac{1000}{5^4}\right]$$

$$= 200 + 40 + 8 + 1$$

$$= 249,$$

故 1000! 的末尾共有 249 个零.

例9.18 设 n 为任一正整数，试证：$2^n \mid (n+1)(n+2)\cdots(2n-1)(2n)$.

证明： 因为

$$(n+1)(n+2)\cdots(2n-1)(2n) = \frac{(2n)!}{n!},$$

而 $(2n)!$ 的标准分解式中 2 的指数为

$$h_1 = \sum_{k=1}^{\infty}\left[\frac{2n}{2^k}\right] = \sum_{k=1}^{\infty}\left[\frac{n}{2^{k-1}}\right] = n + \sum_{k=1}^{\infty}\left[\frac{n}{2^k}\right],$$

$n!$ 的标准分解式中 2 的指数为

$$h_2 = \sum_{k=1}^{\infty}\left[\frac{n}{2^k}\right],$$

所以 $(n+1)(n+2)\cdots(2n-1)(2n)$ 的标准分解式中 2 的指数是

$$h_1 - h_2 = n.$$

也就是说，$2^n \mid (n+1)(n+2)\cdots(2n-1)(2n)$.

例9.19 设 m,n 都是正整数，试证：$m!n!(m+n)! \mid (2m)!(2n)!$.

证明： 易知，对一切实数 α,β，有

$$[2\alpha] + [2\beta] \geqslant [\alpha] + [\beta] + [\alpha+\beta].$$

取 $\alpha = \dfrac{m}{p^k}, \beta = \dfrac{n}{p^k}$，这里 p 为素数，得

$$\left[\frac{2m}{p^k}\right] + \left[\frac{2n}{p^k}\right] \geqslant \left[\frac{m}{p^k}\right] + \left[\frac{n}{p^k}\right] + \left[\frac{m+n}{p^k}\right].$$

从而对任一素数 p，有

$$\sum_{k=1}^{\infty}\left(\left[\frac{2m}{p^k}\right] + \left[\frac{2n}{p^k}\right]\right) \geqslant \sum_{k=1}^{\infty}\left(\left[\frac{m}{p^k}\right] + \left[\frac{n}{p^k}\right] + \left[\frac{m+n}{p^k}\right]\right).$$

这说明，$m!n!(m+n)! \mid (2m)!(2n)!$.

例9.20 若正整数有八个正因数，且这八个正因数之和为3240，则称这个正整数为"好数"，例如，2006 是好数，因为其因数有 1，2，17，34，59，118，1003，2006 的和为 3240. 求最小的好数.

解： 设 $n = p_1^{\alpha_1}p_2^{\alpha_2}\cdots p_k^{\alpha_k}$. 则 $(1+\alpha_1)(1+\alpha_2)\cdots(1+\alpha_k) = 8$.

故当 $k = 1$ 时，$\alpha_1 = 7$；

当 $k = 2$ 时，$\alpha_1 = 1, \alpha_2 = 3$ 或 $\alpha_1 = 3, \alpha_2 = 1$；

当 $k = 3$ 时，$\alpha_1 = \alpha_2 = \alpha_3 = 1$；

当 $k \geqslant 4$ 时，无解.

(1) 若 $n = p^7$（p 为素数），则

$$\sum_{i=0}^{7} p^i = 3240.$$

(i) 若 $p \geqslant 3$，则

$$\sum_{j=0}^{7} p^j \geqslant \sum_{j=0}^{7} 3^j = \frac{3^8 - 1}{2} = 3280 > 3240.$$

（ii）若 $p = 2$，则

$$\sum_{j=0}^{7} p^j = 2^8 - 1 = 511 \neq 3240.$$

因而，当 $n = p^7$ 时，无满足题意的解.

（2）若 $n = p^3 q$（p, q 为素数，且 $p \neq q$），则有正因数之和

$$3240 = (1 + p + p^2 + p^3)(1 + q) = (1 + p)(1 + p^2)(1 + q) = 2^3 \times 3^4 \times 5.$$

由 $q \geqslant 2$，得 $1 + q \geqslant 3$. 从而

$$1 + p + p^2 + p^3 \leqslant 1080.$$

故 $p \leqslant 7$.

（i）若 $p = 7$，则

$$1 + p^2 = 50 \times 3240,$$

矛盾.

（ii）若 $p = 5$，则

$$1 + p^2 = 26 \times 3240,$$

矛盾.

（iii）若 $p = 3$，则

$$1 + q = \frac{3240}{40} = 81, 从而, q = 80.$$

矛盾.

（iv）若 $p = 2$，则

$$1 + q = \frac{3240}{15} = 216, 从而, q = 215.$$

亦矛盾.

（3）若 $n = pqr$（$p < q < r$），则

$$(1 + p)(1 + q)(1 + r) = 3240 = 2^3 \times 3^4 \times 5.$$

（i）若 $p = 2$，则 $(1 + q)(1 + r) = 2^3 \times 3^3 \times 5$.

注意到要让 pqr 尽量小，则 qr 尽量小. 于是由

$$qr = (1 + q)(1 + r) - (q + 1) - (r + 1) + 1,$$

知 $(q + 1)(r + 1)$ 应尽量大.

结合 $(1 + q)(1 + r) = 2^3 \times 3^3 \times 5$，可知 $q + 1$ 应尽量小，$r + 1$ 应尽量大.

故取 $q = 3, r = 269$. 此时，

$$n_{\min} = 2 \times 3 \times 269 = 1614.$$

（ii）若 $p \geqslant 3$，则

$$\frac{1 + p}{2} \cdot \frac{1 + q}{2} \cdot \frac{1 + r}{2} = 3^4 \times 5.$$

由于 p, q, r 的大小关系，故只有两种情形：

$$\frac{1 + p}{2} = 3, \frac{1 + q}{2} = 3^2, \frac{1 + r}{2} = 3 \times 5;$$

及

$$\frac{1+p}{2} = 3, \frac{1+q}{2} = 5, \frac{1+r}{2} = 3^3.$$

易解出对应的 $(p,q,r) = (5,17,29)$ 或 $(5,9,53)$.

又因为 9 不是素数，所以，$(p,q,r) = (5,17,29)$.

因此，$n = 2465$.

综上，满足题意的 $n_{\min} = 1614$.

例 9.21 求所有的正整数 n，满足 n 为合数，且其所有大于 1 的因数可以放在一个圆上，使得任意两个相邻的因数都不是互素的.

解： 若 $n = pq$（p,q 为不同的素数），则其大于 1 的因数 p,q,pq 无论怎样放于圆周上，p,q 总会相邻，不合题意.

若 $n = p^m$（p 为素数，正整数 $m \geq 2$），则无论怎样将 n 的大于 1 的因数放在一个圆上，任意两个相邻的因数都不是互素，满足题意.

若 $n = p_1^{m_1} p_2^{m_2} \cdots p_k^{m_k}$，其中，素数 $p_1 < p_2 < \cdots < p_k, m_1, m_2, \cdots, m_k$ 为正整数，且当 $k > 2$ 或 $k = 2$ 时，$\max\{m_1, m_2\} > 1$.

设集合 $D_n = \{d \mid d \mid n$ 且 $d > 1\}$.

下面说明可找到满足题意的方法.

首先，将 $n, p_1 p_2, p_2 p_3, \cdots, p_{k-1} p_k$ 按顺时针放在圆上.

在 n 和 $p_1 p_2$ 之间依任意的顺序放入 D_n 中所有以 p_1 为最小素因数的正整数（不包括 $p_1 p_2$）；在 $p_1 p_2$ 和 $p_2 p_3$ 之间，依任意的次序放入 D_n 中所有以 p_2 为最小素因数的正整数（不包括 $p_2 p_3$）；……在 $p_{k-1} p_k$ 和 n 之间，依任意的次序放入 $p_k, p_k^2, \cdots, p_k^{m_k}$. 于是，$D_n$ 中的所有元素恰被放在圆周上一次，且任意相邻的两个数有一个公共的素因数.

例 9.22 求自然数 N，使它能被 5 和 49 整除，并且包括 1 和 N 在内，它共有 10 个约数.

解： 设 $N = 2^{a_1} \times 3^{a_2} \times 5^{a_3} \times 7^{a_4} \times \ldots \times p_n^{a_n}$（$a_i \in N, i = 1,2,\ldots,n$）. 则

$$(a_1 + 1)(a_2 + 1)\ldots(a_n + 1) = 10.$$

由于 $5 \mid N, 7^2 \mid N$，则 $a_3 + 1 \geq 2, a_4 + 1 \geq 3$. 故 $a_1, a_2, a_5, a_6, \ldots, a_n$ 必然均为 0. 进而解得 2 $N = 5 \times 7^4$.

例 9.23 证明：如果 $a^2 \mid b^2$，那么 $a \mid b$.

证明： 不妨假设 a, b 均是正整数，且

$$a = p_1^{\alpha_1} p_2^{\alpha_2} \ldots p_k^{\alpha_k}, b = p_1^{\beta_1} p_2^{\beta_2} \ldots p_k^{\beta_k},$$

其中 $\alpha_i \geq 0, \beta_i \geq 0 (1 \leq i \leq k)$. 因为 $a^2 \mid b^2$，所以存在 $q \in Z$，使得 $b^2 = a^2 q$，即

$$p_1^{2\beta_1} p_2^{2\beta_2} \ldots p_k^{2\beta_k} = p_1^{2\alpha_1} p_2^{2\alpha_2} \ldots p_k^{2\alpha_k} q.$$

由此可见，$2\alpha_i \leq 2\beta_i (1 \leq i \leq k)$，即 $a_i \leq \beta_i (1 \leq i \leq k)$，因此 $a \mid b$.

标准分解式也可用于刻画最大公因数和最小公倍数.

例 9.24 设

$$A = \frac{1 \times 3 \times 5 \times \cdots \times (2n - 1)}{2 \times 4 \times 6 \times \cdots \times (2n)},$$

其中 n 为正整数. 试证明数列 $A, 2A, 2^2 A, \cdots, 2^k A, \cdots$, 从某一项开始都是整数.

分析： 这是一个与阶乘数有关的问题. 有时为了书写简单, 常把 A 的分子写成 $(2n-1)!!$, 分母写成 $(2n)!!$, 称为双阶乘. 如果能将这组数适当地变形, 使其与组合数建立起联系, 再利用组合数 C_n^k 是整数这个特点, 问题就可解决.

证明： 首先将分子的双阶乘变为阶乘. 在 A 的分子分母上同时乘以 $2 \times 4 \times 6 \times \cdots \times (2n-2)$, 可得

$$A = \frac{1 \times 2 \times 3 \times 4 \times \cdots \times (2n - 1)}{\left[2 \times 4 \times 6 \times \cdots \times (2n - 2)\right]^2 \times 2n}$$

$$= \frac{1}{2^{2n-1}} \cdot \frac{(2n - 1)!}{\left[(n-1)!\right]^2 \times n}$$

$$= \frac{1}{2^{2n-1}} \cdot \frac{(2n - 1)!}{n!(n-1)!}$$

$$= \frac{1}{2^{2n-1}} C_{2n-1}^{n-1},$$

因为组合数 C_{2n-1}^{n-1} 是整数, 所以 $2^{2n-1} A$ 及以后的项都是整数.

例 9.25 证明：设 $a = p_1^{\alpha_1} p_2^{\alpha_2} \cdots p_k^{\alpha_k}, b = p_1^{\beta_1} p_2^{\beta_2} \cdots p_k^{\beta_k}$, 其中 $\alpha_i \geq 0, \beta_i \geq 0 (1 \leq i \leq k)$, 则

$$(a, b) = p_1^{s_1} p_2^{s_2} \cdots p_k^{s_k}, s_i = \min\{\alpha_i, \beta_i\} (1 \leq i \leq k);$$

$$[a, b] = p_1^{t_1} p_2^{t_2} \cdots p_k^{t_k}, s_i = \max\{\alpha_i, \beta_i\} (1 \leq i \leq k);$$

证明： 直接由定义即得.

例如, 当 $a = 72 = 2^3 \times 3^2, b = 100 = 2^2 \times 5^2$ 时,

有 $(a, b) = 2^3 \times 3^2 \times 5^2$.

9.3.2 典型练习题

练习 9.1 求 300! 的标准分解式中素因数 7 的指数.

练习 9.2 求 30! 的标准分解式.

练习 9.3 在 $(2000!)^3$ 的末尾共有多少个零?

练习 9.4 试证：$\dfrac{(2n)!}{(n!)^2}$ 是偶数.

练习 9.5 已知 N 为正整数, 恰有 2005 个正整数有序对 (x, y) 满足 $\dfrac{1}{x} + \dfrac{1}{y} = \dfrac{1}{N}$. 证明 N 是完全平方数.

[提示：应注意到 $x, y > N$. 否则, $\dfrac{1}{x}$ 或 $\dfrac{1}{y}$ 之一将大于 $\dfrac{1}{N}$. 则 $\dfrac{1}{x} + \dfrac{1}{y} = \dfrac{1}{N} \Rightarrow y = \dfrac{N^2}{x - N} + N$. 这样, N^2 的每个正因子 d 均唯一对应着一组解 $(x, y) = \left(d + N, \dfrac{N^2}{d} + N\right)$. 令 $N = P_1^{a_1} p_2^{a_2} \cdots p_n^{a_n}$ 则 $N^2 = p_1^{2a_1} p_2^{2a_2} \cdots p_n^{2a_n}$. 由题意知 $(2a_1 + 1)(2a_2 + 1) \cdots (2a_n + 1) = 2005$. 因

为 2005 的正因子均模 4 余 1，所以，a_i 必为偶数，故 N 是完全平方数]

练习 9.6 如果 p, q 是两个素数，并且 $q = p + 2$，证明：$p^q + q^p$ 能被 $p + q$ 整除.

[提示：对 $p^q + q^p$ 进行变形有 $p^q + q^p = (p^p + q^p) + (p^{p+2} - p^p)$. 则需证明 $p + q \mid (p^{p+2} - p^p)$. 而 $p + q = 2(p + 1)$，故只需证明 $2(p + 1) \mid (p^{p+2} - p^p)$]

第 10 章　勾股数组

最合宜的是祭品，最明智的是数目，最优美的是和谐，最有力的是知识，最美好的是幸福，最佳的能力是医术，最真的话语是"人是邪恶的"．

——毕达哥拉斯

问：2 个平方数的和可能等于平方数吗？

答：有可能．比如：$9 + 16 = 25$，$25 + 144 = 169$ 等等．事实上，这样的数我们中国古代就开始研究了，有图文为证：

"句广三，股修四，径隅五．"（《周髀算经》上卷二，公元前 5 ~ 2 世纪）

图 10.1

《九章算术》中，赵爽描述此图："勾股各自乘，并之为玄实．开方除之，即玄．案玄图有可以勾股相乘为朱实二，倍之为朱实四．以勾股之差自相乘为中黄实．加差实亦成玄实．以差实减玄实，半其余．以差为从法，开方除之，复得勾矣．加差于勾即股．凡并勾股之实，即成玄实．或矩于内，或方于外．形诡而量均，体殊而数齐．勾实之矩以股玄差为广，股玄并为袤．而股实方其里．减矩勾之实于玄实，开其余即股．倍股在两边为从法，开矩勾之角，即股玄差．加股为玄．以差除勾实得股玄并．以并除勾实亦得股玄差．令并自乘与勾实为实．倍并为法，所得亦玄．勾实减并自乘，如法为股．股实之矩以勾玄差为广，勾玄并为袤．而勾实方其里，减矩股之实于玄实，开其余即勾．倍勾在两边为从法，开矩股之角，即勾玄差．加勾为玄．以差除股实得勾玄并．以并除股实亦得勾玄差．令并自乘与股实为实．倍并为法，所得亦玄．股实减并自乘如法为勾，两差相乘倍而开之，所得以股玄差增之为勾．以勾玄差增之为股．两差增之为玄．倍玄实列勾股差实，见并实者，以图考之，倍玄实满外大方而多黄实．黄实之多，即勾股差实．以差实减之，开其余，得外大方．大方之面，即勾股并也．令并自乘，倍玄实乃减之，开其余，得中黄方．黄方之面，即勾股差．以差减并而半之为勾，加差于并而半之为股．其倍玄为广袤合．令勾股见者自乘为其实．四实以减之，开其余，所得为差．以差减合半其余为广．减广于玄即所求也．"

用现代数学语言描述就是黄实的面积等于大正方形的面积减去四个朱实的面积，若将其用适当的字母进行标记，将得到：

$$c^2 = \left(\frac{1}{2}ab\right) \times 4 + (b - a)^2 = a^2 + b^2$$

2002 年第 24 届国际数学家大会（ICM）的会标即为该图.

10.1 平方数

平方数，也叫完全平方数，或正方形数，如图 10.2，指的是如 1，4，9，16，25，36，49……这样能够写成一个整数的平方形式的非负数.

$$1 = 1 \times 1 = 1^2$$
$$4 = 2 \times 2 = 2^2$$
$$9 = 3 \times 3 = 3^2$$
$$16 = 4 \times 4 = 4^2$$
$$25 = 5 \times 5 = 5^2$$
$$36 = 6 \times 6 = 6^2$$
$$49 = 7 \times 7 = 7^2$$

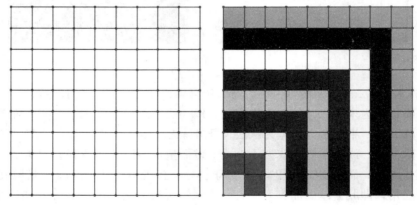

图 10.2

由图 10.2 左图可以看出：若 n 是一个平方数，则将 n 个点排成矩形，可以排成一个正方形.

由图 10.2 右图可以看出：若 n 是一个平方数，则可将 n 个单位正方形拼成一个正方形.

由平方数的定义可以知道，最小的平方数是 0，最小的正平方数是 1. 由九九乘法表容易看出，平方数的个位数字只看是 0，1，4，5，6，9. 事实上，我们有：

若一个数的个位数是 0，则其平方数的个位数也为 0，且其他数字也构成一个平方数.

若一个数的个位数是 1 或 9，则其平方数的个位数为 1，且其他数字构成的数能被 4 整除.

若一个数的个位数是 2 或 8，则其平方数的个位数为 4，且其他数字构成一个偶数.

若一个数的个位数是 3 或 7，则其平方数的个位数为 9，且其他数字构成的数能被 4 整除.

若一个数的个位数是 4 或 6，则其平方数的个位数为 6，且其他数字构成一个奇数.

若一个数的个位数是 5，则其平方数的个位数也为 5.

平方数具有以下性质：

性质 1　$n^2 = 1 + 3 + \cdots + (2n - 1)$，例如 $2^2 = 1 + 3$，$6^2 = 1 + 3 + 5 + 7 + 9 + 11$ 等.

性质 2　任何偶数的平方一定能被 4 整除；任何奇数的平方被 4（或 8）除余 1，即被 4 除余 2 或 3 的数一定不是完全平方数.

性质 3　完全平方数的个位数字是奇数时，其十位上的数字必为偶数. 完全平方数的个位数字是 6 时，其十位数字必为奇数.

性质 4　凡个位数字是 5 但末两位数字不是 25 的自然数不是完全平方数；末尾只有奇数个 0 的自然数不是完全平方数；个位数字是 1，4，9 而十位数字为奇数的自然数不是完全平方数.

性质 5　除 1 外，一个完全平方数分解质因数后，各个质因数的指数都是偶数，如果一个数质分解后，各个指数都为偶数，那么它肯定是个平方数. 完全平方数的所有因数的总个数是奇数个. 因数个数为奇数的自然数一定是完全平方数.

性质 6　若质数 p 整除完全平方数 a，则 $p^2 \mid a$. 如果质数 p 能整除 a，但 p 的平方不能整除 a，则 a 不是完全平方数.

性质 7　如果 a, b 是平方数，$a = bc$，那么 c 也是完全平方数.

性质 8　两个连续自然数的乘积一定不是平方数，两个连续自然数的平方数之间不再有平方数. 在两个相邻的整数的平方数之间的所有整数都不是完全平方数.

性质 9　如果十位数字是奇数，则它的个位数字一定是 6，反之也成立. 这意味着：如果一个数的十位数字是奇数，而个位数字不是 6，那么这个数一定不是完全平方数. 如果一个完全平方数的个位数字不是 6，则它的十位数字是偶数.

性质 10　偶数的平方是 4 的倍数；奇数的平方是 4 的倍数加 1.

性质 11　奇数的平方是 $8n + 1$ 型；偶数的平方为 $8n$ 或 $8n + 4$ 型.（奇数：n 为那个所乘的数 -1；偶数：n 为那个所乘的数 -2）.

性质 12　平方数的形式必为下列两种之一：$3k, 3k + 1$.

性质 13　不是 5 的因数或倍数的数的平方为 $5k \pm 1$ 型，是 5 的因数或倍数的数为 $5k$ 型.

性质 14　平方数的形式具有下列形式之一：$16m, 16m + 1, 16m + 4, 16m + 9$.

性质 15　一个正整数 n 是完全平方数的充分必要条件是 n 有奇数个因数（包括 1 和 n 本身）.

要得到一个平方数是很简单的，只需要用一个整数进行自乘，也即平方运算即可. 同理，要判断一个数是不是平方数就需要进行开平方运算，若开平方后其结果为整数，则此数为平方数，否则就是非平方数. 只是这样的开平方有时候并没有那么容易进行，有一种较为容易操作的方法是将这个数进行因数分解，也即求出其所有最简因数，若所有互异因数的个数都是偶数，则该数是平方数，否则若存在某因数的个数是奇数，则为非平方数.

要判断某数不是平方数，有以下一些结论：

（1）个位数是 2，3，7，8 的整数一定不是完全平方数；

（2）个位数和十位数都是奇数的整数一定不是完全平方数；

（3）个位数是 6，十位数是偶数的整数一定不是完全平方数；

（4）形如 $3n + 2$ 型的整数一定不是完全平方数；

（5）形如 $4n + 2$ 和 $4n + 2$ 型的整数一定不是完全平方数；

（6）形如 $5n \pm 2$ 型的整数一定不是完全平方数；

（7）形如 $8n + 2, 8n + 3, 8n + 5, 8n + 6, 8n + 7$ 型的整数一定不是完全平方数；

（8）数字和是 2，3，5，6，8 的整数一定不是完全平方数；

（9）四平方和定理：每个正整数均可表示为 4 个整数的平方和；

（10）完全平方数的因数个数一定是奇数.

例 10.1 设正整数 d 不等于 2，5，13，求证在集合 $\{2, 5, 13, d\}$ 中可以找到两个不同的元素 a, b，使得 $ab - 1$ 不是完全平方数.（1986 年第 27 届 IMO 试题）

解：显然

$$2 \times 5 - 1 = 9$$
$$2 \times 13 - 1 = 25$$
$$5 \times 13 - 1 = 64$$

都为完全平方数.

假设 $2d - 1$ 为完全平方数，注意到 d 是正整数，所以 $2d - 1$ 为奇数，不妨假设 $2d - 1 = (2n - 1)^2$ 得：

$$d = 2n^2 - 2n + 1.$$

此时 $5d - 1 = 10n^2 - 10n + 4$ 不是完全平方数.

同理假设 $5d - 1, 13d - 1$ 为完全平方数，可以分 d 为奇偶去证明.

10.2 勾股定理

勾股定理，又称为商高定理、毕达哥拉斯定理、百牛定理等. 它是平面几何中一个很重要的定理.

有关勾股定理的研究，最早的记录是中国古代（公元前 11 世纪）的《周髀算经》[①]. 其中记载了一段可称为数学史上具有特殊地位的对话：

周公：天没有阶梯可以攀登，地没有尺子可以测量，请问怎么求得天之高地之广呢?

商高：按勾三股四弦五的比例去算.

随后《周髀算经》将"勾三股四弦五"推广为以下商高定理：

"求斜至日者，以日下为勾，日高为股，勾股各自乘，并以开方除之，得斜至日."

也就是说：要求直角三角形斜边的长度，首先将两直角边分别进行平方后相加，再将此和开平方，开平方得到的算术平方根即为所求的斜边长. 用现代数学的语言表述就是下面的

---

① 周，指的是成书年代，也即周朝. 髀，就是股骨. 周髀指的是周朝测量日光影子长度的一种工具，长 8 尺. 这种记录值得推敲，因为即使是牛的大腿骨，也没有这么长，可能是几根牛的或者是人的大腿骨捆接在一起做成的.

勾股定理.

勾股定理：在任意一个直角三角形中，两条直角边 a, b 长度平方之和等于斜边 c 长度的平方，一般用公式表示为：

$$a^2 + b^2 = c^2$$

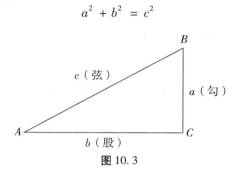

图 10.3

因此，在直角三角形的三边 a, b, c 中，任意给定两条边的长度，另一边就可以通过计算得到：

$$c = \sqrt{a^2 + b^2}$$
$$a = \sqrt{c^2 - b^2}$$
$$b = \sqrt{c^2 - a^2}$$

由于 a, b, c 表示的是长度，所以只要大于 0 即可.

可惜的是，我国的商高和《周髀算经》都没有给出勾股定理的具体证明，直到三国时代的赵爽在公元 3 世纪初才给出了一个高水平证明，号称无字证明，也即前文说到的"朱实黄实"的证明.

而在此之前，约公元前 6 世纪，古希腊数学家毕达哥拉斯第一个给出了勾股定理的证明，也正因此，西方人习惯的称该定理为毕达哥拉斯定理.

此后，公元前 4 世纪，欧几里得将该定理收录在其《几何原本》中，并给出了一个证明，史称"新娘的椅子"的证明. 1876 年，美国第 20 任总统加菲尔德在《新英格兰教育日志》给出了他的证明，史称"总统证法". 时至今日，勾股定理的证明方法已有五百多种，可以说是数学界证明方法最多的定理之一，在现行义务教育数学课程里，一般引用了16 种有代表性的证明. 感兴趣的读者不妨自己查阅相关资料，抑或给出你独特的证明，为勾股定理添加有意义的一笔.

勾股定理从中国的计算数学开始，证明是论证数学的开端. 勾股定理是历史上第一个把数与形联系起来的定理，是数形结合的典型代表，首次将代数和几何联系了起来. 勾股定理还引起了第一次数学危机，极大的加深了人们对数的认识与理解. 勾股定理还是欧几里得几何的基础定理，有巨大的实用价值. 勾股定理是历史上第一个给出完全解答的不定方程，它引出了费马大定理.

10.3　勾股数组及其存在性

问：若用 a, b 表示直角三角形的两条直角边，用 c 表示直角三角形的斜边. 是否存在

a,b,c 都是整数的直角三角形?

答: 存在, 比如 $a = 3, b = 4, c = 5$ 时就是.

这是从几何的角度进行研究, 如果换一个角度, 从数论的角度来提问, 那就变成了以下的问答.

问: 在 $a^2 + b^2 = c^2$ 中, a, b, c 会不会同时是整数?

答: 有可能, 比如三元数组 $(3, 4, 5)$, 还有:

$$5^2 = 3^2 + 4^2$$
$$13^2 = 5^2 + 12^2$$
$$17^2 = 8^2 + 15^2$$
$$53^2 = 28^2 + 45^2$$

像上面这些可以构成一个直角三角形三条边的 3 个正整数称为一个勾股数组, 也叫毕达哥拉斯三元数组.

用现代数学的观点来描述该问题的话, 那就是: 对于不定方程 $z^2 = x^2 + y^2$ 是否存在整数解?

10.4 勾股数组的个数

容易看出, 对于任意整数 p 和勾股数组 (a, b, c), 必然有 (pa, pb, pc) 也同时为勾股数组, 事实上, 由于 (a, b, c) 是一个勾股数组, 所以

$$c^2 = a^2 + b^2$$

在等式两边同时乘以整数 p, 可得:

$$(pc)^2 = (pa)^2 + (pb)^2$$

于是 (pa, pb, pc) 也同时为勾股数组.

从几何的角度来看, 说的是给定一个直角三角形, 对两条边施行延伸 p 倍的变化, 则通过对另一条施行同样的变化可以得到一个直角三角形, 该直角三角形的边长变化导致了周长变化和面积变化, 但直角三角形其本身特性没有变化, 得出的新的直角三角形与原直角三角形构成相似三角形的关系, 见图 10.4.

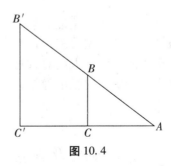

图 10.4

我们甚至可以求出, 周长变化的倍数是 3 倍, 而面积变化的倍数是 p^2.

易知, $a = 3, b = 4, c = 5$ 是一组勾股数组, 所以对 $(3, 4, 5)$ 两边同时乘以任意一个

整数得到的都是勾股数组，而整数个数是无限的，所以勾股数组的个数也是无限的.

只不过这样以公因数 p 构建出的勾股数组意义不大，也不太能吸引人. 于是我们要进一步探讨本原勾股数组.

10.5　本原勾股数组

本原勾股数组指的是没有公因数的勾股数组，如（3，4，5），（5，12，13），（7，24，25），（8，15，17），（9，40，41），（11，60，61），（12，35，37），（16，63，65），（20，21，29），（28，45，53），（33，56，65）等都是本原勾股数组.

考察不定方程 $x^2 + y^2 = z^2$，易知（0，0，0）是其一组解，$(0, -a, a)$ 和 $(a, 0, -a)$ 也是它的解，其中的 a 可以是任何的整数. 在数论中，称有零的解为平凡解，由于平凡解较为容易，我们主要探讨其非平凡解，于是一般假定所求的整数解为正整数解.

结合本原勾股数组，我们一般需要探讨的是不定方程 $x^2 + y^2 = z^2$ 的本原解. 也即没有公因数且满足不定方程的 (x, y, z).

事实上，若 (x, y, z) 是不定方程 $x^2 + y^2 = z^2$ 的本原解，那么

$$(x, y) = 1$$
$$(x, z) = 1$$
$$(y, z) = 1$$

否则的话，不妨假定 $(x, y) = d \neq 1$，则由 $x^2 + y^2 = z^2$ 将可得到 $d \mid z$ 的结论，而这和 (x, y, z) 是的本原解矛盾.

将这一结论和 $d = 2$ 应用进去，将得到：

一个本原勾股数组只可能是一个偶数和两个奇数的组合. 也即不可能存在有两个数是偶数的本原勾股数组.

进一步，我们还可以通过初等方法、几何方法或高斯整数法得到方程的本原解.[①]

满足不定方程 $x^2 + y^2 = z^2$ 的本原解可以表示为：

$$x = m^2 - n^2$$
$$y = 2mn$$
$$z = m^2 + n^2$$

其中 m, n 是正整数，满足条件：

（1）$(m, n) = 1$；

（2）$m > n$；

（3）m 和 n 中有一个是偶数另一个是奇数.

反过来，若 m 和 n 满足上述条件，则 $x = m^2 - n^2, y = 2mn, z = m^2 + n^2$ 是不定方程 $x^2 + y^2 = z^2$ 的本原解.

于是我们可以得到一种构建本原勾股数组 (a, b, c) 的方法，令：

---

① 张顺燕. 数学的源与流[M]. 北京：高等教育出版社，2005：142-149.

$$a = \frac{1}{2}y = mn$$

$$b = \frac{1}{2}x = \frac{m^2 - n^2}{2}$$

$$c = \frac{1}{2}z = \frac{m^2 + n^2}{2}$$

只要满足 $m > n \geqslant 1$ 是没有公因数的奇数即可. 表 10.1 列出了当 $m \leqslant 9$ 时的所有可能的本原勾股数组.

表 10.1

m	n	a	b	c
3	1	3	4	5
5	1	5	12	13
5	3	15	8	17
7	1	7	24	25
7	3	21	20	29
7	5	35	12	37
9	1	9	40	41
9	5	45	28	53
9	7	63	16	65

由本原解可以构建出的直角三角形称为本原三角形. 从解三角形的角度出发,我们有以下问题需要解决:已知边 x 求解本原三角形、已知边 y 求解本原三角形、已知边 z 求解本原三角形,下面依次讨论.

1. 已知边 x 求解本原三角形

根据上述不定方程 $x^2 + y^2 = z^2$ 的本原解的线性表示,我们可以得到:

$$x = m^2 - n^2 = (m + n)(m - n)$$

所以 $(m + n)$, $(m - n)$ 是互素的,于是不妨假设 $m + n = a$, $m - n = b$,有:

$$x = ab, a > b, (a,b) = 1$$

由此,可以得到

$$m = \frac{1}{2}(a + b), n = \frac{1}{2}(a - b)$$

则 m 和 n 满足 $(m,n) = 1$,$m > n$,m 和 n 中有一个是偶数另一个是奇数. 这告诉我们可以从 x 的分解式出发去求解本原三角形.

例 10.2 设本原三角形的 $x = 15$,求解该本原三角形.

解:因为 $x = 15$ 可以分解为

$$x = 15 \times 1$$

于是可以由此分解出

$$m = \frac{1}{2}(a + b) = 8$$

$$n = \frac{1}{2}(a - b) = 7$$

由此可求出

$$x = 15, y = 112, z = 113$$

$x = 15$ 还可以分解为

$$x = 5 \times 3$$

于是可以由此分解出

$$m = \frac{1}{2}(a + b) = 4$$

$$n = \frac{1}{2}(a - b) = 1$$

由此可求出

$$x = 15, y = 8, z = 17$$

2. 已知边 y 求解本原三角形

因为 m 和 n 中有一个是偶数另一个是奇数，故由

$$y = 2mn$$

可知，y 必然被 4 整除.

把 $\frac{y}{2}$ 分解为两个互素的因数的乘积，并使其中一个是奇数，一个是偶数，就可以把较大的一个取作 m，较小的一个取作 n，由此来进行求解.

例 10.3　设本原三角形的 $y = 24$，求解该本原三角形.

解：因为 $y = 24$，所以可以得到

$$\frac{y}{2} = 12 \times 1$$

于是可以由此分解出

$$m = 12$$

$$n = 1$$

由此可求出

$$x = 143, y = 24, z = 145$$

$\frac{y}{2}$ 还可以分解为

$$\frac{y}{2} = 4 \times 3$$

于是可以由此分解出

$$m = 4$$

$$n = 3$$

由此可求出

$$x = 7, y = 24, z = 25$$

3. 已知边 z 求解本原三角形

根据上述不定方程 $x^2 + y^2 = z^2$ 的本原解的线性表示，我们可以得到：

$$z = m^2 + n^2$$

即 z 是满足条件的两个数 m 和 n 的平方和. 这将引发下一章对费马大定理的讨论，在此先埋个伏笔.

例 10.4 设本原三角形的 $z = 41$，求解该本原三角形.

解： 因为 $z = 41$ 可以表示为：

$$z = 5^2 + 4^2$$

所以

$$m = 5$$
$$n = 4$$

由此可求出

$$x = 9, y = 40, z = 41$$

例 10.5 设本原三角形的 $z = 1105$，求解该本原三角形.

解： 因为 $z = 1105$ 可以分解为：

$$z = 1105 = 5 \times 13 \times 17$$

所以 $z = 1105$ 可以表示为：

$$z = 33^2 + 4^2$$

于是可以得到

$$m = 33$$
$$n = 4$$

由此可求出

$$x = 1073, y = 264, z = 1105$$

因为 $z = 1105$ 还可以表示为

$$z = 32^2 + 9^2$$

于是可以得到

$$m = 32$$
$$n = 9$$

由此可求出

$$x = 943, y = 576, z = 1105$$

因为 $z = 1105$ 还可以表示为：

$$z = 31^2 + 12^2$$

于是可以得到

$$m = 31$$
$$n = 12$$

由此可求出
$$x = 817, y = 744, z = 1105$$

因为 $z = 1105$ 还可以表示为：
$$z = 24^2 + 23^2$$

于是可以得到
$$m = 24$$
$$n = 23$$

由此可求出
$$x = 47, y = 1104, z = 1105$$

4. 已知面积求解本原三角形

由于直角三角形的面积 S 等于两直角边乘积的一半，也即：
$$S = \frac{1}{2}xy = mn(m + n)(m - n)$$

式中 4 个因数，有 3 个是奇数，4 个因数都是两两互素的．所以，为了求出所有可能的 m 和 n 的值，可以挑选 S 的两个互素的奇因数 $k, l(k > l)$，并令：
$$m + n = k$$
$$m - n = l$$

也即：
$$m = \frac{1}{2}(k + l)$$

$$n = \frac{1}{2}(k - l)$$

然后，再验证这些值是否确实满足本原三角形的三个条件，以确定有没有这样的三角形．

例 10.6　是否存在面积 $S = 360$ 的本原三角形．

解：因为 $S = 360$ 可以分解为：
$$S = 360 = 2^3 \times 3^2 \times 5$$

所以把 S 写成 4 个两两互素的因数的乘积的唯一方法是：
$$S = 8 \times 1 \times 5 \times 9$$

所以一定有
$$m + n = 9$$

假设
$$m = 8$$
$$n = 1$$

则
$$m - n = 7 \nmid S$$

假设
$$m = 1$$

$$n = 8$$

则不满足 $m > n$ 的条件，故没有这样的本原三角形.

例 10.7 是否存在面积 $S = 360$ 的直角三角形.

解： 由例 10.6 可知，不存在面积为 360 的本原三角形，那么需要考虑的就是面积为 360 的非本原三角形，也即在直角三角形的三边存在公因数时候的情况，不妨假设其最大公因数为 d，于是直角三角形的三边分别为：

$$dx, dy, dz$$

此时的 x, y, z 两两互素，由于 $S = 360$，所以：

$$S = 360 = \frac{1}{2}(dx)(dy) = \frac{1}{2}d^2xy = d^2mn(m - n)(m + n)$$

于是可知 d^2 是 S 的因数. 且由于 d 是最大公因数，所以

$$S_0 = \frac{S}{d^2} = mn(m - n)(m + n)$$

一定是一个本原三角形的面积.

考虑 360 所有可能是平方数的因数，分别有

$$d_1 = 4, d_2 = 9, d_3 = 36$$

依次进行验证，可以分别得到：

$$S_1 = \frac{S}{d_1^2} = 90 = 2 \times 3^2 \times 5$$

$$S_2 = \frac{S}{d_2^2} = 40 = 2^3 \times 5$$

$$S_3 = \frac{S}{d_3^2} = 10 = 2 \times 5$$

也即 40 和 10 都不可能表示为 4 个两两互素的因数的乘积，而对于 90 有且仅有一种表示方法：

$$90 = 2 \times 3^2 \times 5 = 1 \times 2 \times 5 \times 9$$

因为 9 是最大的因数，所以取 $m + n = 9$，由此可得所有可能是：

$$m = 1, n = 8$$
$$m = 2, n = 7$$
$$m = 5, n = 4$$

对这三种情况来说仅有 $m = 5, n = 4$ 满足条件 $m > n$，但是在这种情况下：

$$mn(m - n)(m + n) \neq 90$$

所以可以得到：没有一个本原三角形使得其面积为 90. 于是进一步可以得到：没有一个直角三角形其面积是 360.

第 11 章　费马大定理

只要一门科学分支能提出大量的问题，它就充满着生命力，而问题缺乏则预示着独立发展的终止或衰亡.

——希尔伯特

　　勾股定理回答了 2 个平方数之和能否等于平方数的问题，那自然会继续考虑 2 个立方数之和可能等于立方数吗？2 个 4 次方数之和可能是 4 次方数吗？以此类推考虑更一般的情况又如何呢？

11.1　来　源

　　费马大定理（Fermat's Last Theorem），又被称为"费马最后的定理"，由 17 世纪法国数学家 Pierre de Fermat[①] 提出．

　　古希腊亚历山大时期数学家 Diophantus[②]（丢番图）的著作《算术》的拉丁文译本传到费马手中并进行研究时，他曾在第 11 卷第 8 命题旁写道："将一个立方数分成两个立方数之和，或一个四次幂分成两个四次幂之和，或者一般地将一个高于二次的幂分成两个同次幂之和，这是不可能的．关于此，我确信已发现了一种美妙的证法，可惜这里空白的地方太小，写不下．"

　　在第 10 章我们讨论了不定方程 $z^2 = x^2 + y^2$ 的整数解和本原解的情况．那自然的，我们就会进一步考虑

$$z^3 = x^3 + y^3$$
$$z^4 = x^4 + y^4$$
$$\cdots\cdots$$

乃至更高次幂时不定方程的整数解的情况，这就是著名的费马大定理．

11.2　费马大定理

　　当 $n > 2$ 时，不定方程 $z^n = x^n + y^n$ 没有正整数解．

　　也就是说，不存在正整数 x, y, z，可以使得上述方程成立．如 $z^3 = x^3 + y^3$，$z^4 = x^4 + y^4$ 等都没有正整数解．

---

　　① 　Pierre de Fermat，皮埃尔·德·费马，法国律师和业余数学家．他在数学上的成就不比职业数学家差，他似乎对数论最有兴趣，亦对现代微积分的建立有所贡献，被誉为"业余数学家之王"．费马，是当今常见译法，20 世纪 80 年代的书籍文章也多见译为"费尔玛"的情况，但"费玛"则少见．他生前从未发表过一篇相关论文，他的著作是在死后由其子萨缪尔整理文章、信件以及对丢番图《算术》一书的批注后发表的．费马在数学上的成就较多，如他与帕斯卡共同创立了概率论．他独立发现了解析几何的原理．他曾在微积分建立中做出了极大的贡献，让牛顿和莱布尼兹都承认其画切线方法为微积分创立的起因．在数论方面，他发现了 17396 和 18416 是一对亲和数，证明了 26 是唯一一个夹在一个平方数和一个立方数之间的数，费马小定理和费马大定理更是引领数学发展数百年的"一只会下金蛋的鹅"．

　　② 　丢番图（Diophantus）是古希腊亚历山大大学后期的重要学者和数学家（约公元 246—330 年，据推断和计算而知）．丢番图是代数学的创始人之一，对算术理论有深入研究，他完全脱离了几何形式，在希腊数学中独树一帜．

11.3　有关证明

毕竟费马没有写下证明，而他的其他猜想对数学贡献良多，由此激发了许多数学家对这一猜想的兴趣．数学家们的有关工作丰富了数论的内容，推动了数论的发展．

11.3.1　欧　拉

欧拉（Leonhard Euler，1707—1783），瑞士数学家、自然科学家．1707 年 4 月 15 日出生于瑞士的巴塞尔，1783 年 9 月 18 日于俄国圣彼得堡去世．欧拉出生于牧师家庭，自幼受父亲的影响．13 岁时入读巴塞尔大学，15 岁大学毕业，16 岁获得硕士学位．他的主要成就是创立函数的符号，创立分析力学，解决了柯尼斯堡七桥问题，给出各种欧拉公式等．欧拉是 18 世纪数学界最杰出的人物之一，他不但为数学界做出贡献，更把整个数学推至物理的领域．他是数学史上最多产的数学家，平均每年写出八百多页的论文，还写了大量的力学、分析学、几何学、变分法等的课本，其《无穷小分析引论》《微分学原理》《积分学原理》等都成为数学界中的经典著作．欧拉对数学的研究如此之广泛，因此在许多数学的分支中也可经常见到以他的名字命名的重要常数、公式和定理．此外欧拉还涉及建筑学、弹道学、航海学等领域．瑞士教育与研究国务秘书 Charles Kleiber 曾表示："没有欧拉的众多科学发现，今天的我们将过着完全不一样的生活．"法国数学家拉普拉斯则说："读读欧拉，他是所有人的老师．"2007 年，为庆祝欧拉诞辰 300 周年，瑞士政府、中国科学院及中国教育部于 2007 年 4 月 23 日下午在北京的中国科学院文献情报中心共同举办纪念活动，回顾欧拉的生平、工作以及对现代生活的影响．

欧拉证明费马大定理的想法是：

在 n 个方程中，先证明其中一个方程没有整数解，然后再把这个结果推广到其他方程上去．

基于这个想法，欧拉在重新研究费马所研究的《算术》时采用了无穷递降法，证明了当 $n = 4$ 时费马大定理是正确的．也就是：不定方程 $z^4 = x^4 + y^4$ 没有整数解．

他曾用同样的无穷递降法证明当 $n = 3$ 时费马大定理的正确性，但是并未获得成功．随后又在证明过程中利用了虚数，证明了此时费马大定理是成立的．也就是：不定方程 $z^3 = x^3 + y^3$ 没有整数解．

欧拉用 $n = 3,4$ 时的两个特例迈出了费马大定理证明道路上的第一步．

11.3.2　热尔曼

索菲·热尔曼（Sophie Germain，1776—1831 年）是法国的女数学家．她出身巴黎一个殷实的商人家庭，从小热爱数学，但不为家庭所鼓励．她对拉格朗日的教学很有兴趣，但由于当时的女子不得接近大学，于是以拉白朗（Augusts Antoine Le Blanc）之名提交课业及论文等．拉格朗日要求见其人一面，于是热尔曼便说明一切，后来拉格朗日成为了热尔曼的导师．在拉格朗日的指导下，热尔曼进步更快了，后来她成为法国历史上最有名的女数学家．1830 年，在高斯的推荐下，哥廷根大学颁发了荣誉学位予热尔曼．

　　热尔曼对费马大定理的证明思路跟欧拉是完全不一样的, 她没有从方程的特殊性入手, 而是打算一次性证明多种情形. 为此, 热尔曼找出了一些特殊的素数, 这些素数被后人称为热尔曼素数. 其指的是这样的一类素数: 素数 p, 它的 2 倍加 1, 也就是 p 和 $2p + 1$ 同时为素数的这一类素数. 例如 $p = 2$ 是素数, $2p + 1 = 5$ 也是素数. $p = 3$ 是素数, $2p + 1 = 7$ 也是素数. $p = 5$ 是素数, $2p + 1 = 11$ 也是素数. $p = 11$ 是素数, $2p + 1 = 23$ 也是素数. 这些就是热尔曼素数. 但是当 $p = 7, 13$ 等时, $2p + 1 = 15, 27$ 等并不是素数, 不在热尔曼素数范围内.

　　热尔曼证明了当费马大定理的方程中的 n 是热尔曼素数时, 方程 "可能" 不存在整数解. 所谓 "可能", 说的是若有解, 则需对解进行很严格的限定.

　　由于高斯的兴趣从纯数学转向应用数学, 热尔曼和他的通信中断, 热尔曼提出这一想法后, 逐渐将兴趣也转移到了弹性力学方面, 最终导致未能完成最后的证明.

　　直到 1825 年, 狄利克雷和勒让德才分别独立地用热尔曼的方法证明了当 $n = 5$ 时费马大定理是正确的. 也就是: 不定方程 $z^5 = x^5 + y^5$ 没有整数解.

　　1839 年, 法国力学家拉梅在热尔曼的基础上作了改进后证明了当 $n = 7$ 时费马大定理是正确的. 也就是: 不定方程 $z^7 = x^7 + y^7$ 没有整数解.

11.3.3　库默尔

　　库默尔 (Ernst Eduard Kummer, 1810—1893), 德国数学家. 库默尔在数论、几何学、函数论、数学分析、方程论等方面都有较大的贡献, 但最主要的是在函数论、数论和几何三个方面. 在函数论方面. 他研究了超几何级数, 首次对这些级数的单值群的代换进行计算. 他发明的级数变换法是相当有名的, 在级数的数值计算中有广泛的应用. 在几何方面. 他研究了一般射线系统, 并用纯代数方法构作了一个四次曲面. 它有 16 个孤立的二重点, 16 个奇异切平面, 称之为库默尔曲面. 在数论方面. 库默尔花的时间最多, 贡献也最大. 他创立了理想数理论, 这让费马大定理的证明工作取得了空前的进展 (除 $n = 37, 59$, 67 外, 证明了费马大定理当 $2 < n < 100$ 时都成立).

11.3.4　沃尔夫斯凯尔

　　沃尔夫斯凯尔 (Paul Wolfskehl) 是一个很有意思的数论爱好者而非数学家, 他的真实身份是德国一个实业家. 他曾因热恋遭受拒绝而轻生, 其轻生也是非常具有喜剧色彩的, 他自己给自己制定了非常详细的计划, 定好了离世前所要做的每一件事和完成时间, 甚至定好了自杀的日期, 决定于当日零时告别人世. 由于其计划制定较保守和工作效率高的原因, 他提前几个小时完成了该做的所有事情.

　　为了打发时间, 沃尔夫斯凯尔就到图书馆借了一本关于数论的数来看, 他看到的正是刘维尔在 1847 年宣读库默尔的信的事. 于是, 他很快就被库默尔的工作吸引而进入了推导, 在推导中他发现了库默尔的一个问题: 库默尔提出了一个假设, 但这个假设的依据却并未阐述. 他不知道是自己错了还是库默尔错了, 于是决定要把事情弄清楚. 毕竟, 如果是库默尔错了的话, 那费马大定理的证明就不会那么复杂了, 这一发现是很重要的.

　　于是, 沃尔夫斯凯尔针对库默尔所做的工作, 尤其是那个假设的问题开始自己进行证明, 等他证明完成的时候天已大亮. 当然, 库默尔并没有错, 不过自此他因发现自己在数学方面的信心而决定不再自杀，直至 1908 年去世，他重新立了遗嘱，决定从其遗产中拿出相当于 1500 万人民币的钱来奖励证明费马大定理的人．这里要特别说明的一点是，像沃尔夫斯凯尔这样，因费马大定理而痴迷甚至改变一生命运的人在历史上来说有许许多多，也因此，费马大定理甚至被后人趣称为"一只会下金蛋的鹅"．

11.3.5　哥德尔

哥德尔（Kurt Gödel）（1906—1978）是位数学家、逻辑学家和哲学家．哥德尔发展了冯·诺伊曼和伯奈斯等人的工作，其主要贡献在逻辑学和数学基础方面，最杰出的贡献是哥德尔不完全性定理．

20 世纪初，他证明了形式数论（即算术逻辑）系统的"不完全性定理"：即使把初等数论形式化之后，在这个形式的演绎系统中也总可以找出一个合理的命题来，在该系统中既无法证明它为真，也无法证明它为假．这一著名结果发表在 1931 年的论文中．他还致力于连续统假设的研究，在 1930 年采用一种不同的方法得到了选择公理的相容性证明．3 年以后又证明了（广义）连续统假设的相容性定理，并于 1940 年发表．他的工作对公理集合论有重要影响，而且直接导致了集合和序数上的递归论的产生．

事实上，哥德尔对费马大定理并没有什么实质性贡献，但是他的思想却使得整个数学的基础都产生了混乱，尤其是在 1931 年其所著的《〈数学原理〉及有关系统中的形式不可判定命题》中证明了：相容性永远不可能被证明，也就是说，人们永远不可能确定，他们选择的公理不会导致矛盾的出现．相当于告诉人们：也许费马大定理是不可判定的，其证明也是根本不可能的．根据他的思想，人们得出的结论是：费马大定理可能是对的，但是可能并不存在证明费马大定理的方法．

11.3.6　谷山丰和志村五郎

1955 年，日本谷山丰偶然发现，一个具体的模形式①的 M – 序列中的前几项与某一个椭圆方程的 E – 序列中列出的前几项的数是完全一样的．随后他进行了一系列的验证，发现模形式和椭圆方程这两种数学对象之间存在着某种基本的联系．这给人们一个启示：如果知道了某个椭圆方程的 E – 序列，也就可以通过这种方式判定而知道该椭圆方程对应的模形式的 M – 序列．

当时和谷山丰一起研究的还有志村五郎，在绝大多数数学家对谷山丰的发现采取一种偶然巧合的判定的时候，谷山丰于两年后自杀了．志村五郎坚信谷山丰的证明，于是他停止了一切其他工作，全力以赴的证明谷山丰的想法，获得丰富的成果．

后来这个问题也逐渐为数学界认可，并被称为谷山 – 志村猜想．该猜想主要是建立了椭圆曲线这一代数几何对象和模形式这一数论中周期性全纯函数之间的重要联系，猜想最后由安德鲁·怀尔斯、ChristopheBreuil、BrianConrad、FredDiamond 和理查·泰勒证明．

---

①　模形式是 19 世纪发现的一种数学基本运算，特点是具有无限的对称性．简单说就是模形式可以通过无限多种方式做平移、旋转、反射和交换而保持不变，被认为是最对称的数学研究对象．

11.3.7　弗　雷

1986 年，德国数学家弗雷（Gerhard Frey）提出，如果谷山 – 志村猜想能得到证明，那么费马大定理自动得到证明．其逻辑是：如果费马大定理是错的，那么至少存在一个解：

$$A^N = B^N + C^N$$

经过一系列复杂的推导，弗雷将具有这个假设存在的解的费马大定理的方程变成：

$$y^2 = x^3 + (A^N - B^N) x^2 - A^N B^N$$

可以证明，这个方程是一个椭圆方程，然而这是一个特殊的椭圆方程，它不可能跟任何一个模形式有关．而根据谷山 – 志村猜想，每一个椭圆方程一定与一个模形式有关．这就是说，弗雷方程的存在否定了谷山 – 志村猜想，这样就把费马大定理和谷山 – 志村猜想联系了起来．

归纳起来，弗雷的逻辑是：如果谷山 – 志村猜想能被证明是正确的，则任何椭圆方程都一定可以模形式化，则弗雷的椭圆方程不可能存在，于世费马大定理不可能有解，于世费马大定理得到证明．

弗雷的这一猜想最后由美国加利福尼亚大学伯克利分校的肯·李贝特教授证明．

于是，证明费马大定理就变成了证明谷山 – 志村猜想．

11.3.8　怀尔斯

安德鲁·怀尔斯（Andrew Wiles）是英国著名数学家、牛津大学默顿学院教授．

怀尔斯对数学的最大贡献是证明了历时 350 多年的著名的费尔马大定理．在此之前，他于 1977 年和科茨（Coates）共同证明了椭圆曲线中最重要的猜想——伯奇 – 斯温耐顿 – 代尔（Birch-Swinnerton-Dyer）猜想的特殊情形（即对于具有复数乘法的椭圆曲线）；1984 年和马祖尔（Mazur）一起证明了岩泽理论中的主猜想．在这些工作的基础上，他于 1994 年通过证明半稳定的椭圆曲线的谷山 – 志村 – 韦伊猜想，从而完全证明了费马最后定理．

因为证明了费马大定理，他获得了如下各种奖项和荣誉：

表 11.1

时　间	荣　誉	评选（颁奖）机构及备注
1995 年	Schock 数学奖	瑞典皇家学会
1996 年 3 月	沃尔夫奖	沃尔夫基金会
1996 年 6 月	美国国家科学院外籍院士、该科学院数学奖	美国国家科学院
1996 年	瑞典科学院舍克奖	瑞典科学院
	英国皇家学会皇家奖章	英国皇家学会
	奥斯特洛夫斯基奖	瑞士奥斯特洛夫斯基基金会
	费马奖	法国
1997 年	科尔奖	美国数学会
1997 年 6 月	沃尔夫斯科尔（Wolfskehl）奖（10 万马克奖金）	1908 年为解决费马猜想而设置

续　表

时　　间	荣　　誉	评选（颁奖）机构及备注
1998 年 8 月	国际数学联盟特别奖（菲尔兹奖银质奖章）	国际数学联盟（第 23 届国际数学家大会）颁特别奖而非菲尔兹奖的原因是他当年已经超过 40 岁
1999 年	首届克莱数学研究奖（Clay Research Award）	美国克莱数学研究所
2000 年	爵士	英国女王
2005 年	邵逸夫数学科学奖（Shaw Prize），奖金 100 万美金	邵逸夫奖基金会
2016 年 3 月	阿贝尔奖	挪威科学与文学院

11.3.9　其他突出贡献者

除了上述的一些学者对费马大定理的证明做出的贡献外，还有很多学者都曾致力于费马大定理的研究和证明，并为其最后证明提供了有益的帮助，为数论乃至数学的发展产生了一定的影响和促进作用，记录较为完善的主要有以下这些.

表 11.2

姓　　名	年　　代
费马（Pierre de Fermat）	1601—1665
欧拉（Leonhard Euler）	1707—1783
热尔曼（Sophie Germain）	1776—1831
高斯（Carl Friedrich Gauss）	1777—1855
拉梅（Gabriel Lame）	1795—1870
狄利克雷（Peter Gustav Lejeune Dirichlet）	1805—1859
刘维尔（Joseph Liouville）	1809—1882
库默尔（Ernst Eduard Kummer）	1810—1893
樊迪维尔（Harry Schultz Vandiver）	1882—1973
志村五郎	1926—
谷山丰	1927—1958
弗雷（Gerhard Frey）	—
法尔廷斯（Gerd Faltings）	—
吕贝特（Kenneth A. Ribet）	—
怀尔斯（Andrew J. Wiles）	—

第 12 章　哥德巴赫猜想

在学习中要敢于做减法，就是减去前人已经解决的部分，看看还有那些问题没有解决，需要我们去探索解决.

——华罗庚

关于素数的研究很容易开始，却很难结束．容易开始，是因为其表述简洁，基本上一个小学阶段正常智力的孩子就可以进行相关学习和研究．很难结束是因为关于素数的研究方向、方法等众多，研究主题各异，一个人、一代人乃至几代人都未必能够研究完．如果再遇到一些类似哥德巴赫猜想这样的看上去非常简单，而证明却非常难的题目，那就更加突出了．

12.1　来　源

1742 年 6 月 7 日，德国数学家哥德巴赫在写给著名数学家欧拉的一封求助信中，提出了两个大胆的猜想：

（1）任何不小于 4 的偶数，都可以是两个质数之和；

（2）任何不小于 7 的奇数，都可以是三个质数之和．

这就是数学史上著名的"哥德巴赫猜想"．显然，第二个猜想是第一个猜想的推论，因此，只需在两个猜想中证明一个就足够了．

乍一看，这一猜想的研究基础非常浅显，无外乎就是偶数奇数和质数的加法表示而已，如，对于第一个猜想来说：

$$4 = 2 + 2$$
$$6 = 3 + 3$$
$$8 = 3 + 5$$
$$10 = 5 + 5$$
$$12 = 5 + 7$$
$$14 = 7 + 7$$
$$16 = 5 + 11$$
$$18 = 7 + 11$$
$$\cdots\cdots$$

这样的加法算式每个人都可以写出很多很多，但是却不能以此来说明猜想的正确性，因为偶数有无穷多个，要这样的证明需要将其全部列举完，而这是不可能的．

对于第二个猜想来说：

$$9 = 3 + 3 + 3$$
$$11 = 2 + 2 + 7$$
$$13 = 3 + 5 + 5$$
$$15 = 5 + 5 + 5$$
$$17 = 5 + 5 + 7$$
$$19 = 5 + 7 + 7$$
$$21 = 7 + 7 + 7$$
$$23 = 5 + 7 + 11$$
$$\cdots\cdots$$

这里遇到了跟第一个猜想一样的问题，无法用列举法完成证明，但是这样的表达式却如此简单，那么究竟应当如何在这些表达式中找到证明，亦或者找到哪怕一点可能的突破口呢？有许许多多知名的、不知名的、专业的、业余的数学家或爱好者投入了毕生精力来完成或完善这一猜想的研究.

12.2　谁来摘取"数学王冠上的明珠"

由于其表述的简洁和证明的难度，数学家们将哥德巴赫猜想比喻为"数学王冠上的明珠"，研究者们都期待能够摘取摘取这一明珠，历代研究者们都做了极大的贡献，却很难将其破解.

1742 年 6 月 30 日，欧拉在给哥德巴赫的回信中，明确表示他深信哥德巴赫的这两个猜想都是正确的定理，但是欧拉至死也没给出证明.

1900 年，20 世纪最伟大的数学家希尔伯特，在国际数学会议上把"哥德巴赫猜想"列为 23 个数学难题之一. 此后，20 世纪的数学家们在世界范围内"联手"进攻"哥德巴赫猜想"堡垒，终于取得了辉煌的成果.

20 世纪的数学家们研究哥德巴赫猜想所采用的主要方法，是筛法、圆法、密率法和三角和法等等高深的数学方法. 解决这个猜想的思路，就像"缩小包围圈"一样，逐步逼近最后的结果.

1920 年，挪威数学家布朗证明了定理"9 + 9"，由此划定了进攻"哥德巴赫猜想"的"大包围圈". 这个"9 + 9"是怎么回事呢？所谓"9 + 9"，翻译成数学语言就是："任何一个足够大的偶数，都可以表示成其他两个数之和，而这两个数中的每个数，都是 9 个奇质数之积." 从这个"9 + 9"开始，全世界的数学家集中力量"缩小包围圈"，当然最后的目标就是"1 + 1"了.

1924 年，德国数学家雷德马赫证明了定理"7 + 7". 很快，"6 + 6""5 + 5""4 + 4"和"3 + 3"逐一被攻陷. 1957 年，中国数学家王元证明了"2 + 3". 1962 年，中国数学家潘承洞证明了"1 + 5"，同年又和王元合作证明了"1 + 4". 1965 年，苏联数学家证明了"1 + 3".

1966 年，中国著名数学家陈景润攻克了"1 + 2"，也就是："任何一个足够大的偶数，都可以表示成两个数之和，而这两个数中的一个就是奇质数，另一个则是两个奇质数的积." 这个定理被世界数学界称为"陈氏定理".

由于陈景润的贡献，人类距离哥德巴赫猜想的最后结果"1 + 1"仅有一步之遥了. 但为了实现这最后的一步，也许还要历经一个漫长的探索过程. 有许多数学家认为，要想证明"1 + 1"，必须通过创造新的数学方法，以往的路很可能都是走不通的.

附件一　50000 以内的质数表

1～100：2，3，5，7，11，13，17，19，23，29，31，37，41，43，47，53，59，61，67，71，73，79，83，89，97

101～200：101，103，107，109，113，127，131，137，139，149，151，157，163，167，173，179，181，191，193，197，199

201～300：211，223，227，229，233，239，241，251，257，263，269，271，277，281，283，293

301～400：307，311，313，317，331，337，347，349，353，359，367，373，379，383，389，397

401～500：401，409，419，421，431，433，439，443，449，457，461，463，467，479，487，491，499

501～600：503，509，521，523，541，547，557，563，569，571，577，587，593，599

601～700：601，607，613，617，619，631，641，643，647，653，659，661，673，677，683，691

701～800：701，709，719，727，733，739，743，751，757，761，769，773，787，797

801～900：809，811，821，823，827，829，839，853，857，859，863，877，881，883，887

901～1000：907，911，919，929，937，941，947，953，967，971，977，983，991，997

1001～1100：1009，1013，1019，1021，1031，1033，1039，1049，1051，1061，1063，1069，1087，1091，1093，1097

1101～1200：1103，1109，1117，1123，1129，1151，1153，1163，1171，1181，1187，1193

1201～1300：1201，1213，1217，1223，1229，1231，1237，1249，1259，1277，1279，1283，1289，1291，1297

1301～1400：1301，1303，1307，1319，1321，1327，1361，1367，1373，1381，1399

1401～1500：1409，1423，1427，1429，1433，1439，1447，1451，1453，1459，1471，1481，1483，1487，1489，1493，1499

1501～1600：1511，1523，1531，1543，1549，1553，1559，1567，1571，1579，1583，1597

1601～1700：1601，1607，1609，1613，1619，1621，1627，1637，1657，1663，1667，1669，1693，1697，1699

1701～1800：1709，1721，1723，1733，1741，1747，1753，1759，1777，1783，1787，1789

1801～1900：1801，1811，1823，1831，1847，1861，1867，1871，1873，1877，1879，1889

1901～2000：1901，1907，1913，1931，1933，1949，1951，1973，1979，1987，1993，1997，1999

2001 ~ 2100：2003，2011，2017，2027，2029，2039，2053，2063，2069，2081，2083，2087，2089，2099

2101 ~ 2200：2111，2113，2129，2131，2137，2141，2143，2153，2161，2179

2201 ~ 2300：2203，2207，2213，2221，2237，2239，2243，2251，2267，2269，2273，2281，2287，2293，2297

2301 ~ 2400：2309，2311，2333，2339，2341，2347，2351，2357，2371，2377，2381，2383，2389，2393，2399

2401 ~ 2500：2411，2417，2423，2437，2441，2447，2459，2467，2473，2477

2501 ~ 2600：2503，2521，2531，2539，2543，2549，2551，2557，2579，2591，2593

2601 ~ 2700：2609，2617，2621，2633，2647，2657，2659，2663，2671，2677，2683，2687，2689，2693，2699

2701 ~ 2800：2707，2711，2713，2719，2729，2731，2741，2749，2753，2767，2777，2789，2791，2797

2801 ~ 2900：2801，2803，2819，2833，2837，2843，2851，2857，2861，2879，2887，2897

2901 ~ 3000：2903，2909，2917，2927，2939，2953，2957，2963，2969，2971，2999

3001 ~ 3100：3001，3011，3019，3023，3037，3041，3049，3061，3067，3079，3083，3089

3101 ~ 3200：3109，3119，3121，3137，3163，3167，3169，3181，3187，3191

3201 ~ 3300：3203，3209，3217，3221，3229，3251，3253，3257，3259，3271，3299

3301 ~ 3400：3301，3307，3313，3319，3323，3329，3331，3343，3347，3359，3361，3371，3373，3389，3391

3401 ~ 3500：3407，3413，3433，3449，3457，3461，3463，3467，3469，3491，3499

3501 ~ 3600：3511，3517，3527，3529，3533，3539，3541，3547，3557，3559，3571，3581，3583，3593

3601 ~ 3700：3607，3613，3617，3623，3631，3637，3643，3659，3671，3673，3677，3691，3697

3701 ~ 3800：3701，3709，3719，3727，3733，3739，3761，3767，3769，3779，3793，3797

3801 ~ 3900：3803，3821，3823，3833，3847，3851，3853，3863，3877，3881，3889

3901 ~ 4000：3907，3911，3917，3919，3923，3929，3931，3943，3947，3967，3989

4001 ~ 4100：4001，4003，4007，4013，4019，4021，4027，4049，4051，4057，4073，4079，4091，4093，4099

4101 ~ 4200：4111，4127，4129，4133，4139，4153，4157，4159，4177

4201 ~ 4300：4201，4211，4217，4219，4229，4231，4241，4243，4253，4259，4261，4271，4273，4283，4289，4297

4301 ~ 4400：4327，4337，4339，4349，4357，4363，4373，4391，4397

4401 ~ 4500：4409，4421，4423，4441，4447，4451，4457，4463，4481，4483，4493

4501 ~ 4600：4507，4513，4517，4519，4523，4547，4549，4561，4567，4583，

4591, 4597

4601~4700：4603, 4621, 4637, 4639, 4643, 4649, 4651, 4657, 4663, 4673, 4679, 4691

4701~4800：4703, 4721, 4723, 4729, 4733, 4751, 4759, 4783, 4787, 4789, 4793, 4799

4801~4900：4801, 4813, 4817, 4831, 4861, 4871, 4877, 4889

4901~5000：4903, 4909, 4919, 4931, 4933, 4937, 4943, 4951, 4957, 4967, 4969, 4973, 4987, 4993, 4999

5001~5100：5003, 5009, 5011, 5021, 5023, 5039, 5051, 5059, 5077, 5081, 5087, 5099

5101~5200：5101, 5107, 5113, 5119, 5147, 5153, 5167, 5171, 5179, 5189, 5197

5201~5300：5209, 5227, 5231, 5233, 5237, 5261, 5273, 5279, 5281, 5297

5301~5400：5303, 5309, 5323, 5333, 5347, 5351, 5381, 5387, 5393, 5399

5401~5500：5407, 5413, 5417, 5419, 5431, 5437, 5441, 5443, 5449, 5471, 5477, 5479, 5483

5501~5600：5501, 5503, 5507, 5519, 5521, 5527, 5531, 5557, 5563, 5569, 5573, 5581, 5591

5601~5700：5623, 5639, 5641, 5647, 5651, 5653, 5657, 5659, 5669, 5683, 5689, 5693

5701~5800：5701, 5711, 5717, 5737, 5741, 5743, 5749, 5779, 5783, 5791

5801~5900：5801, 5807, 5813, 5821, 5827, 5839, 5843, 5849, 5851, 5857, 5861, 5867, 5869, 5879, 5881, 5897

5901~6000：5903, 5923, 5927, 5939, 5953, 5981, 5987

6001~6100：6007, 6011, 6029, 6037, 6043, 6047, 6053, 6067, 6073, 6079, 6089, 6091

6101~6200：6101, 6113, 6121, 6131, 6133, 6143, 6151, 6163, 6173, 6197, 6199

6201~6300：6203, 6211, 6217, 6221, 6229, 6247, 6257, 6263, 6269, 6271, 6277, 6287, 6299

6301~6400：6301, 6311, 6317, 6323, 6329, 6337, 6343, 6353, 6359, 6361, 6367, 6373, 6379, 6389, 6397

6401~6500：6421, 6427, 6449, 6451, 6469, 6473, 6481, 6491

6501~6600：6521, 6529, 6547, 6551, 6553, 6563, 6569, 6571, 6577, 6581, 6599

6601~6700：6607, 6619, 6637, 6653, 6659, 6661, 6673, 6679, 6689, 6691

6701~6800：6701, 6703, 6709, 6719, 6733, 6737, 6761, 6763, 6779, 6781, 6791, 6793

6801~6900：6803, 6823, 6827, 6829, 6833, 6841, 6857, 6863, 6869, 6871, 6883, 6899

6901~7000：6907, 6911, 6917, 6947, 6949, 6959, 6961, 6967, 6971, 6977, 6983, 6991, 6997

7001～7100：7001，7013，7019，7027，7039，7043，7057，7069，7079

7101～7200：7103，7109，7121，7127，7129，7151，7159，7177，7187，7193

7201～7300：7207，7211，7213，7219，7229，7237，7243，7247，7253，7283，7297

7301～7400：7307，7309，7321，7331，7333，7349，7351，7369，7393

7401～7500：7411，7417，7433，7451，7457，7459，7477，7481，7487，7489，7499

7501～7600：7507，7517，7523，7529，7537，7541，7547，7549，7559，7561，7573，7577，7583，7589，7591

7601～7700：7603，7607，7621，7639，7643，7649，7669，7673，7681，7687，7691，7699

7701～7800：7703，7717，7723，7727，7741，7753，7757，7759，7789，7793

7801～7900：7817，7823，7829，7841，7853，7867，7873，7877，7879，7883

7901～8000：7901，7907，7919，7927，7933，7937，7949，7951，7963，7993

8001～8100：8009，8011，8017，8039，8053，8059，8069，8081，8087，8089，8093

8101～8200：8101，8111，8117，8123，8147，8161，8167，8171，8179，8191

8201～8300：8209，8219，8221，8231，8233，8237，8243，8263，8269，8273，8287，8291，8293，8297

8301～8400：8311，8317，8329，8353，8363，8369，8377，8387，8389

8401～8500：8419，8423，8429，8431，8443，8447，8461，8467

8501～8600：8501，8513，8521，8527，8537，8539，8543，8563，8573，8581，8597，8599

8601～8700：8609，8623，8627，8629，8641，8647，8663，8669，8677，8681，8689，8693，8699

8701～8800：8707，8713，8719，8731，8737，8741，8747，8753，8761，8779，8783

8801～8900：8803，8807，8819，8821，8831，8837，8839，8849，8861，8863，8867，8887，8893

8901～9000：8923，8929，8933，8941，8951，8963，8969，8971，8999

9001～9100：9001，9007，9011，9013，9029，9041，9043，9049，9059，9067，9091

9101～9200：9103，9109，9127，9133，9137，9151，9157，9161，9173，9181，9187，9199

9201～9300：9203，9209，9221，9227，9239，9241，9257，9277，9281，9283，9293

9301～9400：9311，9319，9323，9337，9341，9343，9349，9371，9377，9391，9397

9401～9500：9403，9413，9419，9421，9431，9433，9437，9439，9461，9463，9467，9473，9479，9491，9497

9501～9600：9511，9521，9533，9539，9547，9551，9587

9601～9700：9601，9613，9619，9623，9629，9631，9643，9649，9661，9677，9679，9689，9697

9701～9800：9719，9721，9733，9739，9743，9749，9767，9769，9781，9787，9791

9801～9900：9803，9811，9817，9829，9833，9839，9851，9857，9859，9871，

9883，9887

9901 ~ 10000：9901, 9907, 9923, 9929, 9931, 9941, 9949, 9967, 9973

10001 ~ 11000：10007, 10009, 10037, 10039, 10061, 10067, 10069, 10079, 10091, 10093, 10099, 10103, 10111, 10133, 10139, 10141, 10151, 10159, 10163, 10169, 10177, 10181, 10193, 10211, 10223, 10243, 10247, 10253, 10259, 10267, 10271, 10273, 10289, 10301, 10303, 10313, 10321, 10331, 10333, 10337, 10343, 10357, 10369, 10391, 10399, 10427, 10429, 10433, 10453, 10457, 10459, 10463, 10477, 10487, 10499, 10501, 10513, 10529, 10531, 10559, 10567, 10589, 10597, 10601, 10607, 10613, 10627, 10631, 10639, 10651, 10657, 10663, 10667, 10687, 10691, 10709, 10711, 10723, 10729, 10733, 10739, 10753, 10771, 10781, 10789, 10799, 10831, 10837, 10847, 10853, 10859, 10861, 10867, 10883, 10889, 10891, 10903, 10909, 10937, 10939, 10949, 10957, 10973, 10979, 10987, 10993

11001 ~ 12000：11003, 11027, 11047, 11057, 11059, 11069, 11071, 11083, 11087, 11093, 11113, 11117, 11119, 11131, 11149, 11159, 11161, 11171, 11173, 11177, 11197, 11213, 11239, 11243, 11251, 11257, 11261, 11273, 11279, 11287, 11299, 11311, 11317, 11321, 11329, 11351, 11353, 11369, 11383, 11393, 11399, 11411, 11423, 11437, 11443, 11447, 11467, 11471, 11483, 11489, 11491, 11497, 11503, 11519, 11527, 11549, 11551, 11579, 11587, 11593, 11597, 11617, 11621, 11633, 11657, 11677, 11681, 11689, 11699, 11701, 11717, 11719, 11731, 11743, 11777, 11779, 11783, 11789, 11801, 11807, 11813, 11821, 11827, 11831, 11833, 11839, 11863, 11867, 11887, 11897, 11903, 11909, 11923, 11927, 11933, 11939, 11941, 11953, 11959, 11969, 11971, 11981, 11987

12001 ~ 13000：12007, 12011, 12037, 12041, 12043, 12049, 12071, 12073, 12097, 12101, 12107, 12109, 12113, 12119, 12143, 12149, 12157, 12161, 12163, 12197, 12203, 12211, 12227, 12239, 12241, 12251, 12253, 12263, 12269, 12277, 12281, 12289, 12301, 12323, 12329, 12343, 12347, 12373, 12377, 12379, 12391, 12401, 12409, 12413, 12421, 12433, 12437, 12451, 12457, 12473, 12479, 12487, 12491, 12497, 12503, 12511, 12517, 12527, 12539, 12541, 12547, 12553, 12569, 12577, 12583, 12589, 12601, 12611, 12613, 12619, 12637, 12641, 12647, 12653, 12659, 12671, 12689, 12697, 12703, 12713, 12721, 12739, 12743, 12757, 12763, 12781, 12791, 12799, 12809, 12821, 12823, 12829, 12841, 12853, 12889, 12893, 12899, 12907, 12911, 12917, 12919, 12923, 12941, 12953, 12959, 12967, 12973, 12979, 12983

13001 ~ 14000：13001, 13003, 13007, 13009, 13033, 13037, 13043, 13049, 13063, 13093, 13099, 13103, 13109, 13121, 13127, 13147, 13151, 13159, 13163, 13171, 13177, 13183, 13187, 13217, 13219, 13229, 13241, 13249, 13259, 13267, 13291, 13297, 13309, 13313, 13327, 13331, 13337, 13339, 13367, 13381, 13397, 13399, 13411, 13417, 13421, 13441, 13451, 13457, 13463, 13469, 13477, 13487, 13499, 13513, 13523, 13537, 13553, 13567, 13577, 13591, 13597, 13613, 13619,

13627，13633，13649，13669，13679，13681，13687，13691，13693，13697，13709，
13711，13721，13723，13729，13751，13757，13759，13763，13781，13789，13799，
13807，13829，13831，13841，13859，13873，13877，13879，13883，13901，13903，
13907，13913，13921，13931，13933，13963，13967，13997，13999

14001～15000：14009，14011，14029，14033，14051，14057，14071，14081，
14083，14087，14107，14143，14149，14153，14159，14173，14177，14197，14207，
14221，14243，14249，14251，14281，14293，14303，14321，14323，14327，14341，
14347，14369，14387，14389，14401，14407，14411，14419，14423，14431，14437，
14447，14449，14461，14479，14489，14503，14519，14533，14537，14543，14549，
14551，14557，14561，14563，14591，14593，14621，14627，14629，14633，14639，
14653，14657，14669，14683，14699，14713，14717，14723，14731，14737，14741，
14747，14753，14759，14767，14771，14779，14783，14797，14813，14821，14827，
14831，14843，14851，14867，14869，14879，14887，14891，14897，14923，14929，
14939，14947，14951，14957，14969，14983

15001～16000：15013，15017，15031，15053，15061，15073，15077，15083，
15091，15101，15107，15121，15131，15137，15139，15149，15161，15173，15187，
15193，15199，15217，15227，15233，15241，15259，15263，15269，15271，15277，
15287，15289，15299，15307，15313，15319，15329，15331，15349，15359，15361，
15373，15377，15383，15391，15401，15413，15427，15439，15443，15451，15461，
15467，15473，15493，15497，15511，15527，15541，15551，15559，15569，15581，
15583，15601，15607，15619，15629，15641，15643，15647，15649，15661，15667，
15671，15679，15683，15727，15731，15733，15737，15739，15749，15761，15767，
15773，15787，15791，15797，15803，15809，15817，15823，15859，15877，15881，
15887，15889，15901，15907，15913，15919，15923，15937，15959，15971，
15973，15991

16001～17000：16001，16007，16033，16057，16061，16063，16067，16069，
16073，16087，16091，16097，16103，16111，16127，16139，16141，16183，16187，
16189，16193，16217，16223，16229，16231，16249，16253，16267，16273，16301，
16319，16333，16339，16349，16361，16363，16369，16381，16411，16417，16421，
16427，16433，16447，16451，16453，16477，16481，16487，16493，16519，16529，
16547，16553，16561，16567，16573，16603，16607，16619，16631，16633，16649，
16651，16657，16661，16673，16691，16693，16699，16703，16729，16741，16747，
16759，16763，16787，16811，16823，16829，16831，16843，16871，16879，16883，
16889，16901，16903，16921，16927，16931，16937，16943，16963，16979，16981，
16987，16993

17001～18000：17011，17021，17027，17029，17033，17041，17047，17053，
17077，17093，17099，17107，17117，17123，17137，17159，17167，17183，17189，
17191，17203，17207，17209，17231，17239，17257，17291，17293，17299，17317，
17321，17327，17333，17341，17351，17359，17377，17383，17387，17389，17393，

17401, 17417, 17419, 17431, 17443, 17449, 17467, 17471, 17477, 17483, 17489, 17491, 17497, 17509, 17519, 17539, 17551, 17569, 17573, 17579, 17581, 17597, 17599, 17609, 17623, 17627, 17657, 17659, 17669, 17681, 17683, 17707, 17713, 17729, 17737, 17747, 17749, 17761, 17783, 17789, 17791, 17807, 17827, 17837, 17839, 17851, 17863, 17881, 17891, 17903, 17909, 17911, 17921, 17923, 17929, 17939, 17957, 17959, 17971, 17977, 17981, 17987, 17989

18001 ~ 19000：18013, 18041, 18043, 18047, 18049, 18059, 18061, 18077, 18089, 18097, 18119, 18121, 18127, 18131, 18133, 18143, 18149, 18169, 18181, 18191, 18199, 18211, 18217, 18223, 18229, 18233, 18251, 18253, 18257, 18269, 18287, 18289, 18301, 18307, 18311, 18313, 18329, 18341, 18353, 18367, 18371, 18379, 18397, 18401, 18413, 18427, 18433, 18439, 18443, 18451, 18457, 18461, 18481, 18493, 18503, 18517, 18521, 18523, 18539, 18541, 18553, 18583, 18587, 18593, 18617, 18637, 18661, 18671, 18679, 18691, 18701, 18713, 18719, 18731, 18743, 18749, 18757, 18773, 18787, 18793, 18797, 18803, 18839, 18859, 18869, 18899, 18911, 18913, 18917, 18919, 18947, 18959, 18973, 18979

19001 ~ 20000：19001, 19009, 19013, 19031, 19037, 19051, 19069, 19073, 19079, 19081, 19087, 19121, 19139, 19141, 19157, 19163, 19181, 19183, 19207, 19211, 19213, 19219, 19231, 19237, 19249, 19259, 19267, 19273, 19289, 19301, 19309, 19319, 19333, 19373, 19379, 19381, 19387, 19391, 19403, 19417, 19421, 19423, 19427, 19429, 19433, 19441, 19447, 19457, 19463, 19469, 19471, 19477, 19483, 19489, 19501, 19507, 19531, 19541, 19543, 19553, 19559, 19571, 19577, 19583, 19597, 19603, 19609, 19661, 19681, 19687, 19697, 19699, 19709, 19717, 19727, 19739, 19751, 19753, 19759, 19763, 19777, 19793, 19801, 19813, 19819, 19841, 19843, 19853, 19861, 19867, 19889, 19891, 19913, 19919, 19927, 19937, 19949, 19961, 19963, 19973, 19979, 19991, 19993, 19997

20001 ~ 30000：20011, 20021, 20023, 20029, 20047, 20051, 20063, 20071, 20089, 20101, 20107, 20113, 20117, 20123, 20129, 20143, 20147, 20149, 20161, 20173, 20177, 20183, 20201, 20219, 20231, 20233, 20249, 20261, 20269, 20287, 20297, 20323, 20327, 20333, 20341, 20347, 20353, 20357, 20359, 20369, 20389, 20393, 20399, 20407, 20411, 20431, 20441, 20443, 20477, 20479, 20483, 20507, 20509, 20521, 20533, 20543, 20549, 20551, 20563, 20593, 20599, 20611, 20627, 20639, 20641, 20663, 20681, 20693, 20707, 20717, 20719, 20731, 20743, 20747, 20749, 20753, 20759, 20771, 20773, 20789, 20807, 20809, 20849, 20857, 20873, 20879, 20887, 20897, 20899, 20903, 20921, 20929, 20939, 20947, 20959, 20963, 20981, 20983, 21001, 21011, 21013, 21017, 21019, 21023, 21031, 21059, 21061, 21067, 21089, 21101, 21107, 21121, 21139, 21143, 21149, 21157, 21163, 21169, 21179, 21187, 21191, 21193, 21211, 21221, 21227, 21247, 21269, 21277, 21283, 21313, 21317, 21319, 21323, 21341, 21347, 21377, 21379, 21383, 21391, 21397, 21401, 21407, 21419, 21433, 21467, 21481, 21487, 21491, 21493, 21499, 21503,

21517，21521，21523，21529，21557，21559，21563，21569，21577，21587，21589，
21599，21601，21611，21613，21617，21647，21649，21661，21673，21683，21701，
21713，21727，21737，21739，21751，21757，21767，21773，21787，21799，21803，
21817，21821，21839，21841，21851，21859，21863，21871，21881，21893，21911，
21929，21937，21943，21961，21977，21991，21997，22003，22013，22027，22031，
22037，22039，22051，22063，22067，22073，22079，22091，22093，22109，22111，
22123，22129，22133，22147，22153，22157，22159，22171，22189，22193，22229，
22247，22259，22271，22273，22277，22279，22283，22291，22303，22307，22343，
22349，22367，22369，22381，22391，22397，22409，22433，22441，22447，22453，
22469，22481，22483，22501，22511，22531，22541，22543，22549，22567，22571，
22573，22613，22619，22621，22637，22639，22643，22651，22669，22679，22691，
22697，22699，22709，22717，22721，22727，22739，22741，22751，22769，22777，
22783，22787，22807，22811，22817，22853，22859，22861，22871，22877，22901，
22907，22921，22937，22943，22961，22963，22973，22993，23003，23011，23017，
23021，23027，23029，23039，23041，23053，23057，23059，23063，23071，23081，
23087，23099，23117，23131，23143，23159，23167，23173，23189，23197，23201，
23203，23209，23227，23251，23269，23279，23291，23293，23297，23311，23321，
23327，23333，23339，23357，23369，23371，23399，23417，23431，23447，23459，
23473，23497，23509，23531，23537，23539，23549，23557，23561，23563，23567，
23581，23593，23599，23603，23609，23623，23627，23629，23633，23663，23669，
23671，23677，23687，23689，23719，23741，23743，23747，23753，23761，23767，
23773，23789，23801，23813，23819，23827，23831，23833，23857，23869，3873，
23879，23887，23893，23899，23909，23911，23917，23929，23957，23971，23977，
23981，23993，24001，24007，24019，24023，24029，24043，24049，24061，24071，
24077，24083，24091，24097，24103，24107，24109，24113，24121，24133，24137，
24151，24169，24179，24181，24197，24203，24223，24229，24239，24247，24251，
24281，24317，24329，24337，24359，24371，24373，24379，24391，24407，24413，
24419，24421，24439，24443，24469，24473，24481，24499，24509，24517，24527，
24533，24547，24551，24571，24593，24611，24623，24631，24659，24671，24677，
24683，24691，24697，24709，24733，24749，24763，24767，24781，24793，24799，
24809，24821，24841，24847，24851，24859，24877，24889，24907，24917，24919，
24923，24943，24953，24967，24971，24977，24979，24989，25013，25031，25033，
25037，25057，25073，25087，25097，25111，25117，25121，25127，25147，25153，
25163，25169，25171，25183，25189，25219，25229，25237，25243，25247，25253，
25261，25301，25303，25307，25309，25321，25339，25343，25349，25357，25367，
25373，25391，25409，25411，25423，25439，25447，25453，25457，25463，25469，
25471，25523，25537，25541，25561，25577，25579，25583，25589，25601，25603，
25609，25621，25633，25639，25643，25657，25667，25673，25679，25693，25703，
25717，25733，25741，25747，25759，25763，25771，25793，25799，25801，25819，

25841，25847，25849，25867，25873，25889，25903，25913，25919，25931，25933，
25939，25943，25951，25969，25981，25997，25999，26003，26017，26021，26029，
26041，26053，26083，26099，26107，26111，26113，26119，26141，26153，26161，
26171，26177，26183，26189，26203，26209，26227，26237，26249，26251，26261，
26263，26267，26293，26297，26309，26317，26321，26339，26347，26357，26371，
26387，26393，26399，26407，26417，26423，26431，26437，26449，26459，26479，
26489，26497，26501，26513，26539，26557，26561，26573，26591，26597，26627，
26633，26641，26647，26669，26681，26683，26687，26693，26699，26701，26711，
26713，26717，26723，26729，26731，26737，26759，26777，26783，26801，26813，
26821，26833，26839，26849，26861，26863，26879，26881，26891，26893，26903，
26921，26927，26947，26951，26953，26959，26981，26987，26993，27011，27017，
27031，27043，27059，27061，27067，27073，27077，27091，27103，27107，27109，
27127，27143，27179，27191，27197，27211，27239，27241，27253，27259，27271，
27277，27281，27283，27299，27329，27337，27361，27367，27397，27407，27409，
27427，27431，27437，27449，27457，27479，27481，27487，27509，27527，27529，
27539，27541，27551，27581，27583，27611，27617，27631，27647，27653，27673，
27689，27691，27697，27701，27733，27737，27739，27743，27749，27751，27763，
27767，27773，27779，27791，27793，27799，27803，27809，27817，27823，27827，
27847，27851，27883，27893，27901，27917，27919，27941，27943，27947，27953，
27961，27967，27983，27997，28001，28019，28027，28031，28051，28057，28069，
28081，28087，28097，28099，28109，28111，28123，28151，28163，28181，28183，
28201，28211，28219，28229，28277，28279，28283，28289，28297，28307，28309，
28319，28349，28351，28387，28393，28403，28409，28411，28429，28433，28439，
28447，28463，28477，28493，28499，28513，28517，28537，28541，28547，28549，
28559，28571，28573，28579，28591，28597，28603，28607，28619，28621，28627，
28631，28643，28649，28657，28661，28663，28669，28687，28697，28703，28711，
28723，28729，28751，28753，28759，28771，28789，28793，28807，28813，28817，
28837，28843，28859，28867，28871，28879，28901，28909，28921，28927，28933，
28949，28961，28979，29009，29017，29021，29023，29027，29033，29059，29063，
29077，29101，29123，29129，29131，29137，29147，29153，29167，29173，29179，
29191，29201，29207，29209，29221，29231，29243，29251，29269，29287，29297，
29303，29311，29327，29333，29339，29347，29363，29383，29387，29389，29399，
29401，29411，29423，29429，29437，29443，29453，29473，29483，29501，29527，
29531，29537，29567，29569，29573，29581，29587，29599，29611，29629，29633，
29641，29663，29669，29671，29683，29717，29723，29741，29753，29759，29761，
29789，29803，29819，29833，29837，29851，29863，29867，29873，29879，29881，
29917，29921，29927，29947，29959，29983，29989

30001～40000：30011，30013，30029，30047，30059，30071，30089，30091，
30097，3010330109，30113，30119，30133，30137，30139，30161，30169，30181，

30187，30197，30203，30211，30223，30241，30253，30259，30269，30271，30293，
30307，30313，30319，30323，30341，30347，30367，30389，30391，30403，30427，
30431，30449，30467，30469，30491，30493，30497，30509，30517，30529，30539，
30553，30557，30559，30577，30593，30631，30637，30643，30649，30661，30671，
30677，30689，30697，30703，30707，30713，30727，30757，30763，30773，30781，
30803，30809，30817，30829，30839，30841，30851，30853，30859，30869，30871，
30881，30893，30911，30931，30937，30941，30949，30971，30977，30983，31013，
31019，31033，31039，31051，31063，31069，31079，31081，31091，31121，31123，
31139，31147，31151，31153，31159，31177，31181，31183，31189，31193，31219，
31223，31231，31237，31247，31249，31253，31259，31267，31271，31277，31307，
31319，31321，31327，31333，31337，31357，31379，31387，31391，31393，31397，
31469，31477，31481，31489，31511，31513，31517，31531，31541，31543，31547，
31567，31573，31583，31601，31607，31627，31643，31649，31657，31663，31667，
31687，31699，31721，31723，31727，31729，31741，31751，31769，31771，31793，
31799，31817，31847，31849，31859，31873，31883，31891，31907，31957，31963，
31973，31981，31991，2003，32009，32027，32029，32051，32057，32059，32063，
32069，32077，32083，32089，32099，32117，32119，32141，32143，32159，32173，
32183，32189，32191，32203，32213，32233，32237，32251，32257，32261，32297，
32299，32303，32309，32321，32323，32327，32341，32353，32359，32363，32369，
32371，32377，32381，32401，32411，32413，32423，32429，32441，32443，32467，
32479，32491，32497，32503，32507，32531，32533，32537，32561，32563，32569，
32573，32579，32587，32603，32609，32611，32621，32633，32647，32653，32687，
32693，32707，32713，32717，32719，32749，32771，32779，32783，32789，32797，
32801，32803，32831，32833，32839，32843，32869，32887，32909，32911，32917，
32933，32939，32941，32957，32969，32971，32983，32987，32993，32999，33013，
33023，33029，33037，33049，33053，33071，33073，33083，33091，33107，33113，
33119，33149，33151，33161，33179，33181，33191，33199，33203，33211，33223，
33247，33287，33289，33301，33311，33317，33329，33331，33343，33347，33349，
33353，33359，33377，33391，33403，33409，33413，33427，33457，33461，33469，
33479，33487，33493，33503，33521，33529，33533，33547，33563，33569，33577，
33581，33587，33589，33599，33601，33613，33617，33619，33623，33629，33637，
33641，33647，33679，33703，33713，33721，33739，33749，33751，33757，33767，
33769，33773，33791，33797，33809，33811，33827，33829，33851，33857，33863，
33871，33889，33893，33911，33923，33931，33937，33941，33961，33967，33997，
34019，34031，34033，34039，34057，34061，34123，34127，34129，34141，34147，
34157，34159，34171，34183，34211，34213，34217，34231，34253，34259，34261，
34267，34273，34283，34297，34301，34303，34313，34319，34327，34337，34351，
34361，34367，34369，34381，34403，34421，34429，34439，34457，34469，34471，
34483，34487，34499，34501，34511，34513，34519，34537，34543，34549，34583，

34589, 34591, 34603, 34607, 34613, 34631, 34649, 34651, 34667, 34673, 34679,
34687, 34693, 34703, 34721, 34729, 34739, 34747, 34757, 34759, 34763, 34781,
34807, 34819, 34841, 34843, 34847, 34849, 34871, 34877, 34883, 34897, 34913,
34919, 34939, 34949, 34961, 34963, 34981, 35023, 35027, 35051, 35053, 35059,
35069, 35081, 35083, 35089, 35099, 35107, 35111, 35117, 35129, 35141, 35149,
35153, 35159, 35171, 35201, 35221, 35227, 35251, 35257, 35267, 35279, 35281,
35291, 35311, 35317, 35323, 35327, 35339, 35353, 35363, 35381, 35393, 35401,
35407, 35419, 35423, 35437, 35447, 35449, 35461, 35491, 35507, 35509, 35521,
35527, 35531, 35533, 35537, 35543, 35569, 35573, 35591, 35593, 35597, 35603,
35617, 35671, 35677, 35729, 35731, 35747, 35753, 35759, 35771, 35797, 35801,
35803, 35809, 35831, 35837, 35839, 35851, 35863, 35869, 35879, 35897, 35899,
35911, 35923, 35933, 35951, 35963, 35969, 35977, 35983, 35993, 35999, 36007,
36011, 36013, 36017, 36037, 36061, 36067, 36073, 36083, 36097, 36107, 36109,
36131, 36137, 36151, 36161, 36187, 36191, 36209, 36217, 36229, 36241, 36251,
36263, 36269, 36277, 36293, 36299, 36307, 36313, 36319, 36341, 36343, 36353,
36373, 36383, 36389, 36433, 36451, 36457, 36467, 36469, 36473, 36479, 36493,
36497, 36523, 36527, 36529, 36541, 36551, 36559, 36563, 36571, 36583, 36587,
36599, 36607, 36629, 36637, 36643, 36653, 36671, 36677, 36683, 36691, 36697,
36709, 36713, 36721, 36739, 36749, 36761, 36767, 36779, 36781, 36787, 36791,
36793, 36809, 36821, 36833, 36847, 36857, 36871, 36877, 36887, 36899, 36901,
36913, 36919, 36923, 36929, 36931, 36943, 36947, 36973, 36979, 36997, 37003,
37013, 37019, 37021, 37039, 37049, 37057, 37061, 37087, 37097, 37117, 37123,
37139, 37159, 37171, 37181, 37189, 37199, 37201, 37217, 37223, 37243, 37253,
37273, 37277, 37307, 37309, 37313, 37321, 37337, 37339, 37357, 37361, 37363,
37369, 37379, 37397, 37409, 37423, 37441, 37447, 37463, 37483, 37489, 37493,
37501, 37507, 37511, 37517, 37529, 37537, 37547, 37549, 37561, 37567, 37571,
37573, 37579, 37589, 37591, 37607, 37619, 37633, 37643, 37649, 37657, 37663,
37691, 37693, 37699, 37717, 37747, 37781, 37783, 37799, 37811, 37813, 37831,
37847, 37853, 37861, 37871, 37879, 37889, 37897, 37907, 37951, 37957, 37963,
37967, 37987, 37991, 37993, 37997, 38011, 38039, 38047, 38053, 38069, 38083,
38113, 38119, 38149, 38153, 38167, 38177, 38183, 38189, 38197, 38201, 38219,
38231, 38237, 38239, 38261, 38273, 38281, 38287, 38299, 38303, 38317, 38321,
38327, 38329, 38333, 38351, 38371, 38377, 38393, 38431, 38447, 38449, 38453,
38459, 38461, 38501, 38543, 38557, 38561, 38567, 38569, 38593, 38603, 38609,
38611, 38629, 38639, 38651, 38653, 38669, 38671, 38677, 38693, 38699, 38707,
38711, 38713, 38723, 38729, 38737, 38747, 38749, 38767, 38783, 38791, 38803,
38821, 38833, 38839, 38851, 38861, 38867, 38873, 38891, 38903, 38917, 38921,
38923, 38933, 38953, 38959, 38971, 38977, 38993, 39019, 39023, 39041, 39043,
39047, 39079, 39089, 39097, 39103, 39107, 39113, 39119, 39133, 39139, 39157,

39161，39163，39181，39191，39199，39209，39217，39227，39229，39233，39239，
39241，39251，39293，39301，39313，39317，39323，39341，39343，39359，39367，
39371，39373，39383，39397，39409，39419，39439，39443，39451，39461，39499，
39503，39509，39511，39521，39541，39551，39563，39569，39581，39607，39619，
39623，39631，39659，39667，39671，39679，39703，39709，39719，39727，39733，
39749，39761，39769，39779，39791，39799，39821，39827，39829，39839，39841，
39847，39857，39863，39869，39877，39883，39887，39901，39929，39937，39953，
39971，39979，39983，39989

40001～50000：40009，40013，40031，40037，40039，40063，40087，40093，
40099，40111，40123，40127，40129，40151，40153，40163，40169，40177，40189，
40193，40213，40231，40237，40241，40253，40277，40283，40289，40343，40351，
40357，40361，40387，40423，40427，40429，40433，40459，40471，40483，40487，
40493，40499，40507，40519，40529，40531，40543，40559，40577，40583，40591，
40597，40609，40627，40637，40639，40693，40697，40699，40709，40739，40751，
40759，40763，40771，40787，40801，40813，40819，40823，40829，40841，40847，
40849，40853，40867，40879，40883，40897，40903，40927，40933，40939，40949，
40961，40973，40993，41011，41017，41023，41039，41047，41051，41057，41077，
41081，41113，41117，41131，41141，41143，41149，41161，41177，41179，41183，
41189，41201，41203，41213，41221，41227，41231，41233，41243，41257，41263，
41269，41281，41299，41333，41341，41351，41357，41381，41387，41389，41399，
41411，41413，41443，41453，41467，41479，41491，41507，41513，41519，41521，
41539，41543，41549，41579，41593，41597，41603，41609，41611，41617，41621，
41627，41641，41647，41651，41659，41669，41681，41687，41719，41729，41737，
41759，41761，41771，41777，41801，41809，41813，41843，41849，41851，41863，
41879，41887，41893，41897，41903，41911，41927，41941，41947，41953，41957，
41959，41969，41981，41983，41999，42013，42017，42019，42023，42043，42061，
42071，42073，42083，42089，42101，42131，42139，42157，42169，42179，42181，
42187，42193，42197，42209，42221，42223，42227，42239，42257，42281，42283，
42293，42299，42307，42323，42331，42337，42349，42359，42373，42379，42391，
42397，42403，42407，42409，42433，42437，42443，42451，42457，42461，42463，
42467，42473，42487，42491，42499，42509，42533，42557，42569，42571，42577，
42589，42611，42641，42643，42649，42667，42677，42683，42689，42697，42701，
42703，42709，42719，42727，42737，42743，42751，42767，42773，42787，42793，
42797，42821，42829，42839，42841，42853，42859，42863，42899，42901，42923，
42929，42937，42943，42953，42961，42967，42979，42989，43003，43013，43019，
43037，43049，43051，43063，43067，43093，43103，43117，43133，43151，43159，
43177，43189，43201，43207，43223，43237，43261，43271，43283，43291，43313，
43319，43321，43331，43391，43397，43399，43403，43411，43427，43441，43451，
43457，43481，43487，43499，43517，43541，43543，43573，43577，43579，43591，

43597, 43607, 43609, 43613, 43627, 43633, 43649, 43651, 43661, 43669, 43691,
43711, 43717, 43721, 43753, 43759, 43777, 43781, 43783, 43787, 43789, 43793,
43801, 43853, 43867, 43889, 43891, 43913, 43933, 43943, 43951, 43961, 43963,
43969, 43973, 43987, 43991, 43997, 44017, 44021, 44027, 44029, 44041, 44053,
44059, 44071, 44087, 44089, 44101, 44111, 44119, 44123, 44129, 44131, 44159,
44171, 44179, 44189, 44201, 44203, 44207, 44221, 44249, 44257, 44263, 44267,
44269, 44273, 44279, 44281, 44293, 44351, 44357, 44371, 44381, 44383, 44389,
44417, 44449, 44453, 44483, 44491, 44497, 44501, 44507, 44519, 44531, 44533,
44537, 44543, 44549, 44563, 44579, 44587, 44617, 44621, 44623, 44633, 44641,
44647, 44651, 44657, 44683, 44687, 44699, 44701, 44711, 44729, 44741, 44753,
44771, 44773, 44777, 44789, 44797, 44809, 44819, 44839, 44843, 44851, 44867,
44879, 44887, 44893, 44909, 44917, 44927, 44939, 44953, 44959, 44963, 44971,
44983, 44987, 45007, 45013, 45053, 45061, 45077, 45083, 45119, 45121, 45127,
45131, 45137, 45139, 45161, 45179, 45181, 45191, 45197, 45233, 45247, 45259,
45263, 45281, 45289, 45293, 45307, 45317, 45319, 45329, 45337, 45341, 45343,
45361, 45377, 45389, 45403, 45413, 45427, 45433, 45439, 45481, 45491, 45497,
45503, 45523, 45533, 45541, 45553, 45557, 45569, 45587, 45589, 45599, 45613,
45631, 45641, 45659, 45667, 45673, 45677, 45691, 45697, 45707, 45737, 45751,
45757, 45763, 45767, 45779, 45817, 45821, 45823, 45827, 45833, 45841, 45853,
45863, 45869, 45887, 45893, 45943, 45949, 45953, 45959, 45971, 45979, 45989,
46021, 46027, 46049, 46051, 46061, 46073, 46091, 46093, 46099, 46103, 46133,
46141, 46147, 46153, 46171, 46181, 46183, 46187, 46199, 46219, 46229, 46237,
46261, 46271, 46273, 46279, 46301, 46307, 46309, 46327, 46337, 46349, 46351,
46381, 46399, 46411, 46439, 46441, 46447, 46451, 46457, 46471, 46477, 46489,
46499, 46507, 46511, 46523, 46549, 46559, 46567, 46573, 46589, 46591, 46601,
46619, 46633, 46639, 46643, 46649, 46663, 46679, 46681, 46687, 46691, 46703,
46723, 46727, 46747, 46751, 46757, 46769, 46771, 46807, 46811, 46817, 46819,
46829, 46831, 46853, 46861, 46867, 46877, 46889, 46901, 46919, 46933, 46957,
46993, 46997, 47017, 47041, 47051, 47057, 47059, 47087, 47093, 47111, 47119,
47123, 47129, 47137, 47143, 47147, 47149, 47161, 47189, 47207, 47221, 47237,
47251, 47269, 47279, 47287, 47293, 47297, 47303, 47309, 47317, 47339, 47351,
47353, 47363, 47381, 47387, 47389, 47407, 47417, 47419, 47431, 47441, 47459,
47491, 47497, 47501, 47507, 47513, 47521, 47527, 47533, 47543, 47563, 47569,
47581, 47591, 47599, 47609, 47623, 47629, 47639, 47653, 47657, 47659, 47681,
47699, 47701, 47711, 47713, 47717, 47737, 47741, 47743, 47777, 47779, 47791,
47797, 47807, 47809, 47819, 47837, 47843, 47857, 47869, 47881, 47903, 47911,
47917, 47933, 47939, 47947, 47951, 47963, 47969, 47977, 47981, 48017, 48023,
48029, 48049, 48073, 48079, 48091, 48109, 48119, 48121, 48131, 48157, 48163,
48179, 48187, 48193, 48197, 48221, 48239, 48247, 48259, 48271, 48281, 48299,

48311，48313，48337，48341，48353，48371，48383，48397，48407，48409，48413，
48437，48449，48463，48473，48479，48481，48487，48491，48497，48523，48527，
48533，48539，48541，48563，48571，48589，48593，48611，48619，48623，48647，
48649，48661，48673，48677，48679，48731，48733，48751，48757，48761，48767，
48779，48781，48787，48799，48809，48817，48821，48823，48847，48857，48859，
48869，48871，48883，48889，48907，48947，48953，48973，48989，48991，49003，
49009，49019，49031，49033，49037，49043，49057，49069，49081，49103，49109，
49117，49121，49123，49139，49157，49169，49171，49177，49193，49199，49201，
49207，49211，49223，49253，49261，49277，49279，49297，49307，49331，49333，
49339，49363，49367，49369，49391，49393，49409，49411，49417，49429，49433，
49451，49459，49463，49477，49481，49499，49523，49529，49531，49537，49547，
49549，49559，49597，49603，49613，49627，49633，49639，49663，49667，49669，
49681，49697，49711，49727，49739，49741，49747，49757，49783，49787，49789，
49801，49807，49811，49823，49831，49843，49853，49871，49877，49891，49919，
49921，49927，49937，49939，49943，49957，49991，49993，49999

附件二　亲和数

	亲和数	因数之积	因数之和
A	220	$2 \times 2 \times 5 \times 11$	$1 + 2 + 4 + 5 + 10 + 11 + 20 + 22 + 44 + 55 + 110 = 284$
B	284	$2 \times 2 \times 71$	$1 + 2 + 4 + 71 + 142 = 220$
A	1184	$2 \times 2 \times 2 \times 2 \times 2 \times 37$	$1 + 2 + 4 + 8 + 16 + 32 + 37 + 74 + 148 + 296 + 592 = 1210$
B	1210	$2 \times 5 \times 11 \times 11$	$1 + 2 + 5 + 10 + 11 + 22 + 55 + 110 + 121 + 242 + 605 = 1184$
A	2620	$2 \times 2 \times 5 \times 13$	$1 + 2 + 4 + 5 + 10 + 20 + 131 + 262 + 524 + 655 + 1310 = 2924$
B	2924	$2 \times 2 \times 17 \times 43$	$1 + 2 + 4 + 17 + 34 + 43 + 68 + 86 + 172 + 731 + 1462 = 2620$
A	5020	$2 \times 2 \times 2 \times 5 \times 251$	$1 + 2 + 4 + 5 + 10 + 20 + 1004 + 502 + 251 + 1255 + 2510 = 5564$
B	5564	$2 \times 2 \times 13 \times 107$	$1 + 2 + 4 + 13 + 26 + 52 + 107 + 214 + 428 + 1391 + 2782 = 5020$
A	6232	$2 \times 2 \times 2 \times 19 \times 41$	$1 + 2 + 4 + 8 + 19 + 38 + 41 + 76 + 82 + 152 + 164 + 328 + 779 + 1558 + 3116 = 6368$
B	6368	$2 \times 2 \times 2 \times 2 \times 2 \times 199$	$1 + 2 + 4 + 8 + 16 + 32 + 199 + 398 + 796 + 1592 + 3184 = 6232$
A	10744	$2 \times 2 \times 2 \times 17 \times 79$	$1 + 2 + 4 + 8 + 17 + 34 + 68 + 136 + 79 + 158 + 316 + 632 + 1343 + 2686 + 5372 = 10856$
B	10856	$2 \times 2 \times 2 \times 23 \times 59$	$1 + 2 + 4 + 8 + 23 + 46 + 92 + 184 + 59 + 118 + 236 + 472 + 1357 + 2714 + 5428 = 10744$
A	12285	$3 \times 3 \times 3 \times 5 \times 7 \times 13$	$1 + 3 + 9 + 27 + 5 + 15 + 45 + 135 + 91 + 273 + 819 + 2457 + 455 + 1365 + 4095 = 14595$
B	14595	$3 \times 5 \times 7 \times 139$	$1 + 3 + 5 + 7 + 15 + 21 + 35 + 105 + 139 + 417 + 695 + 973 + 2085 + 2919 + 4865 = 12285$
A	17296	$2 \times 2 \times 2 \times 2 \times 23 \times 47$	$1 + 2 + 4 + 8 + 16 + 23 + 46 + 47 + 92 + 94 + 184 + 188 + 368 + 376 + 752 + 1081 + 2162 + 4324 + 8648 = 18416$
B	18416	$2 \times 2 \times 2 \times 2 \times 1151$	$1 + 2 + 4 + 8 + 16 + 1151 + 2302 + 4604 + 9208 = 17296$
A	63020	$2 \times 2 \times 5 \times 23 \times 137$	$1 + 2 + 4 + 5 + 10 + 20 + 23 + 46 + 92 + 115 + 137 + 230 + 274 + 460 + 548 + 685 + 1370 + 2740 + 3151 + 6302 + 12604 + 15755 + 31510 = 76084$
B	76084	$2 \times 2 \times 23 \times 827$	$1 + 2 + 4 + 23 + 46 + 92 + 827 + 1654 + 3308 + 19021 + 38042 = 63020$
A	66928	$2 \times 2 \times 2 \times 2 \times 47 \times 89$	$1 + 2 + 4 + 8 + 16 + 47 + 89 + 94 + 178 + 188 + 356 + 376 + 712 + 752 + 1424 + 4183 + 8366 + 16732 + 33464 = 66992$
B	66992	$2 \times 2 \times 2 \times 2 \times 53 \times 79$	$1 + 2 + 4 + 8 + 16 + 53 + 79 + 106 + 158 + 212 + 316 + 424 + 632 + 848 + 1264 + 4187 + 8374 + 16748 + 33496 = 66928$

续　表

亲和数		因数之积	因数之和
A	67095	$3 \times 3 \times 3 \times 5 \times 7 \times 71$	$1 + 3 + 5 + 7 + 9 + 15 + 21 + 27 + 35 + 45 + 63 + 71 + 105 + 135 + 189 + 213 + 315 + 355 + 497 + 639 + 945 + 1065 + 1494 + 1917 + 2485 + 3195 + 4473 + 7455 + 9585 + 13419 + 22365 = 71145$
B	71145	$3 \times 3 \times 3 \times 5 \times 17 \times 31$	$1 + 3 + 5 + 9 + 15 + 17 + 27 + 31 + 45 + 51 + 85 + 93 + 135 + 153 + 155 + 255 + 179 + 459 + 465 + 527 + 765 + 837 + 1395 + 1581 + 2295 + 2635 + 4185 + 4743 + 7905 + 14229 + 23715 = 67095$
A	69615	$3 \times 3 \times 5 \times 7 \times 13 \times 17$	$1 + 3 + 5 + 7 + 9 + 13 + 15 + 17 + 21 + 35 + 39 + 45 + 51 + 63 + 65 + 85 + 91 + 105 + 117 + 119 + 153 + 195 + 221 + 255 + 273 + 315 + 357 + 455 + 585 + 595 + 663 + 765 + 819 + 1071 + 1105 + 1365 + 1547 + 1785 + 1989 + 3315 + 4095 + 4641 + 5355 + 7735 + 9945 + 13923 + 23205 = 87633$
B	87633	$3 \times 3 \times 7 \times 13 \times 107$	$1 + 3 + 7 + 9 + 13 + 21 + 39 + 63 + 91 + 107 + 117 + 273 + 321 + 749 + 819 + 963 + 1391 + 2247 + 4173 + 6741 + 9737 + 12519 + 29211 = 69615$
A	79750	$2 \times 5 \times 5 \times 5 \times 11 \times 29$	$1 + 2 + 5 + 10 + 11 + 22 + 25 + 50 + 55 + 58 + 110 + 125 + 145 + 250 + 275 + 290 + 319 + 550 + 638 + 725 + 1375 + 1450 + 1595 + 2750 + 3190 + 3625 + 7250 + 7975 + 15950 + 39875 = 88730$
B	88730	$2 \times 5 \times 19 \times 467$	$1 + 2 + 5 + 10 + 19 + 38 + 95 + 190 + 467 + 934 + 2335 + 4670 + 8873 + 17746 + 44365 = 79750$
A	100485	$3 \times 3 \times 5 \times 7 \times 11 \times 29$	$1 + 3 + 5 + 7 + 9 + 11 + 15 + 21 + 29 + 33 + 35 + 45 + 55 + 63 + 77 + 87 + 99 + 105 + 145 + 165 + 203 + 231 + 261 + 315 + 319 + 385 + 435 + 495 + 609 + 693 + 957 + 1015 + 1155 + 1305 + 1595 + 1827 + 2233 + 2871 + 3045 + 3465 + 4785 + 6699 + 9135 + 11165 + 14355 + 20097 + 33495 = 124155$
B	124155	$3 \times 3 \times 5 \times 31 \times 89$	$1 + 3 + 5 + 9 + 15 + 31 + 45 + 89 + 93 + 155 + 267 + 279 + 445 + 465 + 801 + 1335 + 1395 + 2759 + 4005 + 8277 + 13795 + 24831 + 41385 = 100485$
A	122368	$2 \times 2 \times 2 \times 2 \times 2 \times 2 \times 2 \times 2 \times 2 \times 239$	$1 + 2 + 4 + 8 + 16 + 32 + 64 + 128 + 239 + 256 + 478 + 512 + 956 + 1912 + 3824 + 7648 + 15296 + 30592 + 61184 = 123152$
B	123152	$2 \times 2 \times 2 \times 2 \times 43 \times 179$	$1 + 2 + 4 + 8 + 16 + 43 + 86 + 172 + 179 + 344 + 358 + 688 + 716 + 1432 + 2864 + 7697 + 15394 + 30788 + 61576 = 122368$

续 表

	亲和数	因数之积	因数之和
A	141664	$2 \times 2 \times 2 \times 2 \times 2 \times 19 \times 233$	$1 + 2 + 4 + 8 + 16 + 19 + 32 + 38 + 76 + 152 + 233 + 304 + 466 + 608 + 932 + 1864 + 3728 + 4427 + 7456 + 8854 + 17708 + 35416 + 70832 = 153176$
B	153176	$2 \times 2 \times 2 \times 41 \times 467$	$1 + 2 + 4 + 8 + 41 + 82 + 164 + 328 + 467 + 934 + 1868 + 3736 + 19147 + 38294 + 76588 = 141664$
A	142310	$2 \times 5 \times 7 \times 19 \times 107$	$1 + 2 + 5 + 7 + 10 + 14 + 19 + 35 + 38 + 70 + 95 + 107 + 133 + 190 + 214 + 266 + 535 + 665 + 749 + 1070 + 1330 + 1498 + 2033 + 3745 + 4066 + 7490 + 10165 + 14231 + 20330 + 27462 + 71155 = 168730$
B	168730	$2 \times 5 \times 47 \times 359$	$1 + 2 + 5 + 10 + 47 + 94 + 235 + 359 + 470 + 718 + 1795 + 3590 + 16873 + 33746 + 84365 = 142310$
A	171856	$2 \times 2 \times 2 \times 2 \times 23 \times 467$	$1 + 2 + 4 + 8 + 16 + 23 + 46 + 92 + 184 + 368 + 467 + 934 + 1868 + 3736 + 7472 + 10741 + 21482 + 42964 + 85928 = 176336$
B	176336	$2 \times 2 \times 2 \times 2 \times 107 \times 103$	$1 + 2 + 4 + 8 + 16 + 103 + 107 + 206 + 214 + 412 + 428 + 824 + 856 + 1648 + 1712 + 11021 + 22042 + 44084 + 88168 = 171856$
A	176272	$2 \times 2 \times 2 \times 2 \times 23 \times 479$	$1 + 2 + 4 + 16 + 23 + 46 + 92 + 184 + 368 + 479 + 958 + 1916 + 3832 + 7664 + 11017 + 22034 + 44068 + 88136 = 180848$
B	180848	$2 \times 2 \times 2 \times 2 \times 89 \times 127$	$1 + 2 + 4 + 8 + 16 + 89 + 127 + 178 + 254 + 356 + 508 + 712 + 1016 + 1424 + 2032 + 11303 + 22606 + 45212 + 90424 = 176272$
A	185368	$2 \times 2 \times 2 \times 17 \times 29 \times 47$	$1 + 2 + 4 + 8 + 17 + 29 + 34 + 47 + 58 + 68 + 94 + 116 + 136 + 188 + 232 + 376 + 493 + 799 + 986 + 1363 + 1598 + 1972 + 2726 + 3196 + 3944 + 5452 + 6392 + 10904 + 23171 + 46342 + 92684 = 203432$
B	203432	$2 \times 2 \times 2 \times 59 \times 431$	$1 + 2 + 4 + 8 + 59 + 118 + 236 + 431 + 472 + 862 + 1724 + 3448 + 25429 + 50858 + 101716 = 185368$
A	196724	$2 \times 2 \times 11 \times 17 \times 263$	$1 + 2 + 4 + 11 + 17 + 22 + 34 + 44 + 68 + 187 + 263 + 374 + 526 + 748 + 1052 + 2893 + 4471 + 5786 + 8942 + 11572 + 17884 + 49181 + 98362 = 202444$
B	202444	$2 \times 2 \times 11 \times 43 \times 107$	$1 + 2 + 4 + 11 + 22 + 43 + 44 + 86 + 107 + 172 + 214 + 428 + 473 + 946 + 1177 + 1892 + 2354 + 4601 + 4708 + 9202 + 18404 + 50611 + 101222 = 196724$

续 表

亲和数		因数之积	因数之和
A	280540	$2 \times 2 \times 5 \times 13 \times 13 \times 83$	$1 + 2 + 4 + 5 + 10 + 13 + 20 + 26 + 52 + 65 + 83 + 130 + 166 + 169 + 260 + 332 + 338 + 415 + 676 + 830 + 845 + 1079 + 1660 + 1690 + 2158 + 3380 + 4316 + 5395 + 10790 + 14027 + 21580 + 28054 + 56108 + 70135 + 140270 = 365084$
B	365084	$2 \times 2 \times 107 \times 853$	$1 + 2 + 4 + 107 + 214 + 428 + 853 + 1706 + 3412 + 91271 + 182542 = 280540$
A	308620	$2 \times 2 \times 5 \times 13 \times 1187$	$1 + 2 + 4 + 5 + 10 + 13 + 20 + 26 + 52 + 65 + 130 + 260 + 1187 + 2374 + 4748 + 5935 + 11870 + 15431 + 23740 + 30862 + 61724 + 77155 + 154310 = 389924$
B	389924	$2 \times 2 \times 43 \times 2267$	$1 + 2 + 4 + 43 + 86 + 172 + 2267 + 4534 + 9068 + 97481 + 194962 = 308620$
A	319550	$2 \times 5 \times 5 \times 7 \times 11 \times 83$	$1 + 2 + 5 + 7 + 10 + 11 + 14 + 22 + 25 + 35 + 50 + 55 + 70 + 77 + 83 + 110 + 154 + 166 + 175 + 275 + 350 + 385 + 415 + 550 + 581 + 770 + 830 + 913 + 1162 + 1826 + 1925 + 2075 + 2905 + 3850 + 4150 + 4565 + 5810 + 6391 + 9130 + 12782 + 14525 + 22825 + 29050 + 31955 + 45650 + 63910 + 159775 = 430402$
B	430402	$2 \times 7 \times 71 \times 433$	$1 + 2 + 7 + 14 + 71 + 142 + 433 + 497 + 866 + 994 + 3631 + 6062 + 30743 + 61486 = 319550$

附件三　相关研究论文

Zhao and Wang *Journal of Inequalities and Applications* (2018) 2018:160
https://doi.org/10.1186/s13660-018-1753-4

 Journal of Inequalities and Applications
a SpringerOpen Journal

RESEARCH　　　　　　　　　　　　　　　　　　　　　　　**Open Access**

CrossMark

One kind hybrid character sums and their upper bound estimates

Jianhong Zhao[1] and Xiao Wang[2*]

*Correspondence:
wangxiao_0606@stumail.nwu.edu.cn
[2] School of Mathematics, Northwest
University, Xi'an, P.R. China
Full list of author information is
available at the end of the article

Abstract

The main purpose of this paper is applying the analysis method, the properties of Lucas polynomials and Gauss sums to study the estimation problems of some kind hybrid character sums. In the end, we obtain several sharp upper bound estimates for them. As some applications, we prove some new and interesting combinatorial identities.

MSC: 11T24

Keywords: The hybrid character sums; Lucas polynomials; Upper bound estimate; Gauss sums; Combinatorial identity

1 Introduction

As usual, let $q \geq 3$ be an integer, χ denotes any Dirichlet character mod q. The classical Gauss sum $\tau(\chi)$ is defined by

$$\tau(\chi) = \sum_{a=1}^{q} \chi(a) e\left(\frac{a}{q}\right),$$

where $e(y) = e^{2\pi i y}$.

We know that this sum plays a very important role in analytic number theory; plenty of number theory problems (such as Dirichlet L-functions and distribution of primes) are closely related to it. Concerning the various elementary properties of $\tau(\chi)$, some authors also studied it and obtained a series of interesting results, some conclusions can be found in Refs. [1] and [2]. For example, if χ is a primitive character mod q, then, for any integer n, one has $|\tau(\chi)| = \sqrt{q}$ and the identity

$$\sum_{a=1}^{q} \chi(a) e\left(\frac{na}{q}\right) = \overline{\chi}(n) \tau(\chi).$$

From the Euler formula we know that $e(x) = \cos(2\pi x) + i \sin(2\pi x)$. So scholars will naturally ask, for any positive integer n, whether there exists a similar estimate for

$$\left| \sum_{a=1}^{q} \chi(a) \cos^{n}\left(\frac{2\pi a}{q}\right) \right| \quad \text{and} \quad \left| \sum_{a=1}^{q} \chi(a) \sin^{n}\left(\frac{2\pi a}{q}\right) \right|. \tag{1}$$

Zhao and Wang *Journal of Inequalities and Applications* (2018) 2018:160

The estimates or calculations for (1) are significant, because they are closely related to the famous Gauss sums, so they have many interesting applications in analytic number theory, especially various estimates for hybrid character sums and generalized Klooster-man sums.

As far as we know, it seems that nobody has studied the estimate for (1), at least we have not seen related papers before. In this paper, we shall use the analytic method, the properties of Lucas polynomials and Gauss sums to do research on these problems, and obtain some sharp upper bound estimates for them. Our main idea is to put the nth power $\sin^n(x)$ (or $\cos^n(x)$) into a combination of $\sin(kx)$ (or $\cos(kx)$), where $1 \leq k \leq n$. Then using the estimate for classical Gauss sums we give some sharp upper bound estimates for (1). As some applications of our main results, we also give some new and interesting combinatorial identities.

2 Main results and discussion

The main results in this paper are detailed in the following.

Theorem 1 *If q is an integer with $q > 2$, then, for any positive integer n and primitive character χ mod q, we have the estimate*

(a) $\left| \sum_{a=1}^{q-1} \chi(a) \cos^{2n}\left(\frac{2\pi a}{q}\right) \right| \leq \left(1 - \frac{\binom{2n}{n}}{4^n}\right)\sqrt{q};$

(b) $\left| \sum_{a=1}^{q-1} \chi(a) \cos^{2n-1}\left(\frac{2\pi a}{q}\right) \right| \leq \sqrt{q},$

where $\binom{m}{n} = \frac{m!}{n!(m-n)!}$.

Theorem 2 *If q is an integer with $q > 2$, then, for any positive integer n and primitive character χ mod q, we have the estimate*

(c) $\left| \sum_{a=1}^{q-1} \chi(a) \sin^{2n}\left(\frac{2\pi a}{q}\right) \right| \leq \left(1 - \frac{\binom{2n}{n}}{4^n}\right)\sqrt{q};$

(d) $\left| \sum_{a=1}^{q-1} \chi(a) \sin^{2n-1}\left(\frac{2\pi a}{q}\right) \right| \leq \sqrt{q}.$

From the method of proving Theorem 1 and the orthogonality of characters mod q we can immediately deduce the following combinatorial identity.

Corollary *If n is any positive integer, then we have the identity*

$$\binom{4n}{2n} + \binom{2n}{n}^2 = 2\sum_{k=0}^{n}\binom{2n}{k}^2.$$

Notes Note that $2\cos^2 x = \cos(2x) + 1$ and $|\tau(\chi)| = \sqrt{q}$; we have

$$\left| \sum_{a=1}^{q-1} \chi(a) \cos^2\left(\frac{2\pi a}{q}\right) \right| = \frac{1}{2}\left| \sum_{a=1}^{q-1} \chi(a)\cos\left(\frac{4\pi a}{q}\right) + \sum_{a=1}^{q-1}\chi(a) \right|$$

$$= \frac{1}{4}\left| \sum_{a=1}^{q-1} \chi(a)e\left(\frac{2a}{q}\right) + \sum_{a=1}^{q-1}\chi(a)e\left(\frac{-2a}{q}\right) \right|$$

$$= \frac{1}{4}\sqrt{q}\left| \overline{\chi}(2) + \overline{\chi}(-2) \right|$$

$$= \begin{cases} \frac{1}{2} \cdot \sqrt{q}, & \text{if } \chi(-1) = 1, \\ 0, & \text{if } \chi(-1) = -1. \end{cases}$$

Therefore, if $n = 1$ and χ is an even primitive character mod q, then the equal sign in our theorems holds. So the estimates in our Theorem 1 and Theorem 2 are the best.

3 Several simple lemmas

In order to prove our main results, we first introduce the Fibonacci polynomials and Lucas polynomials as follows.

For integer $n \geq 0$, the famous Fibonacci polynomials $\{F_n(x)\}$ and Lucas polynomials $\{L_n(x)\}$ are defined by $F_0(x) = 0$, $F_1(x) = 1$, $L_0(x) = 2$, $L_1(x) = x$ and $F_{n+2}(x) = xF_{n+1}(x) + F_n(x)$, $L_{n+2}(x) = xL_{n+1}(x) + L_n(x)$ for all $n \geq 0$. In fact the general terms of $F_n(x)$ and $L_n(x)$ are given by

$$F_n(x) = \frac{1}{\sqrt{x^2+4}}\left[\left(\frac{x+\sqrt{x^2+4}}{2} \right)^n - \left(\frac{x-\sqrt{x^2+4}}{2} \right)^n \right]$$

and

$$L_n(x) = \left(\frac{x+\sqrt{x^2+4}}{2} \right)^n + \left(\frac{x-\sqrt{x^2+4}}{2} \right)^n. \tag{2}$$

It is easy to obtain the identities

$$F_{n+1}(x) = \sum_{k=0}^{[\frac{n}{2}]} \binom{n-k}{k} x^{n-2k} \quad \text{and} \quad L_n(x) = \sum_{k=0}^{[\frac{n}{2}]} \frac{n}{n-k}\binom{n-k}{k} x^{n-2k},$$

where $\binom{m}{n} = \frac{m!}{n!(m-n)!}$, and $[x]$ denotes the greatest integer $\leq x$.

Taking $x = 1$, then $\{F_n(x)\}$ becomes the Fibonacci sequences $\{F_n\}$, and $\{L_n(x)\}$ becomes the Lucas sequences $\{L_n\}$. If we take $x = 2$, then $F_n(2) = P_n$, the nth Pell numbers, $P_0 = 0$, $P_1 = 1$, and $P_{n+2} = 2P_{n+1} + P_n$ for all $n \geq 0$.

Since these sequences (or polynomials) occupy a more crucial position in the theory and application of mathematics, many scholars have studied their various elementary properties and obtained a series of important results. See Refs. [3–10]. Here we give some new properties of Lucas polynomials.

Lemma 1 *If k is a non-negative integer, then we have the identities*

$$L_{2k}(2i\sin\theta) = 2 \cdot \cos(2k\theta) \quad \text{and} \quad L_{2k}(2i\cos\theta) = (-1)^k \cdot 2 \cdot \cos(2k\theta);$$

Zhao and Wang *Journal of Inequalities and Applications* (2018) 2018:160

$$L_{2k+1}(2i\sin\theta) = 2i \cdot \sin\big((2k+1)\theta\big),$$

and

$$L_{2k+1}(2i\cos\theta) = (-1)^k \cdot 2i \cdot \cos\big((2k+1)\theta\big),$$

where i is the imaginary unit. That is to say, $i^2 = -1$.

Proof Taking $x = 2i\sin\theta$ in (2), and note that $x^2 + 4 = 4 - 4\sin^2\theta = 4\cos^2\theta$, from the Euler formula we have

$$
\begin{aligned}
L_{2k}(2i\sin\theta) &= \left(\frac{2i\sin\theta + \sqrt{4\cos^2\theta}}{2}\right)^{2k} + \left(\frac{2i\sin\theta - \sqrt{4\cos^2\theta}}{2}\right)^{2k} \\
&= (i\sin\theta + \cos\theta)^{2k} + (i\sin\theta - \cos\theta)^{2k} \\
&= (\cos\theta + i\sin\theta)^{2k} + (\cos\theta - i\sin\theta)^{2k} \\
&= \cos(2k\theta) + i\sin(2k\theta) + \cos(2k\theta) - i\sin(2k\theta) = 2 \cdot \cos(2k\theta)
\end{aligned}
$$

and

$$
\begin{aligned}
L_{2k}(2i\cos\theta) &= \left(\frac{2i\cos\theta + \sqrt{4\sin^2\theta}}{2}\right)^{2k} + \left(\frac{2i\cos\theta - \sqrt{4\sin^2\theta}}{2}\right)^{2k} \\
&= (i\cos\theta + \sin\theta)^{2k} + (i\cos\theta - \sin\theta)^{2k} \\
&= (-1)^k(\cos\theta - i\sin\theta)^{2k} + (-1)^k(\cos\theta + i\sin\theta)^{2k} \\
&= (-1)^k \cdot 2 \cdot \cos(2k\theta).
\end{aligned}
$$

This proves the first and second formulas of Lemma 1.

Similarly, we also have the identities

$$
\begin{aligned}
L_{2k+1}(2i\sin\theta) &= \left(\frac{2i\sin\theta + \sqrt{4\cos^2\theta}}{2}\right)^{2k+1} + \left(\frac{2i\sin\theta - \sqrt{4\cos^2\theta}}{2}\right)^{2k+1} \\
&= (i\sin\theta + \cos\theta)^{2k+1} + (i\sin\theta - \cos\theta)^{2k+1} \\
&= (\cos\theta + i\sin\theta)^{2k+1} - (\cos\theta - i\sin\theta)^{2k+1} \\
&= \cos\big((2k+1)\theta\big) + i\sin\big((2k+1)\theta\big) - \cos\big((2k+1)\theta\big) + i\sin\big((2k+1)\theta\big) \\
&= 2i \cdot \sin\big((2k+1)\theta\big)
\end{aligned}
$$

and

$$
\begin{aligned}
L_{2k+1}(2i\cos\theta) &= \left(\frac{2i\cos\theta + \sqrt{4\sin^2\theta}}{2}\right)^{2k+1} + \left(\frac{2i\cos\theta - \sqrt{4\sin^2\theta}}{2}\right)^{2k+1} \\
&= (i\cos\theta + \sin\theta)^{2k+1} + (i\cos\theta - \sin\theta)^{2k+1} \\
&= i^{2k+1}(\cos\theta - i\sin\theta)^{2k+1} + i^{2k+1}(\cos\theta + i\sin\theta)^{2k+1} \\
&= (-1)^k \cdot 2i \cdot \cos\big((2k+1)\theta\big).
\end{aligned}
$$

This completes the proof of Lemma 1.

Lemma 2 *If n is any non-negative integer, then we have the identities*

$$x^{2n} = \frac{(-1)^n}{2} \binom{2n}{n} \cdot L_0(x) + \sum_{k=1}^{n} (-1)^{n-k} \binom{2n}{n-k} \cdot L_{2k}(x)$$

and

$$x^{2n+1} = \sum_{k=0}^{n} (-1)^{n-k} \binom{2n+1}{n-k} \cdot L_{2k+1}(x).$$

Proof From the definition of $L_n(x)$ we know that $L_{2k}(x)$ is an even function. So we may suppose that

$$x^{2n} = \sum_{k=0}^{n} a_k \cdot L_{2k}(x). \tag{3}$$

Taking $x = 2i\cos\theta$ in (3) and applying Lemma 1, we have

$$(-1)^n 4^n \cos^{2n}\theta = \sum_{k=0}^{n} a_k \cdot L_{2k}(2i\cos\theta) = 2\sum_{k=0}^{n} a_k \cdot (-1)^k \cos(2k\theta). \tag{4}$$

Note that the identities

$$\int_0^\pi 2\cos(m\theta)\cos(n\theta)\,d\theta = \begin{cases} \pi, & \text{if } m = n \neq 0, \\ 0, & \text{if } m \neq n, \\ 2\pi, & \text{if } m = n = 0, \end{cases} \tag{5}$$

and

$$\int_0^\pi \cos^{2n}(\theta)\cos(2k\theta)\,d\theta = \pi \cdot \frac{(2n)!}{(2n-2k)!!(2n+2k)!!} = \frac{\pi}{4^n} \cdot \binom{2n}{n-k},$$

from (4) we have

$$a_k \cdot (-1)^k \pi = (-1)^n 4^n \int_0^\pi \cos^{2n}(\theta)\cos(2k\theta)\,d\theta = (-1)^n \pi \cdot \binom{2n}{n-k}$$

or

$$a_0 = \frac{(-1)^n}{2} \cdot \binom{2n}{n} \quad \text{and} \quad a_k = (-1)^{n-k} \cdot \binom{2n}{n-k}, \quad 1 \le k \le n. \tag{6}$$

Combining identities (3) and (6) we may immediately deduce the first formula of Lemma 2.

Similarly, since $L_{2k+1}(x)$ is an odd function, we can suppose that

$$x^{2n+1} = \sum_{k=0}^{n} b_k \cdot L_{2k+1}(x). \tag{7}$$

Zhao and Wang *Journal of Inequalities and Applications* （2018）2018:160

Taking $x = 2i\cos\theta$ in (7), then applying Lemma 1 we have

$$(-1)^n 4^n \cos^{2n+1}\theta = \sum_{k=0}^{n} b_k \cdot (-1)^k \cos((2k+1)\theta). \tag{8}$$

From (5) and (8) we may immediately deduce that

$$b_k = \frac{2(-1)^{n-k}}{\pi} \cdot 4^n \cdot \int_0^\pi \cos^{2n+1}(\theta)\cos((2k+1)\theta)\,d\theta = (-1)^{n-k}\binom{2n+1}{n-k}. \tag{9}$$

Now the second identity of Lemma 2 follows from (7) and (9).

Lemma 3 *If n is a positive integer, then we have the identity*

$$\sum_{k=1}^{n}\binom{2n}{n-k} = \frac{1}{2}\cdot\left(4^n - \binom{2n}{n}\right) \quad and \quad \sum_{k=0}^{n}\binom{2n+1}{n-k} = 4^n.$$

Proof First applying the binomial theorem we have the identity

$$\sum_{k=0}^{2n}\binom{2n}{2n-k} = \sum_{k=0}^{2n}\binom{2n}{k} = (1+1)^{2n} = 4^n. \tag{10}$$

On the other hand, we also have

$$\begin{aligned}
\sum_{k=0}^{2n}\binom{2n}{k} = (1+1)^{2n} &= \sum_{k=0}^{n}\binom{2n}{k} + \sum_{k=n+1}^{2n}\binom{2n}{k} \\
&= \sum_{k=0}^{n}\binom{2n}{k} + \sum_{k=1}^{n}\binom{2n}{n-k} \\
&= 2\sum_{k=1}^{n}\binom{2n}{n-k} + \binom{2n}{n}.
\end{aligned} \tag{11}$$

Combining (10) and (11) we can deduce

$$\sum_{k=1}^{n}\binom{2n}{n-k} = \frac{1}{2}\cdot\left(4^n - \binom{2n}{n}\right).$$

This proves the first formula of Lemma 3.

Similarly, we can deduce the second formula of Lemma 3.

4 Proofs of the theorems

In this section, we complete the proofs of our main results. First we prove Theorem 1. Let $q \geq 3$ be an integer; χ is any even primitive character mod q. Then taking $x = 2i\cos(\frac{2\pi a}{q})$ in the first formula of Lemma 2, multiplying both sides by $\chi(a)$ and summing over all $1 \leq a \leq q-1$, and noting that

$$\sum_{a=1}^{q-1}\chi(a) = 0 \quad and \quad \sum_{a=1}^{q-1}\chi(a)e\left(\frac{\pm 2ka}{q}\right) = \overline{\chi}(\pm 2k)\tau(\chi),$$

we have

$$(-1)^n 4^n \sum_{a=1}^{q-1} \chi(a) \cos^{2n}\left(\frac{2\pi a}{q}\right)$$

$$= (-1)^n \binom{2n}{n} \sum_{a=1}^{q-1} \chi(a) + 2 \sum_{k=1}^{n} (-1)^{n-k} \binom{2n}{n-k} \sum_{a=1}^{q-1} (-1)^k \chi(a) \cos\left(\frac{4k\pi a}{q}\right)$$

$$= (-1)^n \sum_{k=1}^{n} \binom{2n}{n-k} \sum_{a=1}^{q-1} \chi(a) \left[e\left(\frac{2ka}{q}\right) + e\left(\frac{-2ka}{q}\right) \right]$$

$$= (-1)^n \tau(\chi) \sum_{k=1}^{n} \binom{2n}{n-k} \left(\overline{\chi}(2k) + \overline{\chi}(-2k)\right)$$

$$= 2(-1)^n \overline{\chi}(2) \tau(\chi) \sum_{k=1}^{n} \binom{2n}{n-k} \overline{\chi}(k). \tag{12}$$

Note that $|\tau(\chi)| = \sqrt{q}$, from (12) and Lemma 3 we have the estimate

$$\left| \sum_{a=1}^{q-1} \chi(a) \cos^{2n}\left(\frac{2\pi a}{q}\right) \right| = \frac{2\sqrt{q}}{4^n} \cdot \left| \sum_{k=1}^{n} \binom{2n}{n-k} \overline{\chi}(k) \right|$$

$$\leq \frac{2\sqrt{q}}{4^n} \cdot \sum_{k=1}^{n} \binom{2n}{n-k}$$

$$= \frac{2\sqrt{q}}{4^n} \cdot \frac{4^n - \binom{2n}{n}}{2}$$

$$= \left(1 - \frac{\binom{2n}{n}}{4^n}\right) \cdot \sqrt{q}. \tag{13}$$

Similarly, taking $x = 2i\cos(\frac{2\pi a}{q})$ in the second formula of Lemma 2, then applying Lemma 1 we also have

$$(-1)^n 4^n \sum_{a=1}^{q-1} \chi(a) \cos^{2n+1}\left(\frac{2\pi a}{q}\right)$$

$$= \sum_{k=0}^{n} (-1)^{n-k} \binom{2n+1}{n-k} \sum_{a=1}^{q-1} (-1)^k \chi(a) \cos\left(\frac{2(2k+1)\pi a}{q}\right)$$

$$= \frac{(-1)^n}{2} \cdot \sum_{k=0}^{n} \binom{2n+1}{n-k} \sum_{a=1}^{q-1} \chi(a) \left[e\left(\frac{(2k+1)a}{q}\right) + e\left(\frac{-(2k+1)a}{q}\right) \right]$$

$$= \frac{(-1)^n \tau(\chi)}{2} \cdot \sum_{k=0}^{n} \binom{2n+1}{n-k} \left(\overline{\chi}(2k+1) + \overline{\chi}(-2k-1)\right)$$

$$= (-1)^n \tau(\chi) \sum_{k=0}^{n} \binom{2n+1}{n-k} \overline{\chi}(2k+1). \tag{14}$$

From (14) and Lemma 3 we may immediately deduce the estimate

$$\left| \sum_{a=1}^{q-1} \chi(a) \cos^{2n+1}\left(\frac{2\pi a}{q}\right) \right| = \frac{\sqrt{q}}{4^n} \cdot \left| \sum_{k=0}^{n} \binom{2n+1}{n-k} \overline{\chi}(2k+1) \right|$$

$$\leq \frac{\sqrt{q}}{4^n} \cdot \sum_{k=0}^{n} \binom{2n+1}{n-k} = \sqrt{q}. \tag{15}$$

If χ is an odd primitive character mod q, then it is very easy to prove that

$$\left| \sum_{a=1}^{q-1} \chi(a) \cos^{2n+1}\left(\frac{2\pi a}{q}\right) \right| = \left| \sum_{a=1}^{q-1} \chi(a) \cos^{2n}\left(\frac{2\pi a}{q}\right) \right| = 0. \tag{16}$$

Combining (13), (15), and (16) we may immediately deduce Theorem 1.

Using a very similar method to proving Theorem 1 we can also deduce the estimates in Theorem 2. So it is not repeated here.

Now we prove our corollary. If p is a prime large enough, then, for any fixed integer $n \geq 1$, from [11] we have the identity

$$\sum_{a=0}^{p-1} \cos^{2n}\left(\frac{2\pi a}{p}\right) = \frac{p}{4^n} \cdot \binom{2n}{n}. \tag{17}$$

From the orthogonality of characters mod p and (17) we have

$$\sum_{\chi \bmod p} \left| \sum_{a=1}^{q-1} \chi(a) \cos^{2n}\left(\frac{2\pi a}{q}\right) \right|^2 = (p-1) \sum_{a=1}^{p-1} \cos^{4n}\left(\frac{2\pi a}{p}\right)$$

$$= \frac{p(p-1)}{4^{2n}} \cdot \binom{4n}{2n} - (p-1). \tag{18}$$

On the other hand, from (13), (17), and Lemma 3 we also have

$$\sum_{\chi \bmod p} \left| \sum_{a=1}^{p-1} \chi(a) \cos^{2n}\left(\frac{2\pi a}{p}\right) \right|^2$$

$$= \left| \sum_{a=1}^{p-1} \cos^{2n}\left(\frac{2\pi a}{p}\right) \right|^2 + \sum_{\substack{\chi \bmod p \\ \chi(-1)=1, \chi \neq \chi_0}} \left| \sum_{a=1}^{p-1} \chi(a) \cos^{2n}\left(\frac{2\pi a}{p}\right) \right|^2$$

$$= \left(\frac{p}{4^n} \cdot \binom{2n}{n} - 1 \right)^2 + \frac{4p}{4^{2n}} \sum_{\substack{\chi \bmod p \\ \chi(-1)=1, \chi \neq \chi_0}} \left| \sum_{k=1}^{n} \binom{2n}{n-k} \overline{\chi}(k) \right|^2$$

$$= \left(\frac{p}{4^n} \cdot \binom{2n}{n} - 1 \right)^2 + \frac{4p}{4^{2n}} \cdot \frac{p-1}{2} \sum_{k=1}^{n} \binom{2n}{n-k}^2 - \frac{4p}{4^{2n}} \cdot \left(\sum_{k=1}^{n} \binom{2n}{n-k} \right)^2$$

$$= \left(\frac{p}{4^n} \cdot \binom{2n}{n} - 1 \right)^2 + \frac{4p}{4^{2n}} \cdot \frac{p-1}{2} \sum_{k=1}^{n} \binom{2n}{n-k}^2 - \frac{p}{4^{2n}} \left(4^n - \binom{2n}{n} \right)^2. \tag{19}$$

Combining (18) and (19) we have the identity

$$\frac{p(p-1)}{4^{2n}} \cdot \binom{4n}{2n} - (p-1) = \left(\frac{p}{4^n} \cdot \binom{2n}{n} - 1 \right)^2 + \frac{4p}{4^{2n}} \cdot \frac{p-1}{2} \sum_{k=1}^{n} \binom{2n}{n-k}^2$$
$$- \frac{p}{4^{2n}} \left(4^n - \binom{2n}{n} \right)^2. \tag{20}$$

From (20) we may immediately deduce

$$\binom{4n}{2n} = \binom{2n}{n}^2 + 2 \sum_{k=1}^{n} \binom{2n}{n-k}^2 = 2 \sum_{k=0}^{n} \binom{2n}{k}^2 - \binom{2n}{n}^2.$$

This completes the proofs of our all results.

Acknowledgements
The authors would like to thank the referees for very helpful and detailed comments, which have significantly improved the presentation of this paper.

Funding
This work is supported by the N.S.F. China (Grant No. 11771351).

Competing interests
The authors declare that there are no conflicts of interest regarding the publication of this paper.

Authors' contributions
All authors have equally contributed to this work. All authors read and approved the final manuscript.

Author details
[1]Department of Teachers Education, Lijiang Teachers College, Lijiang, P.R. China. [2]School of Mathematics, Northwest University, Xi'an, P.R. China.

Publisher's Note
Springer Nature remains neutral with regard to jurisdictional claims in published maps and institutional affiliations.

Received: 6 October 2017　Accepted: 23 May 2018　Published online: 04 July 2018

References
1. Apostol, T.M.: Introduction to Analytic Number Theory. Springer, New York (1976)
2. Pan, C.D., Pan, C.B.: Goldbach Conjecture. Science Press, Beijing (2011)
3. Clemente, C.: Identities and generating functions on Chebyshev polynomials. Georgian Math. J. **19**, 427–440 (2012)
4. Lee, C.L., Wong, K.B.: On Chebyshev's polynomials and certain combinatorial identities. Bull. Malays. Math. Sci. Soc. **34**, 279–286 (2011)
5. Doha, E., Bhrawy, A., Ezz-Eldien, S.: Numerical approximations for fractional diffusion equations via a Chebyshev spectral-tau method. Cent. Eur. J. Phys. **11**, 1494–1503 (2013)
6. Ma, R., Zhang, W.P.: Several identities involving the Fibonacci numbers and Lucas numbers. Fibonacci Q. **45**, 164–170 (2007)
7. Ma, Y.K., Lv, X.X.: Several identities involving the reciprocal sums of Chebyshev polynomials. Math. Probl. Eng. **2017**, Article ID 4194579 (2017)
8. Wang, T.T., Zhang, H.: Some identities involving the derivative of the first kind Chebyshev polynomials. Math. Probl. Eng. **2015**, Article ID 146313 (2015)
9. Yi, Y., Zhang, W.P.: Some identities involving the Fibonacci polynomials. Fibonacci Q. **40**, 314–318 (2002)
10. Zhang, W.P., Wang, T.T.: Two identities involving the integral of the first kind Chebyshev polynomials. Bull. Math. Soc. Sci. Math. Roum. **108**, 91–98 (2017)
11. Fonseca, M.C., Glasser, M.L., Kowalenko, V.: Basic trigonometric power sums with applications. Ramanujan J. **42**, 401–428 (2017)

 *symmetry*

Article

Some Symmetric Identities Involving Fubini Polynomials and Euler Numbers

Zhao Jianhong [1] and Chen Zhuoyu [2,*]

[1] Department of Teachers Education, Lijiang Teachers College, Lijiang 674199, China; zjh3004@163.com
[2] School of Mathematics, Northwest University, Xi'an 710127, China
[*] Correspondence: chenzymath@163.com

 check for updates

Received: 30 June 2018; Accepted: 24 July 2018; Published: 1 August 2018

Abstract: The aim of this paper is to use elementary methods and the recursive properties of a special sequence to study the computational problem of one kind symmetric sums involving Fubini polynomials and Euler numbers, and give an interesting computational formula for it. At the same time, we also give a recursive calculation method for the general case.

Keywords: Fubini polynomials; Euler numbers; symmetric identities; elementary method; computational formula

MSC: 11B83; 11B37

1. Introduction

For any integer $n \geq 0$, the Fubini polynomials $\{F_n(y)\}$ are defined by the coefficients of the generating function

$$\frac{1}{1 - y(e^t - 1)} = \sum_{n=0}^{\infty} \frac{F_n(y)}{n!} \cdot t^n, \tag{1}$$

where $F_0(y) = 1$, $F_1(y) = y$, and so on. $F_n(1) = F_n$ are called Fubini numbers. These polynomials and numbers are closely connected with the Stirling numbers. Some contents and properties of Stirling numbers can be found in reference [1]. T. Kim et al. [2] proved the identity

$$F_n(y) = \sum_{k=0}^{n} S_2(n,k) \, k! \, y^k, \ (n \geq 0),$$

where $S_2(n,k)$ are the Stirling numbers of the second kind. It not only associated Fubini polynomials with Stirling numbers, but also stressed the importance of researching Fubini polynomials.

Please note that the identity (see [3,4])

$$\frac{2e^{tx}}{1 + e^t} = \sum_{n=0}^{\infty} \frac{E_n(x)}{n!} \cdot t^n, \tag{2}$$

where $E_n(x)$ signifies the Euler polynomials.

It is distinct that if taking $y = -\frac{1}{2}$ in (1) and $x = 0$ in (2), then from (1) and (2) we can get the identity

$$E_n(0) = F_n\left(-\frac{1}{2}\right), n \in N^*0 \tag{3}$$

Symmetry **2018**, *10*, 303; doi:10.3390/sym10080303   www.mdpi.com/journal/symmetry

Symmetry **2018**, *10*, 303

where $E_n(0) = E_n$ is the Euler number (see [5] for related contents).

On the other hand, two variable Fubini polynomials are defined by means of the following (see [2,6])

$$\frac{e^{xt}}{1 - y(e^t - 1)} = \sum_{n=0}^{\infty} \frac{F_n(x, y)}{n!} \cdot t^n,$$

and $F_n(y) = F_n(0, y)$ for all integers $n \geq 0$. About the properties of $F_n(x, y)$, several scholars have also researched it, especially T. Kim and others have done a large amount of vital works. For instance, they proved a series of identities linked to $F_n(x, y)$ (see [2,7]), one of which is

$$F_n(x, y) = \sum_{l=0}^{n} \binom{n}{l} x^l \cdot F_{n-l}(y), n \in N^*0.$$

These polynomials occupy indispensable positions in the theory and application of mathematics. In particular, they are widely used in combinatorial mathematics. Therefore, several scholars have researched their various properties, and acquired a series of vital results. Some involved contents can be found in references [5,7–17].

The goal of this paper is to use elementary methods and recursive properties of a special sequenc to research the computational problem of the sums

$$\sum_{a_1 + a_2 + \cdots + a_k = n} \frac{F_{a_1}(y)}{(a_1)!} \cdot \frac{F_{a_2}(y)}{(a_2)!} \cdots \frac{F_{a_k}(y)}{(a_k)!}, \tag{4}$$

where the summation is over all k-tuples with non-negative integer coordinates $(a_1, a_2, \cdots, a_k)$ such that $a_1 + a_2 + \cdots + a_k = n$.

About this content, it seems there is no valid method to solve the computational problem of (4). However, this problem is significant, it can reveal the structure of Fubini polynomials itself and its internal relations, at least it can reflect the combination properties of Fubini polynomials.

In this paper, we will take elementary methods and the properties of $F_n(y)$ to obtain a fascinating computational formula for (4). Simultaneously, we can also acquire a recursive calculation method for the general case. That is, we are going to prove the following major result:

Theorem 1. *For any positive integers n and k, we have the identity*

$$\sum_{a_1 + a_2 + \cdots + a_k = n} \frac{F_{a_1}(y)}{(a_1)!} \cdot \frac{F_{a_2}(y)}{(a_2)!} \cdots \frac{F_{a_k}(y)}{(a_k)!}$$

$$= \frac{1}{(k-1)!(y+1)^{k-1}} \cdot \frac{1}{n!} \sum_{i=0}^{k-1} C(k-1, i) F_{n+k-1-i}(y),$$

where the sequence $\{C(k, i)\}$ is defined as follows: For any positive integer k and integers $0 \leq i \leq k$, we define $C(k, 0) = 1$, $C(k, k) = k!$ and

$$C(k+1, i+1) = C(k, i+1) + (k+1)C(k, i), \text{ for all } 0 \leq i < k,$$

providing $C(k, i) = 0$, if $i > k$.

The characteristic of this theorem is to represent a complex sum of Fubini polynomials as a linear combination of a single Fubini polynomial. Of course, our method can also be further generalized, provided a corresponding results for $F_n(x, y)$. It is just that its form is not so pretty, so we are not listing it here. If taking $k = 3, 4$ and 5, then from our theorem we may instantly deduce the following several corollaries:

Symmetry **2018**, 10, 303

Corollary 1. *For any positive integer n, we have the identity*

$$\sum_{a+b+c=n} \frac{F_a(y)}{a!} \cdot \frac{F_b(y)}{b!} \cdot \frac{F_c(y)}{c!} = \frac{1}{2 \cdot n! \cdot (y+1)^2} \left(F_{n+2}(y) + 3F_{n+1}(y) + 2F_n(y) \right).$$

Corollary 2. *For any positive integer n, we have the identity*

$$\sum_{a+b+c+d=n} \frac{F_a(y)}{a!} \cdot \frac{F_b(y)}{b!} \cdot \frac{F_c(y)}{c!} \cdot \frac{F_d(y)}{d!}$$
$$= \frac{1}{6 \cdot n! \cdot (y+1)^3} \left(F_{n+3}(y) + 6F_{n+2}(y) + 11F_{n+1}(y) + 6F_n(y) \right).$$

Corollary 3. *For any positive integer n, we have the identity*

$$\sum_{a+b+c+d+e=n} \frac{F_a(y)}{a!} \cdot \frac{F_b(y)}{b!} \cdot \frac{F_c(y)}{c!} \cdot \frac{F_d(y)}{d!} \cdot \frac{F_e(y)}{e!}$$
$$= \frac{1}{24 \cdot n! \cdot (y+1)^4} \left(F_{n+4}(y) + 10F_{n+3}(y) + 35F_{n+2}(y) + 50F_{n+1}(y) + 24F_n(y) \right).$$

If taking $y = -\frac{1}{2}$ in our theorem, then from (3) we can also infer the following:

Corollary 4. *For any positive integers n and $k \geq 2$, we have the identity*

$$\sum_{a_1+a_2+\cdots+a_k=n} \frac{E_{a_1}}{(a_1)!} \cdot \frac{E_{a_2}}{(a_2)!} \cdots \frac{E_{a_k}}{(a_k)!} = \frac{2^{k-1}}{(k-1)!} \cdot \frac{1}{n!} \sum_{i=0}^{k-1} C(k-1, i) E_{n+k-1-i}.$$

If $n = p$ is an odd prime, then taking $y = 1$ in Corollarys 1 and 2, we also have the following congruences.

Corollary 5. *For any odd prime p, we have the congruence*

$$22F_p \equiv F_{p+2} + 3F_{p+1} \pmod{p}.$$

Corollary 6. *For any odd prime p, we have the congruence*

$$186F_p \equiv F_{p+3} + 6F_{p+2} + 11F_{p+1} \pmod{p}.$$

2. A Simple Lemma

For purpose of proving our theorem, we need a uncomplicated lemma. As a matter of convenience, we first present a new sequence $\{C(k, i)\}$ as follows. For any positive integer k and integers $0 \leq i \leq k$, we define $C(k, 0) = 1$, $C(k, k) = k!$ and

$C(k + 1, i + 1) = C(k, i + 1) + (k + 1)C(k, i)$, $1 \leq i \leq k$, $C(k, i) = 0$, if $i > k$.

For clarity, for $1 \leq k \leq 9$, we list values of $C(k, i)$ in the Table 1.

Table 1. Values of $C(k, i)$.

$C(k,i)$	$i=0$	$i=1$	$i=2$	$i=3$	$i=4$	$i=5$	$i=6$	$i=7$	$i=8$	$i=9$
$k=1$	1	1								
$k=2$	1	3	2							
$k=3$	1	6	11	6						
$k=4$	1	10	35	50	24					
$k=5$	1	15	85	225	274	120				
$k=6$	1	21	175	735	1624	1764	720			
$k=7$	1	28	322	1960	6769	13,132	13,068	5040		
$k=8$	1	36	546	4536	22,449	67,284	118,124	109,584	40,320	
$k=9$	1	45	870	9450	63,273	269,325	723,680	1,172,700	1,026,576	362,880

Obviously, the values of $C(k, i)$ can be easily calculated by using a computer program. Hence, for any positive integer k, the computational problem of (4) can be solved fully.

In this table of numerical values, we also find that for prime $p = 3, 5$ and 7, we have the congruence

$$C(p-1, i) \equiv 0 \,(\mathrm{mod}\, p) \quad \text{for all } 1 \le i \le p-2.$$

For all prime $p > 7$ is true? This is an enjoyable open problem.

If this congruence is true, then we can also deduce that for any positive integer n and odd prime p, one has the congruence

$$F_{n+p-1}(y) + F_n(y) \equiv 0 \,(\mathrm{mod}\ p).$$

Now let function $f(t) = \frac{1}{1-y(e^t-1)}$. Then we have the following

Lemma 1. *For any positive integer k, we have the identity*

$$\sum_{i=0}^{k} C(k, i) f^{(k-i)}(t) = k!(y+1)^k f^{k+1}(t),$$

where $f^{(0)}(t) = f(t)$, $f^{(r)}(t)$ denotes the r-order derivative of $f(t)$ for variable t.

Proof. Now we prove this lemma by induction. From the definition of the derivative we acquire

$$f'(t) = \frac{ye^t}{(1 - y(e^t - 1))^2} = -f(t) + (y+1)f^2(t) \tag{5}$$

or

$$f'(t) + f(t) = (y+1)f^2(t). \tag{6}$$

Please note that $C(1, 0) = 1$ and $C(1, 1) = 1$, so the lemma is true for $k = 1$.
Suppose that the lemma is true for all integer $k \ge 1$. That is,

$$\sum_{i=0}^{k} C(k, i) f^{(k-i)}(t) = k!(y+1)^k f^{k+1}(t). \tag{7}$$

Then take the derivative for t in (7) and applying (5) and (7) we obtain

$$
\begin{aligned}
\sum_{i=0}^{k} C(k, i) f^{(k+1-i)}(t) &= (k+1)!(y+1)^k f^k(t) \cdot f'(t) \\
&= (k+1)!(y+1)^k f^k(t) \cdot \left(-f(t) + (y+1)f^2(t) \right) \\
&= (k+1)!(y+1)^{k+1} f^{k+2}(t) - (k+1)!(y+1)^k f^{k+1}(t) \\
&= (k+1)!(y+1)^{k+1} f^{k+2}(t) - (k+1) \left(\sum_{i=0}^{k} C(k, i) f^{(k-i)}(t) \right).
\end{aligned}
\tag{8}
$$

It is evident that (8) implies

$$
\begin{aligned}
&(k+1)!(y+1)^{k+1}f^{k+2}(t) \\
={}& C(k,0)f^{(k+1)}(t) + \sum_{i=0}^{k-1}\left(C(k,i+1)+(k+1)C(k,i)\right)f^{(k-i)}(t) + (k+1)!f(t) \\
={}& C(k,0)f^{(k+1)}(t) + \sum_{i=0}^{k-1}C(k+1,i+1)f^{(k-i)}(t) + (k+1)!f(t) \\
={}& \sum_{i=0}^{k+1}C(k+1,i)f^{(k+1-i)}(t),
\end{aligned}
\tag{9}
$$

where we have used the identities $C(k,0)=1$ and $C(k,k)=k!$. Now the lemma follows from (9) and mathematical induction. $\square$

3. Proof of the Theorem

In this section, the proof of our theorem will be completed. Firstly, for any positive integer k, from the definition of $f(t)$ and the properties of the power series we obtain

$$
f^{(k)}(t) = \sum_{n=0}^{\infty} \frac{F_{n+k}(y)}{n!}\cdot t^n
\tag{10}
$$

and

$$
\begin{aligned}
f^k(t) ={}& \left(\sum_{a_1=0}^{\infty}\frac{F_{a_1}(y)}{a_1!}\cdot t^{a_2}\right)\left(\sum_{a_2=0}^{\infty}\frac{F_{a_2}(y)}{a_2!}\cdot t^{a_1}\right)\cdots\left(\sum_{a_k=0}^{\infty}\frac{F_{a_k}(y)}{a_k!}\cdot t^{a_k}\right) \\
={}& \left(\sum_{a_1=0}^{\infty}\sum_{a_2=0}^{\infty}\cdots\sum_{a_k=0}^{\infty}\frac{F_{a_1}(y)}{(a_1)!}\cdot\frac{F_{a_2}(y)}{(a_2)!}\cdots\frac{F_{a_k}(y)}{(a_k)!}\cdot t^{a_1+a_2\cdots+a_k}\right) \\
={}& \sum_{n=0}^{\infty}\left(\sum_{a_1+a_2+\cdots+a_k=n}\frac{F_{a_1}(y)}{(a_1)!}\cdot\frac{F_{a_2}(y)}{(a_2)!}\cdots\frac{F_{a_k}(y)}{(a_k)!}\right)\cdot t^n.
\end{aligned}
\tag{11}
$$

From (10), (11) and Lemma we acquire

$$
\begin{aligned}
&\frac{1}{(k-1)!(y+1)^{k-1}}\cdot\sum_{i=0}^{k-1}C(k-1,i)\sum_{n=0}^{\infty}\frac{F_{n+k-1-i}(y)}{n!}\cdot t^n \\
={}& \sum_{n=0}^{\infty}\left(\sum_{a_1+a_2+\cdots+a_k=n}\frac{F_{a_1}(y)}{(a_1)!}\cdot\frac{F_{a_2}(y)}{(a_2)!}\cdots\frac{F_{a_k}(y)}{(a_k)!}\right)\cdot t^n.
\end{aligned}
\tag{12}
$$

Comparing the coefficients of t^n in (12) we have the identity

$$
\begin{aligned}
&\sum_{a_1+a_2+\cdots+a_k=n}\frac{F_{a_1}(y)}{(a_1)!}\cdot\frac{F_{a_2}(y)}{(a_2)!}\cdots\frac{F_{a_k}(y)}{(a_k)!} \\
={}& \frac{1}{(k-1)!(y+1)^{k-1}}\cdot\frac{1}{n!}\sum_{i=0}^{k-1}C(k-1,i)F_{n+k-1-i}(y).
\end{aligned}
$$

This completes the proof of our Theorem.

Author Contributions: Writing-original draft: J.Z.; Writing-review and editing: Z.C.

Funding: This research was funded by the N. S. F. (11771351) of China.

Acknowledgments: The author would like to thank the referees for their very helpful and detailed comments, which have significantly improved the presentation of this paper.

Conflicts of Interest: The authors declare no conflict of interest.

Symmetry **2018**, *10*, 303

References

1. Feng, R.-Q.; Song, C.-W. *Combinatorial Mathematics*; Beijing University Press: Beijing, China, 2015.
2. Kim, T.; Kim, D.S.; Jang, G.-W. A note on degenerate Fubini polynomials. *Proc. Jangjeon Math. Soc.* **2017**, *20*, 521–531.
3. Kim, T. Symmetry of power sum polynomials and multivariate fermionic p-adic invariant integral on Z_p. *Russ. J. Math. Phys.* **2009**, *16*, 93–96. [CrossRef]
4. Kim, D.S.; Park, K.H. Identities of symmetry for Bernoulli polynomials arising from quotients of Volkenborn integrals invariant under S_3. *Appl. Math. Comput.* **2013**, *219*, 5096–5104. [CrossRef]
5. Zhang, W. Some identities involving the Euler and the central factorial numbers. *Fibonacci Q.* **1998**, *36*, 154–157.
6. Kilar, N.; Simesk, Y. A new family of Fubini type numbrs and polynomials associated with Apostol-Bernoulli nujmbers and polynomials. *J. Korean Math. Soc.* **2017**, *54*, 1605–1621.
7. Kim, T.; Kim, D. S.; Jang, G.-W.; Kwon, J. Symmetric identities for Fubini polynomials. *Symmetry* **2018**, *10*, 219. [CrossRef]
8. Kim, T.; Kim, D.S. An identity of symmetry for the degernerate Frobenius-Euler polynomials. *Math. Slovaca* **2018**, *68*, 239–243. [CrossRef]
9. He, Y. Symmetric identities for Calitz's q-Bernoulli numbers and polynomials. *Adv. Differ. Equ.* **2013**, *2013*, 246. [CrossRef]
10. Rim, S.-H.; Jeong, J.-H.; Lee, S.-J.; Moon, E.-J.; Jin, J.-H. On the symmetric properties for the generalized twisted Genocchi polynomials. *ARS Comb.* **2012**, *105*, 267–272.
11. Yi, Y.; Zhang, W. Some identities involving the Fibonacci polynomials. *Fibonacci Q.* **2002**, *40*, 314–318.
12. Ma, R.; Zhang, W. Several identities involving the Fibonacci numbers and Lucas numbers. *Fibonacci Q.* **2007**, *45*, 164–170.
13. Wang, T.; Zhang, W. Some identities involving Fibonacci, Lucas polynomials and their applications. *Bull. Math. Soc. Sci. Math. Roum.* **2012**, *55*, 95–103.
14. Chen, L.; Zhang, W. Chebyshev polynomials and their some interesting applications. *Adv. Differ. Equ.* **2017**, *2017*, 303.
15. Li, X.X. Some identities involving Chebyshev polynomials. *Math. Probl. Eng.* **2015**, *2015*, 950695. [CrossRef]
16. Ma, Y.; Lv, X.-X. Several identities involving the reciprocal sums of Chebyshev polynomials. *Math. Probl. Eng.* **2017**, *2017*, 4194579. [CrossRef]
17. Clemente, C. Identities and generating functions on Chebyshev polynomials. *Georgian Math. J.* **2012**, *19*, 427–440.

IOP Conference Series: Materials Science and Engineering

PAPER • OPEN ACCESS

Elliptic Curve Integral Points on $y^2 = x^3 + 19x + 46$

To cite this article: Jianhong Zhao 2018 IOP Conf. Ser.: Mater. Sci. Eng. 382 052038

View the article online for updates and enhancements.

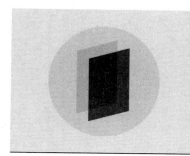

AMIMA 2018   IOP Publishing

IOP Conf. Series: Materials Science and Engineering **382** (2018) 052038   doi:10.1088/1757-899X/382/5/052038

Elliptic Curve Integral Points on $y^2 = x^3 + 19x + 46$

Jianhong Zhao

Department of Teachers and Education, Lijiang Teachers College,Lijiang, Yunnan 674199, China

E-mail:312508050@qq.com

Abstract. By using elementary number theory methods, such as congruence and Legendre Symbol, it can be proved that elliptic curve $y^2 = x^3 + 19x + 46$ has no integer point.

1. Introduction

The positive integer points and integral points of elliptic curves

$$y^2 = (x + a)(x^2 - ax + p), a, p \in Z \qquad (1)$$

are very important in the theory of number and arithmetic algebra, it has a wide range of applications in cryptography and other fields. There are a few results of positive integer points of elliptic curve focused on $a = \pm 2$. When $a = -2$, elliptic curve (2) is equivalent to

$$y^2 = (x - 2)(x^2 + 2x + p), a, p \in Z \qquad (2)$$

On the elliptic curve (2), the result are mainly in paper [1]-[5].

When $a = 2$, elliptic curve (2) is equivalent to

$$y^2 = (x + 2)(x^2 - 2x + p), a, p \in Z \qquad (3)$$

On the elliptic curve (3), the result are mainly in paper [6]-[7].

Up to now, there is no relevant conclusions while $a = 2, p = 23$.

2. Critical lemma

Critical lemma [7] Let D to be a square-free positive integer, then the equation $x^2 - Dy^4 = 1$ will have two sets of positive integer solutions (x, y) at most.

When $D = 2^{4s} \times 1785$, where $s \in \{0,1\}$, we can get that $(x_1, y_1) = (169, 2^{1-s})$ and $(x_2, y_2) = (6525617281, 2^{1-s} \times 6214)$;

Otherwise when $D \neq 2^{4s} \times 1785$, $(x_1, y_1) = (u_1, \sqrt{v_1})$ and $(x_2, y_2) = (u_2, \sqrt{v_2})$, where (u_n, v_n) is a positive integer solution of the Pell equation $U^2 - DV^2 = 1$, if $x^2 - Dy^4 = 1$ has only one set of positive integer solution (x, y) and the positive integer n is suitable for $(x, y^2) = (u_n, v_n)$, then $n = 2$ consequently ; otherwise if n is an even number; otherwise if n is an odd number, then $n = 1$ or p, here p is a prime numbers and $p \equiv 3 \pmod 4$.

3. Proof of main theorem

By using elementary method such as congruence and Legendre Symbol, the integer points on $y^2 = x^3 + 19x + 46$ can be obtained.

3.1. Theorem

Elliptic curve

$$y^2 = (x + 2)(x^2 - 2x + 23) \qquad (4)$$

AMIMA 2018

IOP Publishing

IOP Conf. Series: Materials Science and Engineering **382** (2018) 052038   doi:10.1088/1757-899X/382/5/052038

has no integer point.

3.2. Proof of main theorem

3.2.1. Primary analysis.
Let (x, y) be an integer point of the elliptic curve (4).

$gcd(x + 2, x^2 - 2x + 23) = gcd(x + 2, 31) = 1$ or 31, in other words, the range of this greatest common divisor is $\{1,31\}$. as a result, we have to discuss in two cases of the elliptic curve (4):

  Case I $x + 2 = a^2, x^2 - 2x + 23 = b^2, y = ab, gc\,d(a, b) = 1, a, b \in Z^+$

  Case II $x + 2 = 31a^2, x^2 - 2x + 23 = 31b^2, y = 31ab, gc\,d(a, b) = 1, a, b \in Z^+$

3.2.2. Discusion on Case 1
Obviously $a^2 \equiv 0,1(mod4)$, then $x = a^2 - 2 \equiv 2,3(mod4)$, so $x^2 - 2x + 23 \equiv 2,3(mod4)$.

At the same time $b^2 = x^2 - 2x + 23 \equiv 0,1(mod4)$.Then we will get $2,3(mod4) \equiv 0,1(mod4)$, it is self-contradiction, this shows that (4) has no integer points.

3.2.3. Discusion on Case 2
Divide integers into two categories as $2 \nmid a$ and $2|a$ discuss separately.

First step: Suppose $2 \nmid a$. We will get $a^2 \equiv 1(mod4)$, then $x = 31a^2 - 2 \equiv 1(mod4)$, so $x^2 - 2x + 23 \equiv 2(mod4)$.

At the same time $b^2 \equiv 0,1(mod4)$. Then we will get $31b^2 = x^2 - 2x + 23 \equiv 0,3(mod4)$, it means $2(mod4) \equiv 0,3(mod4)$, it is self-contradiction, this shows that (4) has no integer points.

Second step: Suppose $2|a$. Let $a = 2c, c \in Z$, and $x + 2 = 31a^2$, then $x = 124c^2 - 2$.

From $x^2 - 2x + 23 = 31b^2$, we will get $(124c^2 - 3)^2 + 22 = 31b^2$, then $(12c^2 - 1)^2 + 352c^4 = b^2$,it is equivalent to:

$$(b + 12c^2 - 1)(b - 12c^2 + 1) = 352c^4 \tag{5}$$

Because $2|a$, and $x - 2 = 31a^2$, we will get $2|x$. Go a step further, $2 \nmid x^2 + 2x + 23$, then $2 \nmid 31b^2$, so $2 \nmid b$.Therefore, $2|[b - (12c^2 - 1)]$.

Let $d = \gcd(12c^2 - 1, b)$, then $d|b, d|12c^2 - 1$, for $\gcd(b + 12c^2 - 1, b - 12c^2 + 1) = \gcd(2(12c^2 - 1), b - (12c^2 - 1)) = 2\gcd(12c^2 - 1, b) = 2d$ .So, $d|(b + 12c^2 + 1)$, Therefore, $d|352c^4$. $\gcd(12c^2 - 1, 352c^4) = \gcd(12c^2 - 1, 11) = 1$ or 11.$\gcd(b + 12c^2 - 1, b - 12c^2 + 1) = 2$ or 22.

(i) When $\gcd(b + 12c^2 - 1, b - 12c^2 + 1) = 2$. For $352 = 2^5 \times 11$, (5) is equivalent to:

$$\begin{cases} b + 12c^2 - 1 = 2gm^4 \\ b - 12c^2 + 1 = \frac{146}{g}n^4, \\ c = mn \end{cases} \tag{6}$$

Where $gcd(m, n) = 1, gcd\left(g, \frac{88}{g}\right) = 1, g = 1,11, 2^3, 2^3 \times 11$.

From the first two formulas (6), we will get:

$$12c^2 - 1 = gm^4 - \frac{88}{g}n^4. \tag{7}$$

Making an equivalent of the modulus 4 on (7), we will get:

$$-1 \equiv gm^4 - \frac{88}{g}n^4(mod4) \tag{8}$$

When $g = 1$, (8) is equivalent to:

$$-1 \equiv m^4(mod4) \tag{9}$$

For $m^4 \equiv 0,1(mod4)$. (9) can be $-1 \equiv m^4 \equiv 0,1(mod4)$. It is self-contradiction, this shows that (5) is impossible.

When $g = 2^3$, (8) is equivalent to:

$$-1 \equiv n^4(mod4) \tag{10}$$

Imitating the previous proof of (9) we can get (10) is impossible.

2

AMIMA 2018   IOP Publishing

IOP Conf. Series: Materials Science and Engineering **382** (2018) 052038   doi:10.1088/1757-899X/382/5/052038

When $g = 11$, (7) is equivalent to $12c^2 - 1 = 11m^4 - 8n^4$, from $c = mn$, we will get:
$$12m^2n^2 - 1 = 11m^4 - 8n^4 \tag{11}$$
Making an equivalent of the modulus 8 on (11), we will get:
$$4m^2n^2 - 1 \equiv 3m^4 (mod 8) \tag{12}$$
From (11), we can get $2 \nmid m$, then $m^2 \equiv 1(mod 8)$, so (12) is equivalent to:
$$4n^2 - 1 \equiv 3 (mod 8) \tag{13}$$
From (12), we can get $2 \nmid n$, then $n^2 \equiv 1,9(mod 16)$, $m^2 \equiv 1,9(mod 16)$. Making an equivalent of the modulus 16 on (11), we will get:
$$11 \equiv 3 (mod 16) \tag{14}$$
It shows that (14) is impossible, so we can get (5) is impossible.

When $g = 2^3 \times 11$, the (7) is equivalent to $12c^2 - 1 = 88m^4 - n^4$, from $c = mn$, we will get:
$$(n^2 + 6m^2)^2 = 124n^4 + 1 \tag{15}$$
Let $r = n^2 + 6m^2, r \in Z^+$, (15) is equivalent to:
$$r^2 = 124m^4 + 1 \tag{16}$$
Let $s = 2m$, (16) is equivalent to:
$$r^2 - 31s^2 = 1 \tag{17}$$
We know that (16) has one positive integer point from the Critical lemma at most, suppose (r, m) is the positive integer point, then all of the positive integer point on equation (17) can be represented as: $s_k\sqrt{31} + r_k = (1520 + 273\sqrt{31})^k, k \in Z^+$. Therefore all of the positive integer point of (15) satisfied:
$$(m^2 + 6n^2) + m^2\sqrt{124} = (1520 + 273\sqrt{31})^k, k \in Z^+. \tag{18}$$
So $n^2 + 6m^2 = r_k, 2m^2 = s_k, k \in Z^+$. It is easy to verify the following formula:
$$\therefore s_{k+2} = 3040s_{k+1} - s_k, s_0 = 0, s_1 = 273. \tag{19}$$
Making an equivalent of the modulus 2 on recurrent sequence (19), we will get the residue class sequence $0,1,0,1,\cdots$, cycle for 2.

And just when $k \equiv 1(mod 2)$, $s_k \equiv 1(mod 2)$, just when $k \equiv 0(mod 2)$, $s_k \equiv 0(mod 2)$.

Because $2m^2 = s_k$, s_k must be an even number. It means $k \equiv 0(mod 2)$.

From the Critical lemma, we know that in order to make the (15) set up, $k = 2$ or $2 \nmid k$. So, we will get $n^2 + 6m^2 + m^2\sqrt{124} = m^2 + 6m^2 + 2m^2 = 4260799 + 829920\sqrt{31}$.

Taken together, we will get: $n^2 - 6m^2 = 4260799$, $2m^2 = 829920$, $m^2 = 414960$. All appearance it has no integer points, this shows that equation (15) has no integer points.

(ii) When $\gcd(b + 12c^2 - 1, b - 12c^2 + 1) = 22$. For $352 = 2^5 \times 11$, (5) is equivalent to:
$$\begin{cases} b + 12c^2 - 1 = 22gm^4 \\ b - 12c^2 + 1 = \frac{16}{g}n^4 \\ c = mn \end{cases}, \tag{20}$$
Or:
$$\begin{cases} b - 12c^2 + 1 = 22gm^4 \\ b + 12c^2 - 1 = \frac{16}{g}n^4 \\ c = mn \end{cases}, \tag{21}$$
Where $\gcd(m, n) = 1, \gcd\left(g, \frac{8}{g}\right) = 1, g = 1, 2^3, 11 \nmid m, 11 \nmid n$.

From the first two formulas (20), we will get:
$$12c^2 - 1 = 11gm^4 - \frac{8}{g}n^4. \tag{22}$$
When $g = 1$, from $c = mn$ (22) is equivalent to:
$$12m^2n^2 - 1 = 11m^4 - 8n^4 \tag{23}$$
Imitating the previous proof of (11) we can get (23) is impossible, this shows that (5) has no integer points.

From the first two formulas (21), we will get:

3

AMIMA 2018

IOP Publishing

IOP Conf. Series: Materials Science and Engineering **382** (2018) 052038   doi:10.1088/1757-899X/382/5/052038

$$12c^2 - 1 = \frac{8}{g}n^4 - 11gm^4. \tag{24}$$

When $g = 1$, from $c = mn$ (24) is equivalent to:

$$12c^2 - 1 = n^4 - 11m^4 \tag{25}$$

Making an equivalent of the modulus 4 on (25), we will get:

$$-1 \equiv m^4 (mod 4) \tag{26}$$

Imitating the previous proof of (9) we can get (26) is impossible, so we can get (5) is impossible. When $g = 2^3$, (24) is equivalent to:

$$12c^2 - 1 = n^4 - 88m^4 \tag{27}$$

Making an equivalent of the modulus 4 on (27), we will get: $-1 \equiv n^4 (mod 4)$ .For $n^4 \equiv 0,1(mod 4)$, (28) is equivalent to $-1 \equiv n^4 (mod 4) \equiv 0,1(mod 4)$, it is self-contradiction, this shows that (5) has no integer points. In conclusion, $y^2 = x^3 + 19x + 46$ has no integer point.

4. Conclusion

The positive integer points and integral points of elliptic curves $y^2 = (x + a)(x^2 - ax + p), a, p \in Z$ are very important in the theory of number and arithmetic algebra, it has a wide range of applications in cryptography and other fields. Up to now, there is no relevant conclusions while $a = 2, p = 23$.

In this paper, by using elementary number theory methods, such as congruence and Legendre Symbol, it can be proved that elliptic curve $y^2 = x^3 + 19x + 46$ has no integer point.

Acknowledgment

Supported by Scientific Research Project fund of Education Department of Yunnan Province: 2018JS608.

References

[1]   D. Zagier. Lager Integral Point on Elliptic Curves [J]. Math Comp, 1987,48:425-436.

[2]   Zhu H L, Chen J H. Integral point on $y^2 = x^3 + 27x - 62$ [J]. J Math Study, 2009, 42(2): 117-125.

[3]   Wu H M . Points on the Elliptic Curves $y^2 = x^3 - 27x - 62$ [J] . J Acta Mathematica Curve Sinica,2010,53(1):205-208.

[4]   Li Y Z, Cui B J. Points on the Elliptic $y^2 = x^3 - 21x - 90$ [J]. Journal of Yanan University (Natural Science Edition), 2015, 34(3):14-15.

[5]   Guo J. The Integral Points on the Elliptic Curve $y^2 = x^3 + 27x + 62$ [J]. Journal of Chongqing Normal University (Natural Science), 2016,33(5): 50- 53.

[6]   Guo J. The Positive Integral Points on the Elliptic Curve $y^2 = x^3 - 21x + 90$ [J]. Mathematics in Practice and Theory,2017,47(8):288-291.

[7]   Togbé A.,Voutier P.M.,and Walsh P.G.,Solving a family of Thue equations with an application to the equation $x^2 - dy^2 = 1$.Acta. Arith.,2005,120(1) :39-58.

IOP Conference Series: Earth and Environmental Science

PAPER • OPEN ACCESS

On the Diophantine equations $x^2 - 27y^2 = 1$ and $y^2 - 2^p z^2 = 25$

To cite this article: Jianhong Zhao 2018 *IOP Conf. Ser.: Earth Environ. Sci.* **153** 042001

View the article online for updates and enhancements.

Related content

- On the Diophantine equations x2_7y2 = 1 y2_Dz2 = 9
  Jianhong Zhao and Lixing Yang

- On the Diophantine equations x2_47y2 = 1 and y2_Pz2 = 49
  Jianhong Zhao and Lixing Yang

- A NOTE ON THE DIOPHANTINE EQUATION qx + py = z2
  Lei Lu

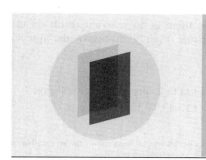

2018 2nd International Workshop on Renewable Energy and Development (IWRED 2018)   IOP Publishing

IOP Conf. Series: Earth and Environmental Science **153** (2018) 042001   doi:10.1088/1755-1315/153/4/042001

On the Diophantine equations $x^2 - 27y^2 = 1$ and $y^2 - 2^p z^2 = 25$

Jianhong Zhao

Department of Teachers and Education, Lijiang Teachers College, Lijiang, Yunnan 674199, China

E-mail:312508050@qq.com

Abstract. In recent years, scholars have paid much attention to the problem of solving the diophantine equations $x^2 - D_1 y^2 = m, (D_1 \in Z^+, m \in Z)$ and $y^2 - D_2 z^2 = n, (D_2 \in Z^+, n \in Z)$.

At present, there are only a few conclusions on $x^2 - D_1 y^2 = 1$ and $y^2 - D_2 z^2 = 1$, and the conclusions mainly concentrated in the number of solutions and the range of it, see Ref [1] and [3].

For odd numbers D_2, the integer solution of $x^2 - D_1 y^2 = 1$ and $y^2 - D_2 z^2 = 4$, see Ref [4] - [6]. For even numbers D_2, the integer solution of (2), see Ref [7] - [11].

Up to now, there is no relevant result on the integer solution of $x^2 - 27y^2 = 1$ and $y^2 - 2^p z^2 = 49$, this paper mainly discusses the integer solution of it.

1. Introduction

In recent years, scholars have paid much attention to the problem of solving the diophantine equations $x^2 - D_1 y^2 = m, (D_1 \in Z^+, m \in Z)$ and $y^2 - D_2 z^2 = n, (D_2 \in Z^+, n \in Z)$. When $m = 1, n = 1$, the equations turns into:

$$x^2 - D_1 y^2 = 1 \text{and } y^2 - D_2 z^2 = 1 \qquad (1)$$

At present, there are only a few conclusions on (1), and the conclusions mainly concentrated in the number of solutions and the range of it, see Ref [1] and [3].

When $m = 1, n = 4$, diophantine equations (1) turns into:

$$x^2 - D_1 y^2 = 1 \text{and } y^2 - D_2 z^2 = 4 \qquad (2)$$

For odd numbers D_2, the integer solution of (2), see Ref [4] - [6]. For even numbers D_2, the integer solution of (2), see Ref [7] - [11].

When $m = 1, n = 25$, the diophantine equations (1) turns into: $x^2 - D_1 y^2 = 1$ and $y^2 - D_2 z^2 = 25$.

In this case, $D_1 = 27, D_2$ can be expressed as 2^p. Up to now, there is no relevant result on the integer solution of $x^2 - 27y^2 = 1$ and $y^2 - 2^p z^2 = 49$, this paper mainly discusses the integer solution of it.

2. Critical lemma

Lemma 1[12]   When $D = 1785, 4 \times 1785, 16 \times 1785$, In addition to 2 sets of integer solution $(x, y) = (13,4), (239,1352) (x, y) = (13,2), (239,676); (x, y) = (13,1), (239,338).$the

Published under licence by IOP Publishing Ltd   1

2018 2nd International Workshop on Renewable Energy and Development (IWRED 2018)　　IOP Publishing
IOP Conf. Series: Earth and Environmental Science 153 (2018) 042001　　doi:10.1088/1755-1315/153/4/042001

indeterminate equations has 1 set of positive integer solution (x_1, y_1), where $x_1{}^2 = x_0$ or $x_1{}^2 = 2x_0{}^2 - 1$, and $\varepsilon = x_0 + y_0\sqrt{D}$ is the basic solution of the Pell equations $x^2 - Dy^2 = 1$.

Lemma 2[12]　　Diophantine equations $x^4 - 27y^2 = 1$ only has common solution $(x, y) = (\pm 1, 0)$.

Proof: For $D = 27$, from Lemma 1 we can get the equation $x^4 - 27y^2 = 1$ has 1 set of positive integer solution at most. $(x_0, y_0) = (26, 3)$ is the basic solution of $x^2 - 27y^2 = 1$, then $x_0 = 26$ is a non square number, $2x_0{}^2 - 1 = 1351$ is a non square number, therefore Diophantine equations $x^4 - 27y^2 = 1$ only has common solution $(x, y) = (\pm 1, 0)$.

Lemma 3[13]　　Let $(x_n, y_n), n \in Z$ be all of the solutions on $x^2 - 27y^2 = 1$, then x_n is a square number if and only if $n = 0$ for any x_n.

Proof: Replacing $x_n = a^2$ into the original equation, we can get $x^4 - 27y^2 = 1$. From Lemma 2 we can get Diophantine equations $x^4 - 27y^2 = 1$ only has common solution $(x, y) = (\pm 1, 0)$, then $x_n = 1$, therefore $n = 0$.

3. Theorem and proof

By using elementary method such as congruence, the integer solution of the diophantine equations on $x^2 - 27y^2 = 1$ and $y^2 - 2^p z^2 = 25$ can be obtained.

3.1. Theorem

Let $p \in Z^+$, then the diophantine equations

$$x^2 - 27y^2 = 1 \text{ and } y^2 - 2^p z^2 = 25 \tag{3}$$

has one and only one common solution $(x, y, z) = (\pm 26, \pm 5, 0)$.

3.2. Proof of main theorem

3.2.1. Primary analysis.

Because $(x_1, y_1) = (26, 5)$ is the basic solution of the Pell equation $x^2 - 27y^2 = 1$, then all solution of the Pell equation $x^2 - 27y^2 = 1$ can be expressed as:

$$x_n + y_n\sqrt{D} = (26 + 5\sqrt{27})^n, n \in Z^+.$$

It is easily shown that

(I) $y_m{}^2 - 25 = y_{m+1}y_{m-1}$;

(II) $y_{2m} = 2x_m y_m$;

(III) $\gcd(x_{2m+1}, y_{2m}) = \gcd(x_{2m+1}, y_{2m+2}) = 26$,
　　　$\gcd(x_{2m}, y_{2m+1}) = \gcd(x_{2m+1}, y_{2m+2}) = 1$.

(IV) $\gcd(x_m, y_m) = 1, \gcd(x_{m+1}, y_{m+1}) = 1, \gcd(x_m, x_{m+1}) = 1, \gcd(y_m, y_{m+1}) = 5$;

(V) $x_{2m} \equiv 1(mod2); y_{2m+1} \equiv 1(mod2)$.

Suppose that $(x, y, z) = (x_m, y_m, z), m \in Z$ is the positive integer solution of the diophantine equation (3), from(I), we can get: $2^p z^2 = y_m{}^2 - 25 = y_{m+1}y_{m-1}$, it is

$$2^p z^2 = y_{m+1}y_{m-1} \tag{4}$$

As a result the equation (4) will be:

Case 1　　　p is an positive odd number.

Case 2　　　p is an positive even number.

3.2.2. Discusion on Case 1

Let $p = 2l - 1, l \in Z$,(4) is equivalent to:

$$2^{2l-1}z^2 = y_{m+1}y_{m-1} \tag{5}$$

1. m is an odd number.

Let $m = 2k - 1, k \in Z$,(5) is equivalent to:

2018 2nd International Workshop on Renewable Energy and Development (IWRED 2018)　　IOP Publishing

IOP Conf. Series: Earth and Environmental Science **153** (2018) 042001　　doi:10.1088/1755-1315/153/4/042001

$$2^{2l-1}z^2 = y_{2(k-1)}y_{2k} \tag{6}$$

From (II), (6) is equivalent to:

$$2^{2l-1}z^2 = 4x_{k-1}y_{k-1}x_ky_k \tag{7}$$

1.1 k is an odd number.

From (III) and (IV), we can get $\gcd(x_{k-1}, y_{k-1}) = \gcd(x_k, y_k) = 1$, $\gcd(x_k, x_{k-1}) = 1$, $\gcd(x_k, y_{k-1}) = 26, \gcd(y_k, y_{k-1}) = 5$, it means $\gcd\left(\frac{y_k}{5}, \frac{y_{k-1}}{5}\right) = 1$. $\gcd\left(\frac{x_k}{26}, \frac{y_{k-1}}{26}\right) = 1$. $x_{k-1}, \frac{y_{k-1}}{52}, \frac{x_k}{26}, \frac{y_k}{5}$ are pairwise coprime.

When $k \neq 1$, from (V), we can get $x_{k-1} \equiv 1 (mod2)$, from Lemma 3, we can get x_{k-1} is a square number if and only if $k = 1$. When $k = 1$, $x_{k-1} = x_0 = 1$, $\frac{x_k}{26} = \frac{x_1}{26} = 1$, $\frac{y_k}{5} = \frac{y_1}{5} = 1$, and $\frac{y_{k-1}}{52} \neq 1$ for any $k \in Z$. So $x_{k-1}, \frac{y_{k-1}}{52}, \frac{x_k}{26}, \frac{y_k}{5}$ can not be 2 times of any square number when $k \neq 1$. It means $4x_{k-1}y_{k-1}x_ky_k = 20^2 \times \frac{x_k}{26} \times \frac{y_k}{5} \times x_{k-1} \times \frac{y_{k-1}}{52}$ can not be 2 times of any square number. Therefore, (5) has no integer solution, the diophantine equation (1) has no integer solution.

When $k = 1$, (7) is equivalent to: $2^{2l-1}z^2 = 4x_0y_0x_1y_1 = 4 \times 26 \times 5 \times 1 \times 0 = 0$, then $z = 0$, Therefore, the diophantine equation (3) has common solution $(x, y, z) = (\pm26, \pm5, 0)$.

1.2 k is an even number.

From (III), we can get $\gcd(x_{k-1}, y_k) = 26$, then $\gcd\left(\frac{x_{k-1}}{26}, \frac{y_k}{26}\right) = 1$. From (IV), we can get $\gcd(x_k, y_k) = 1$, and $\gcd(x_k, x_{k-1}) = 1, \gcd(y_k, y_{k-1}) = 5$. Then $\gcd\left(\frac{y_k}{5}, \frac{y_{k-1}}{5}\right) = 1$, so when k is an even number, $x_k, \frac{y_k}{52}, \frac{x_{k-1}}{26}, \frac{y_{k-1}}{5}$ are pairwise coprime.

When $k = 0$, $x_{k-1} = x_0 = 1$. When $k = 2$, $\frac{x_{k-1}}{26} = \frac{x_1}{26} = 1$, and $\frac{y_{k-1}}{5} = \frac{y_1}{5} = 1$. $\frac{y_k}{52} \neq 1$ for any $k \in Z^+$. So when even number $k \neq 0$, $x_k, \frac{y_k}{52}, \frac{x_{k-1}}{26}, \frac{y_{k-1}}{5}$ can not equal to 1.

From Lemma 3, we can get x_k is a square number if and only if $k = 0$. From (V), we can get $x_k \equiv 1 (mod2)$, So $x_k, \frac{y_k}{52}, \frac{x_{k-1}}{26}, \frac{y_{k-1}}{5}$ can not be 2 times of any square number when $k \neq 0$. It means $4x_{k-1}y_{k-1}x_ky_k = 20^2 \times \frac{y_k}{52} \times \frac{x_{k-1}}{26} \times x_k \times \frac{y_{k-1}}{5}$ can not be 2 times of any square number. Therefore, (5) has no integer solution, the diophantine equation (1) has no integer solution.

When $k = 0$, (5) is equivalent to: $2^{2l-1}z^2 = 4x_0y_0x_{-1}y_{-1} = 0$, then $z = 0$, Therefore, the diophantine equation (1) has common solution $(x, y, z) = (\pm26, \pm5, 0)$.

2. m is an even number.

Let $m = 2k, k \in Z^+$, (2) is equivalent to:

$$2^{2l-1}z^2 = y_{2k-1}y_{2k+1} \tag{8}$$

From (V), we can get $y_{2k-1} \equiv y_{2k+1} \equiv 1 (mod2)$, the power of 2 on the right of (8) should be 0, it is even-power. At the same time, the power of 2 on the left of (8) should be odd-power. Which is contradict with each other. Therefore, (8) has no integer solution when m is an even number, and the Diophantine equation (1) has no integer solution.

3.2.3. Discusion on Case 2

Let $p = 2k, k \in Z^+$, then $D = 2^{2k}$, from $y^2 - 2^pz^2 = 25$, we can get $y^2 - 2^{2k}z^2 = 25$, it is equivalent to:

$$(y + 2^kz)(y - 2^kz) = 25 \tag{9}$$

Solve (9), we can get $y_1 = \pm5, z_1 = 0, y_2 = \pm13, z_2 = \pm3, k = 2$. When $y_2 = \pm13$, from $x^2 - 27y^2 = 1$ we can get $x^2 = 4564$, Obviously it has no integer solution. Therefore, the Diophantine equation (3) has one and only one common solution $(x, y, z) = (\pm26, \pm5, 0)$.

To sum up, the theorem is proved.

4. Conclusion

3

2018 2nd International Workshop on Renewable Energy and Development (IWRED 2018)　　IOP Publishing

IOP Conf. Series: Earth and Environmental Science **153** (2018) 042001　　doi:10.1088/1755-1315/153/4/042001

The integer solution of diophantine equations $x^2 - D_1 y^2 = m, (D_1 \in Z^+, m \in Z)$ and $y^2 - D_2 z^2 = n, (D_2 \in Z^+, n \in Z)$ is a matter of great concern.

By using elementary number theory methods, we solved the common solution and nontrivial solution on the diophantine equation when $m = 1, n = 25, D_1 = 27, D_2$ can be expressed as 2^t, it is the diophantine equations $x^2 - 27y^2 = 1$ and $y^2 - 2^p z^2 = 25$ has one and only one common solution $(x, y, z) = (\pm 26, \pm 5, 0)$.

Acknowledgment

Supported by Scientific Research Project fund of Education Department of Yunnan Province: 2018JS608.

References

[1] Ljunggren W. Litt om Simultane Pellske Ligninger [J]. Norsk Mat. Tidsskr.,1941,23:132-138.

[2] Pan Jia-yu, Zhang Yu-ping, Zou Rong. The Pell Equations $x^2 - ay^2 = 1$ and $y^2 - Dz^2 = 1$ [J]. Chinese Quarterly Journal of Mathematics,1999,14(1):73-77.

[3] Le Mao-hua. On the simultaneous Pell Equations $x^2 - 4D_1 y^2 = 1$ and $y^2 - D_2 z^2 = 1$ [J]. Journal of Foshan University(Natural Science Edition), 2004,22(2)：1-3+9

[4] Chen Yong-gao. Pell Equation $x^2 - 2y^2 = 1$ and $y^2 - Dz^2 = 4$ [J]. Acta Scientia rum Naturahum Universitatis Pekinesis,1994,30(3):298-302

[5] Guan Xun-gui. On the isolution of the Pell Equation $x^2 - 2y^2 = 1$ and $y^2 - Dz^2 = 4$ [J]. Journal of Nanjing Normal University(Natural Science Edition):2014,37(3):44-47.

[6] Guo Jing, Du Xian-cun. On The System of Indefinite Equations $x^2 - 12y^2 = 1$ and $y^2 - Dz^2 = 4$[J]. Mathematics in Practice and Theory.2015,45(9):289-293.

[7] Hu Yong-zhong, Han Qing. On the integer solution of the simultaneous equations $x^2 - 2y^2 = 1$ and $y^2 - Dz^2 = 4$ [J]. Journal of Hua Zhong Normal University (Natural Sciences):2002,36(1):17-19.

[8] Guan Xun-gui. On the integer solution of Pell's equation $x^2 - 2y^2 = 1$ and $y^2 - Dz^2 = 4$ [J]. Journal of Huazhong Normal University (Natural Sciences):2012,46(3):267-269+278.

[9] Du Xian-cun, Li Yu-long, On the system of Diophantine equations $x^2 - 6y^2 = 1$ and $y^2 - Dz^2 = 4$ [J]. Journal of Anhui University(Natural Science Edition):2015.39(6):19-22.

[10] Du Xian-cun, Guan Xun-gui, Yang Hui-zhang, On the system of Diophantine equations $x^2 - 6y^2 = 1$ and $y^2 - Dz^2 = 4$ [J]. Journal of Hua Zhong Normal University (Natural Sciences):2014.48(3):5-8.

[11] Guo Jing, Du Xian-cun. On the System of Pell Equations $x^2 - 30y^2 = 1$ and $y^2 - Dz^2 = 4$[J]. Mathematics in Practice and Theory.2015,45(1):309-314.

[12] Sun Qi, Yuan Ping-zhi, ON THE DIOPHANTINE EQUATION $x^4 - Dy^2 = 1$ [J]. Journal Of Sichuan University (Natural Science Edition),1997:34(3):265-267.

IOP Conference Series: Earth and Environmental Science

PAPER • OPEN ACCESS

On the Diophantine equations $x^2 - 7y^2 = 1$ $y^2 - Dz^2 = 9$

To cite this article: Jianhong Zhao and Lixing Yang 2018 *IOP Conf. Ser.: Earth Environ. Sci.* **128** 012126

View the underline article online for updates and enhancements.

Related content

- On the Diophantine equations x2 27y2 = 1 and y2 2pz2 = 25
  Jianhong Zhao

- A NOTE ON THE DIOPHANTINE EQUATION qx + py = z2
  Lei Lu

- Nonlinear Diophantine equation 11x +13y = z2
  A Sugandha, A Tripena, A Prabowo et al.

IOP ebooks™

Bringing you innovative digital publishing with leading voices to create your essential collection of books in STEM research.

Start exploring the collection - download the first chapter of every title for free.

This content was downloaded from IP address 39.128.230.81 on 29/11/2018 at 03:14

ICEESE 2017

IOP Publishing

IOP Conf. Series: Earth and Environmental Science **128** (2018) 012126   doi:10.1088/1755-1315/128/1/012126

On the Diophantine equations $x^2 - 7y^2 = 1$ and $y^2 - Dz^2 = 9$

Jianhong Zhao[a] Lixing Yang[b]

Department of Teachers and Education, Lijiang Teachers College

Lijiang, Yunnan 674199, China

* Corresponding author

[a] E-mail:312508050@qq.com [b] E-mail: 375734990@qq.com

Abstract. The integer solution of diophantine equations $x^2 - D_1y^2 = m, (D_1 \in Z^+, m \in Z)$ and $y^2 - D_2z^2 = n, (D_2 \in Z^+, n \in Z)$ is a matter of great concern. Researchers study for different m,n and D_1, D_2 , and obtain some correlation results as follows.
When $m = 1, n = 1$, the diophantine equations turns into $x^2 - D_1y^2 = 1$ and $y^2 - D_2z^2 = 1$. At present, there are only a few conclusions on it, see Ref [1] and [2].
When $m = 1, n = 4$, the diophantine equations turns into $x^2 - D_1y^2 = 1$ and $y^2 - D_2z^2 = 4$. For even numbers D_1, D_2, the integer solution see Ref [3] - [9].
When $m = 1, n = 16$, the diophantine equations turns into $x^2 - D_1y^2 = 1$ and $y^2 - D_2z^2 = 16$ .The previous conclusions see Ref [10].
When $m = 1, n = 9$, the diophantine equations turns into $x^2 - D_1y^2 = 1$ and $y^2 - D_2z^2 = 9$. Up to now, there is no relevant result on the integer solution of (5), this paper mainly discusses the integer solution of (5) when $D_1 = 7, D_2$ is an even number.

1.　Introduction

The integer solution of diophantine equations

$$x^2 - D_1y^2 = m, (D_1 \in Z^+, m \in Z) \text{ and } y^2 - D_2z^2 = n, (D_2 \in Z^+, n \in Z) \qquad (1)$$

is a matter of great concern. Researchers study for different m,n and D_1, D_2 , and obtain some correlation results as follows.

When $m = 1, n = 1$, diophantine equations (1) turns into:

$$x^2 - D_1y^2 = 1 \text{ and } y^2 - D_2z^2 = 1 \qquad (2)$$

At present, there are only a few conclusions on (2), see Ref [1] and [2].
When $m = 1, n = 4$, diophantine equations (1) turns into:

$$x^2 - D_1y^2 = 1 \text{ and } y^2 - D_2z^2 = 4 \qquad (3)$$

For even numbers D_1, D_2, the integer solution of (3), see Ref [3] - [9].
When $m = 1, n = 16$, diophantine equations (1) turns into:

$$x^2 - D_1y^2 = 1 \text{ and } y^2 - D_2z^2 = 16 \qquad (4)$$

The previous conclusions on (4), see Ref [10].
When $m = 1, n = 9$, diophantine equations (1) turns into:

$$x^2 - D_1y^2 = 1 \text{ and } y^2 - D_2z^2 = 9 \qquad (5)$$

ICEESE 2017　　　　　　　　　　　　　　　　　　　　　　　　　　IOP Publishing

IOP Conf. Series: Earth and Environmental Science **128** (2018) 012126　　doi:10.1088/1755-1315/128/1/012126

Up to now, there is no relevant result on the integer solution of (5), this paper mainly discusses the integer solution of (5) when　$D_1 = 7, D_2$　is an even number.

2.　Key lemma

Lemma 1[11]　Let p　be an odd prime number, there is no integer solution of the diophantine equation $x^4 - py^2 = 1$　except $p = 5, x = 3, y = 4$　and $p = 29, x = 99, y = 1820$.

　　　Lemma 2[12]　　　There is 1 sets of solutions of the diophantine equation $ax^4 - by^2 = 1$　at most when　a　is a square number which is greater than 1.

　　　Lemma 3[13]　　　Let D　be a square-free positive integer, then the equation　$x^2 - Dy^4 = 1$　has two sets of positive integer solutions　(x, y)　at most. Furthermore, the necessary and sufficient condition of it is $D = 1785$　or $D = 28560$, or $2x_0$　and　y_0　are square numbers where (x_0, y_0)　is the basic solution of　$x^2 - Dy^4 = 1$.

　　　Lemma 4[14]　　　Suppose that all the integer solution on Pell equation　$x^2 - 7y^2 = 1$　could be $(x_n, y_n), n \in Z^+$, let　$m, k \in Z^+$　and　$gcd(m, k) = d$, then the following conclusions are established:

　　　(I)　$gcd(x_m, y_k) = y_d$.

　　　(II)　$gcd(x_m, x_k) = 1$ if $2|\frac{mk}{d^2}$, or else $gcd(x_m, x_k) = x_d$ when $2 \nmid \frac{mk}{d^2}$.

　　　(III)　$gcd(x_k, y_m) = 1$ if $2 \nmid \frac{m}{d}$.

　　　Lemma 5　　　Suppose that all the integer solution on Pell equation　$x^2 - 7y^2 = 1$　could be $(x_n, y_n), n \in Z$, for the arbitrary　$n \in Z$, it has the following properties on　(x_n, y_n):

　　　(I)　x_n　is a square number if and only if　$n = 0$.

　　　(II)$\frac{x_n}{8}$　is a square number if and only if　$n = 1$　or　$n = -1$.

　　　(III)　$\frac{y_n}{3}$　is a square number if and only if　$n = 0$　or　$n = 1$.

　　　Proof: (I) Let　$x_n = a^2$, we will get　$a^4 - 7y^2 = 1$, from Lemma 1 we can get there are only 2 integer solution　$(a, y) = (\pm 1, 0)$　on　$a^4 - 7y^2 = 1$　, so　$x_n = 1, n = 0$. On the contrary, it also holds.

　　　(II) Let　$\frac{x_n}{8} = a^2$, we will get　$64a^4 - 7y^2 = 1$, from Lemma 2 we can get there are only 4 integer solution　$(a, y) = (\pm 1, \pm 3)$　on　$64a^4 - 7y^2 = 1$　, so　$x_n = 8, n = 1$　or　$n = -1$. On the contrary, it also holds.

　　　(III) Let　$\frac{y_n}{3} = b^2$, we will get　$x^2 - 63b^4 = 1$, from Lemma 3 we can get there are only 6 integer solution　$(x, b) = (\pm 1, 0), (\pm 8, \pm 1)$　on　$x^2 - 63b^4 = 1$　, so　$y_n = 0$　or　$y_n = 3$.　$n = 0$　or　$n = 1$. On the contrary, it also holds.

3.　Proof of main theorem

By using elementary method such as congruence, the integer solution of the diophantine equations on $x^2 - 7y^2 = 1$　and　$y^2 - Dz^2 = 9$　can be obtained.

3.1 Theorem

　　　Let　$p_s (1 \leq s \leq 4)$　are diverse odd primes,　$D = 2^k p_1^{a_1} \cdots p_s^{a_s} (a_i = 0\ or\ 1, 1 \leq i \leq 4, k \in Z^+)$, then the diophantine equations

　　　$x^2 - 7y^2 = 1$　and　$y^2 - Dz^2 = 9$　　　　　　　　　　　　　　　　　　　　　　　(6)

　　　(i) has common solution　$(x, y, z) = (\pm 8, \pm 3, 0)$　and nontrivial solution　$(x, y, z) = (\pm 2024, \pm 765, \pm 48)$　when　$D = 2 \times 127$.

　　　(ii) has common solution　$(x, y, z) = (\pm 8, \pm 3, 0)$　and nontrivial solution　$(x, y, z) = (\pm 2024, \pm 765, \pm 24)$　when　$D = 2^3 \times 127$.

　　　(iii) has common solution　$(x, y, z) = (\pm 8, \pm 3, 0)$　and nontrivial solution　$(x, y, z) = (\pm 2024, \pm 765, \pm 12)$　when　$D = 2^5 \times 127$.

　　　(iv) has common solution　$(x, y, z) = (\pm 8, \pm 3, 0)$　and nontrivial solution　$(x, y, z) = (\pm 2024, \pm 765, \pm 6)$　when　$D = 2^7 \times 127$.

2

ICEESE 2017

IOP Publishing

IOP Conf. Series: Earth and Environmental Science **128** (2018) 012126   doi:10.1088/1755-1315/128/1/012126

(v) has common solution $(x, y, z) = (\pm 8, \pm 3, 0)$ and nontrivial solution $(x, y, z) = (\pm 2024, \pm 765, \pm 3)$ when $D = 2^9 \times 127$.

(vi) has only nontrivial solution $(x, y, z) = (\pm 2024, \pm 765, \pm 48)$ when $D \neq 2^\alpha \times 127(\alpha = 1,3,5,7,9)$.

3.2 Proof of main theorem

3.2.1 Primary analysis

It is easily shown that $(x_1, y_1) = (2,1)$ is the minimal solution of the Pell equation $x^2 - 7y^2 = 1$, therefor all integer solution of it will be $x_n + \sqrt{7}y_n = (8 + 3\sqrt{7})^n, n \in Z$. and the following recursive sequence will be established:

$$y_{n+2} = 16y_{n+1} - y_n, y_0 = 0, y_1 = 3 \tag{7}$$
$$x_{n+2} = 16x_{n+1} - x_n, x_0 = 1, x_1 = 8 \tag{8}$$

Using modulo 2 on (7), we will get residue class sequence:0,1,0,1......, and $y_n \equiv 1(mod2)$ only when $n \equiv 1(mod2)$, $y_{2n} \equiv 1(mod2)$ only when $n \equiv 0(mod2)$. as a result $y_{2n} \equiv 0(mod2)$ and $y_{2n+1} \equiv 1(mod2)$.

Using modulo 2 on (8), we will get residue class sequence:1,0,1,0....., and $x_n \equiv 0(mod2)$ only when $n \equiv 1(mod2)$, $y_{2n} \equiv 1(mod2)$ only when $n \equiv 0(mod2)$. as a result $x_{2n} \equiv 1(mod2)$ and $x_{2n+1} \equiv 0(mod2)$.

Suppose $(x, y, z) = (x_{n+1}, y_{n+1}, z), n \in Z$ is the integer solution of (6), then $y_{n+1}^2 - 9 = y_{n+1}^2 - 9(x_{n+1}^2 - 7y_{n+1}^2) = 64y_{n+1}^2 - 9x_{n+1}^2 = (8y_{n+1} + 3x_{n+1})(8y_{n+1} - 3x_{n+1}) = y_n y_{n+2}$, it is equivalent to:

$$Dz^2 = y_n y_{n+2} \tag{9}$$

Obviously $DZ^2 = 0$ when $n = -2$ or $n = 0$, here we will get the common solution $(x, y, z) = (\pm 8, \pm 3, 0)$ on (6).

Because $y_{n+2} \equiv 1(mod2)$, $y_n \equiv y_{n+2} \equiv 1(mod2)$ when n is an positive odd number. Therefor $2(y_n y_{n+2}) = 0$, and $2(D) = 1$, we will get $2(DZ^2)$ is an odd number, it is self-contradiction, this shows that n is an nonnegative even number. Let $n = 2m, m \in Z^+$, (9) is equivalent to:

$$Dz^2 = 4x_m x_{m+1} y_m y_{m+1} \tag{10}$$

As a result the equation (10) will be:

Case 1 m is an positive even number.

Case 2 m is an positive odd number.

3.2.2 Discussion on Case 1

Let $m = 2^t p(t \in Z^+, p$ is an positive odd number), (10) is equivalent to:

$$Dz^2 = 4x_{2^t p} x_{2^t p+1} y_{2^t p} y_{2^t p+1} \tag{11}$$

For $y_{2m} = 2x_m y_m$, (10) is equivalent to:

$$Dz^2 = 2^{2+t} x_{2^t p+1} x_{2^t p} x_{2^{t-1} p} \cdots x_{2p} x_p y_{2^t p+1} y_p \tag{12}$$

From (I) of Lemma 4, we can get $gcd(y_{2^t p+1}, y_p) = y_1 = 3$. From (II) of Lemma 4, we can get $gcd(x_{2^t p+1}, x_p) = x_1 = 8$, so (12) is equivalent to:

$$Dz^2 = 2^{8+t} \cdot 3^2 \cdot x_{2^t p} x_{2^{t-1} p} \cdots x_{2p} \cdot \frac{y_{2^t p+1}}{3} \cdot \frac{y_p}{3} \cdot \frac{x_p}{8} \cdot \frac{x_{2^t p+1}}{8} \tag{13}$$

For $gcd(y_{2^t p+1}, y_p) = 3$, we can get $\left(\frac{y_{2^t p+1}}{3}, \frac{y_p}{3}\right) = 1$, for $gcd(x_{2^t p+1}, x_p) = 8$, we can get $\left(\frac{x_{2^t p+1}}{8}, \frac{x_p}{8}\right) = 1$. From (II) of Lemma 4, we can get $x_{2^t p}, x_{2^{t-1} p}, \cdots, x_{2p}, \cdots x_p$ pairwise coprime. From (III) of Lemma 4, we can get $x_{2^t p+1}, x_{2^t p}, x_{2^{t-1} p}, \cdots, x_{2p}, \cdots x_p$ is coprime with $y_{2^t p+1}$ and y_p. So

3

ICEESE 2017   IOP Publishing

IOP Conf. Series: Earth and Environmental Science **128** (2018) 012126   doi:10.1088/1755-1315/128/1/012126

$\frac{x_{2^t p+1}}{8}, x_{2^t p}, x_{2^{t-1}p}, \cdots, x_{2p}, \frac{x_p}{8}$ is coprime with $\frac{y_{2^t p+1}}{3}$ and $\frac{y_p}{3}$. It means that $x_{2^t p}, x_{2^{t-1}p}, \cdots, x_{2p}, \frac{y_{2^t p+1}}{3}, \frac{y_p}{3}, \frac{x_p}{8}, \frac{x_{2^t p+1}}{8}$ pairwise coprime.

Because $x_{2n} \equiv 1 (mod2), y_{2n+1} \equiv 1(mod2)$, we will get $x_{2^t p}, x_{2^{t-1}p}, \cdots, x_{2p}, y_{2^t p+1}, y_p$ are odd numbers, so $\frac{y_{2^t p+1}}{3}, \frac{y_p}{3}$ are odd numbers. And Because $x_{2n+1} \equiv 8(mod16)$, we will get $\frac{x_p}{8}, \frac{x_{2^t p+1}}{8}$ are odd numbers. $2(D) = 1$, so $2(2D^2)$ is odd number, but $D\left(2^{8+t} \cdot x_{2^t p} \cdot x_{2^{t-1}p} \cdots x_{2p} \cdot \frac{y_{2^t p+1}}{3} \cdot \frac{y_p}{3} \cdot \frac{x_p}{8} \cdot \frac{x_{2^t p+1}}{8}\right) = 8 + t$, t must be positive odd number.

From (ii) of Lemma 5, we can get $\frac{x_p}{8}$ is square number only when $p = 1$ or $p = -1$. From (iii) of Lemma 5, we can get $\frac{y_p}{3}$ is square number only when $p = 1$ or $p = 0$. From Lemma 5, we can get $\frac{x_{2^t p+1}}{8}, x_{2^t p}, x_{2^{t-1}p}, \cdots, x_{2p}, y_{2^t p+1}, \frac{y_{2^t p+1}}{3}$ are non-square numbers for the arbitrary positive odd number p, therefor, $x_{2^t p}, x_{2^{t-1}p}, \cdots, x_{2p}, y_{2^t p+1}, y_p, \frac{x_p}{8}, \frac{x_{2^t p+1}}{8}$ are non-square numbers.

When $p > 1$ is an positive odd number, $x_{2^t p}, x_{2^{t-1}p}, \cdots, x_{2p}, \frac{y_{2^t p+1}}{3}, \frac{y_p}{3}, \frac{x_p}{8}, \frac{x_{2^t p+1}}{8}$ are $t + 4$ odd numbers which is not equal to 1, so they provide $t + 4$ odd prime divisors at least for D. Further more, t is an positive odd number, $t + 4 \geq 5$, it means that the right half part of (13) provide 5 odd prime divisors at least for D, it is self-contradiction.

When $p = 1, t \neq 1$, $\frac{y_p}{3}$ and $\frac{x_p}{8}$ are square numbers, and $\frac{x_p}{8} = \frac{x_1}{8} = 1, \frac{y_p}{3} = \frac{y_1}{3} = 1$, here (13) is equivalent to:

$$Dz^2 = 2^{8+t} \cdot 3^2 \cdot x_{2^t p} x_{2^{t-1}p} \cdots x_{2p} \cdot \frac{y_{2^t p+1}}{3} \cdot \frac{x_{2^t p+1}}{8} \tag{14}$$

At present, $x_{2^t p}, x_{2^{t-1}p}, \cdots, x_{2p}, \frac{y_{2^t p+1}}{3}, \frac{x_{2^t p+1}}{8}$ provide $t + 2$ odd prime divisors at least for D. And $t \neq 1$, t is an positive odd number, so $t \geq 3$, $t + 2 \geq 5$. $x_{2^t p}, x_{2^{t-1}p}, \cdots, x_{2p}, \frac{y_{2^t p+1}}{3}, \frac{x_{2^t p+1}}{8}$ provide 5 odd prime divisors at least for D, it is self-contradiction.

When $p = 1, t = 1$, (14) is is equivalent to:

$$Dz^2 = 2^9 \cdot 3^2 \cdot x_2 \cdot \frac{y_3}{3} \cdot \frac{x_3}{8} = 2^9 \times 3^2 \times 5 \times 11 \times 17 \times 23 \times 127 \tag{15}$$

It shows that the right half part of (14) have 6 different odd prime, it is conflict with topic hypothesis, then (11) is might be wrong, (6) has no integer solution.

3.2.3   Discussion on Case 2

Let $m = 2^t p - 1 (t \in Z^+, p$ is an positive odd number), (10) is equivalent to:

$$Dz^2 = 4x_{2^t p} x_{2^t p-1} y_{2^t p} y_{2^t p-1} \tag{16}$$

It could be proved by imitate 3.2.2 that (16) have nontrivial solution only when $p = 1, t = 1$, here (16) turns into $Dz^2 = 4x_2 x_1 y_2 y_1 = 8x_1^2 x^2 y_1^2 = 2^9 \times 3^2 \times 127$, so $D = 2 \times 127, z = 48$, or $D = 2^3 \times 127, z = 24$, or $D = 2^5 \times 127, z = 12$, or $D = 2^7 \times 127, z = 6$, or $D = 2^9 \times 127, z = 3$, therefor equation (6) have nontrivial solution $(x, y, z) = (\pm 2024, \pm 765, \pm 48)$ when $D = 2 \times 127$, $(x, y, z) = (\pm 2024, \pm 765, \pm 24)$ when $D = 2^3 \times 127$, $(x, y, z) = (\pm 2024, \pm 765, \pm 12)$ when $D = 2^5 \times 127$, $(x, y, z) = (\pm 2024, \pm 765, \pm 6)$ when $D = 2^7 \times 127$, $(x, y, z) = (\pm 2024, \pm 765, \pm 3)$ when $D = 2^9 \times 127$,

In conclusion, the diophantine equations (6) has common solution $(x, y, z) = (\pm 8, \pm 3, 0)$ only when $D = 2^\alpha \times 127 (\alpha = 1,3,5,7,9)$, and nontrivial solution $(x, y, z) = (\pm 2024, \pm 765, \pm 48)$. has when $D = 2 \times 127$. $(x, y, z) = (\pm 2024, \pm 765, \pm 24)$ when $D = 2^3 \times 127$. $(x, y, z) = (\pm 2024, \pm 765, \pm 12)$ when $D = 2^5 \times 127$. $(x, y, z) = (\pm 2024, \pm 765, \pm 6)$ when $D = 2^7 \times 127$.$(x, y, z) = (\pm 2024, \pm 765, \pm 3)$ when $D = 2^9 \times 127$.otherewise it has only nontrivial solution $(x, y, z) = (\pm 2024, \pm 765, \pm 48)$when $D \neq 2^\alpha \times 127 (\alpha = 1,3,5,7,9)$.

4

ICEESE 2017   IOP Publishing

IOP Conf. Series: Earth and Environmental Science **128** (2018) 012126   doi:10.1088/1755-1315/128/1/012126

References

[1] Ljunggren W. Litt om Simultane Pellske Ligninger [J]. Norsk Mat. Tidsskr.,1941,23:132-138.

[2] Pan Jia-yu, Zhang Yu-ping, Zou Rong. The Pell Equations $x^2 - ay^2 = 1$ and $y^2 - Dz^2 = 1$ [J]. Chinese Quarterly Journal of Mathematics,1999,14(1):73-77.

[3] Guan Xun-gui. On the integer solution of Pell's equation $x^2 - 2y^2 = 1$ and $y^2 - Dz^2 = 4$ [J]. Journal of Huazhong Normal University (Natural Sciences):2012,46(3):267-269+278.

[4] Hu Yong-zhong, Han Qing. On the integer solution of the simultaneous equations $x^2 - 2y^2 = 1$ and $y^2 - Dz^2 = 4$ [J]. Journal of Hua Zhong Normal University (Natural Sciences):2002,36(1):17-19.

[5] Du Xian-cun, Guan Xun-gui, Yang Hui-zhang, On the system of Diophantine equations $x^2 - 6y^2 = 1$ and $y^2 - Dz^2 = 4$ [J]. Journal of Hua Zhong Normal University (Natural Sciences):2014.48(3):5-8.

[6] Zhao jian-hong, Wan Fei. On the Common Solution of Pell Equations $x^2 - 3y^2 = 1$ and $y^2 - 2^n y^2 = 16$ [J]. Journal of Hubei University for Nationalities (Natural Science Edition), 2016,2(6):146-148.

[7] He La-rong, Zhang Shu-jing, Yuan Jin. Diophantine Equation $x^2 - 6y^2 = 1$, $y^2 - Dz^2 = 4$ [J]. Journal of Yunnan University of Nationalities (Natural Sciences Edition),2012,21(1):57-58.

[8] Ran Yan-ping. Integer Solution of the Simultaneous Diophantine Equations $x^2 - 10y^2 = 1$ and $y^2 - Dz^2 = 4$ [J]. Journal of Yanan University (Natural Science Edition),2012,31(1):8-10.

[9] Guo Jing, Du Xian-cun. On the System of Pell Equations $x^2 - 30y^2 = 1$ and $y^2 - Dz^2 = 4$[J]. Mathematics in Practice and Theory.2015,45(1):309-314.

[10] Wan Fei, Du Xian-cun. On the Simultaneous Diophantine Equations $x^2 - 5y^2 = 1$ and $y^2 - Dz^2 = 16$ [J]. Journal of Chongqing Normal University (Natural Science),2016,33(4):107-111.

[11] Sun Qi, Yuan Ping-zhi, ON THE DIOPHANTINE EQUATION $x^4 - Dy^2 = 1$ [J]. Journal Of Sichuan University (Natural Science Edition),1997:34(3):265-267.

[12] Le Mao-hua. A Family of Binary Quartic Diophantine Equations [J]. Journal of Yunnan Normal University (Natural Sciences Edition),2010,30(1):12-17.

[13] Walsh G. A note on a theorem of Ljunggren and the Diophantine equations $x^2 - kxy^2 + y^4 = 1$ or 4[J]. Arch. Math.1999,73(2):504-513.

[14] Chen Yong-gao. Pell Equation $x^2 - 2y^2 = 1$ and $y^2 - Dz^2 = 4$ [J]. Acta Scientia rum Naturahum Universitatis Pekinesis,1994,30(3):298-302.

[15] Wang Guan-min, Li Bing-rong. The Integer Solution of The Pell's Equations $x^2 - 6y^2 = 1$ and $y^2 - Dz^2 = 4$ [J]. Journal of ZhangZhou Teachers College，2002,15(4):9-14.

IOP Conference Series: Earth and Environmental Science

PAPER • OPEN ACCESS

Elliptic Curve Integral Points on $y^2 = x^3 + 3x - 14$

To cite this article: Jianhong Zhao 2018 IOP Conf. Ser.: Earth Environ. Sci. **128** 012108

View the article online for updates and enhancements.

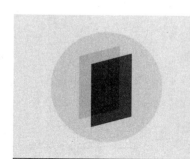

ICEESE 2017

IOP Publishing

IOP Conf. Series: Earth and Environmental Science **128** (2018) 012108   doi:10.1088/1755-1315/128/1/012108

Elliptic Curve Integral Points on $y^2 = x^3 + 3x - 14$

Jianhong Zhao

Department of Teachers and Education, Lijiang Teachers College
Lijiang, 674199 ,Yunnan, China

E-mail:312508050@qq.com

Abstract. The positive integer points and integral points of elliptic curves are very important in the theory of number and arithmetic algebra, it has a wide range of applications in cryptography and other fields. There are some results of positive integer points of elliptic curve $y^2 = x^3 + ax + b, a, b \in Z$ In 1987, D. Zagier submit the question of the integer points on $y^2 = x^3 - 27x + 62$, it count a great deal to the study of the arithmetic properties of elliptic curves.
In 2009, Zhu H L and Chen J H solved the problem of the integer points on $y^2 = x^3 - 27x + 62$ by using algebraic number theory and P-adic analysis method.
In 2010, By using the elementary method, Wu H M obtain all the integral points of elliptic curves $y^2 = x^3 - 27x - 62$.
In 2015, Li Y Z and Cui B J solved the problem of the integer points on $y^2 = x^3 - 21x - 90$ By using the elementary method.
In 2016, Guo J solved the problem of the integer points on $y^2 = x^3 + 27x + 62$ by using the elementary method.
In 2017, Guo J proved that $y^2 = x^3 - 21x + 90$ has no integer points by using the elementary method.
Up to now, there is no relevant conclusions on the integral points of elliptic curves $y^2 = x^3 + 3x - 14$, which is the subject of this paper.
By using congruence and Legendre Symbol, it can be proved that elliptic curve $y^2 = x^3 + 3x - 14$ has only one integer point: $(x, y) = (2,0)$.

1. Introduction

The positive integer points and integral points of elliptic curves are very important in the theory of number and arithmetic algebra, it has a wide range of applications in cryptography and other fields. There are some results of positive integer points of elliptic curve
$$y^2 = x^3 + ax + b, a, b \in Z \qquad (1)$$
In 1987, D. Zagier [1] submit the question of the integer points on elliptic curve (1) while $a = -27, b = 62$, that is $y^2 = x^3 - 27x + 62$, it counts a great deal to the study of the arithmetic properties of elliptic curves.

In 2009, Zhu H L and Chen J H [2] solved the problem what D. Zagier submitted by using algebraic number theory and P-adic analysis method.

In 2010, By using the elementary method, Wu H M [3] obtain all the integral points of elliptic curves $y^2 = x^3 - 27x - 62$.

In 2015, Li Y Z and Cui B J [4] solved the problem of the integer points on $y^2 = x^3 - 21x - 90$ By using the elementary method.

1

ICEESE 2017

IOP Publishing

IOP Conf. Series: Earth and Environmental Science **128** (2018) 012108   doi:10.1088/1755-1315/128/1/012108

In 2016, Guo J [5] solved the problem of the integer points on $y^2 = x^3 + 27x + 62$ by using the elementary method.

In 2017, Guo J [6] proved that $y^2 = x^3 - 21x + 90$ has no integer points by using the elementary method.

Put in a nutshell, Scholars studied the integer points on elliptic curve (1) while $a_1 = -27, b_1 = 62; a_2 = -27, b_2 = -62; a_3 = -21, b_3 = -90; a_4 = 27, b_4 = 62; a_5 = -21, b_5 = 90$. Up to now, there is no relevant conclusions while $a = 3, b = -14$.

2. Key lemma

Key lemma [7] Let D to be a square-free positive integer, then the equation $x^2 - Dy^4 = 1$ will have two sets of positive integer solutions (x, y) at most.

When $D = 2^{4s} \times 1785$, where $s \in \{0,1\}$, we can get that $(x_1, y_1) = (169, 2^{1-s})$ and $(x_2, y_2) = (6525617281, 2^{1-s} \times 6214)$;

Otherwise when $D \neq 2^{4s} \times 1785$, $(x_1, y_1) = (u_1, \sqrt{v_1})$ and $(x_2, y_2) = (u_2, \sqrt{v_2})$, where (u_n, v_n) is a positive integer solution of the Pell equation $U^2 - DV^2 = 1$, if $x^2 - Dy^4 = 1$ has only one set of positive integer solution (x, y) and the positive integer n is suitable for $(x, y^2) = (u_n, v_n)$, then $n = 2$ consequently; otherwise if n is an even number; otherwise if n is an odd number, then $n = 1$ or p, here p is a prime numbers and $p \equiv 3 (mod 4)$.

3. Proof of main theorem

By using elementary method such as congruence and Legendre Symbol, the integer points on $y^2 = x^3 + 27x + 62$ can be obtained.

3.1 Theorem

Elliptic curve

$$y^2 = x^3 + 3x - 14 \qquad (2)$$

has only one integer point $(x, y) = (2,0)$.

3.2 Proof of the main theorem

Elliptic curve (2) is equivalent to

$$y^2 = (x - 2)(x^2 + 2x + 7) \qquad (3)$$

3.2.1 Primary analysis

Suppose (x, y) is a positive integer point of the elliptic curve (3), because $x^2 + 2x + 7 = (x + 1)^2 + 6 \geq 6 > 0$, and $y^2 \geq 0$, we will get $x - 2 \geq 0$ from (3), that is $x \geq 2$. When $x = 2$, the integer points for (3) will be $(x, y) = (2,0)$.

Otherwise we only need to discuss on the situation of the integer point for (3) when $x > 2$.

Let $m = gcd(x - 2, x^2 + 2x + 7) = gcd(x + 2,15)$, then $m \in \{1,3,5,15\}$, as a result, the elliptic curve (3) will be:

Case I $x - 2 = a^2, x^2 + 2x + 7 = b^2, y = ab, gcd(a, b) = 1, a, b \in Z$

Case II $\quad x - 2 = 3a^2, x^2 + 2x + 7 = 3b^2, y = 3ab, gcd(a, b) = 1, a, b \in Z$

Case III $\quad x - 2 = 5a^2, x^2 + 2x + 7 = 5b^2, y = 5ab, gcd(a, b) = 1, a, b \in Z$

Case IV $\quad x - 2 = 15a^2, x^2 + 2x + 7 = 15b^2, y = 15ab, gcd(a, b) = 1, a, b \in Z$

3.2.2 Discussion on Case I

$\because a^2 \equiv 0,1 (mod 4)$.

$\therefore x = a^2 + 2 \equiv 2,3 (mod 4)$.

$\therefore x^2 + 2x + 7 \equiv 2,3 (mod 4)$.

2

ICEESE 2017

IOP Publishing

IOP Conf. Series: Earth and Environmental Science **128** (2018) 012108　　doi:10.1088/1755-1315/128/1/012108

At the same time $b^2 = x^2 + 2x + 7 \equiv 0,1(mod4)$, it means $2,3(mod4) \equiv 0,1(mod4)$, it is self-contradiction, this shows that (3) has no integer points.

3.2.3 Discussion on Case II
∵ $a^2 \equiv 0,1,4(mod8)$.
∴ $x = 3a^2 + 2 \equiv 2,5,6(mod8)$.
∴ $x^2 + 2x + 7 \equiv 2,7(mod8)$.
At the same time $b^2 \equiv 0,1,4(mod8)$.
∴　$3b^2 = x^2 + 2x + 7 \equiv 0,3,4(mod8)$, it means $2,7(mod8) \equiv 0,3,4(mod8)$, it is self-contradiction, this shows that (3) has no integer points.

3.2.4 Discussion on Case III
∵ $a^2 \equiv 0,1(mod4)$.
∴ $x = 5a^2 + 2 \equiv 2,3(mod4)$.
∴ $x^2 + 2x + 7 \equiv 2,3(mod4)$.
At the same time $b^2 \equiv 0,1(mod4)$.
∴　$5b^2 = x^2 + 2x + 7 \equiv 0,1(mod4)$, it means $2,3(mod8) \equiv 0,1(mod4)$, it is self-contradiction, this shows that (3) has no integer points.

3.2.5 Discussion on Case IV
Divide integers into two categories as $2 \nmid a$ and $2 | a$, discuss separately.
First step: $2 \nmid a$.
∵ $2 \nmid a$.
∴ $a^2 \equiv 1(mod4)$.
∴ $x = 15a^2 + 2 \equiv 1(mod4)$.
∴ $x^2 + 2x + 7 \equiv 2(mod4)$.
At the same time $b^2 \equiv 0,1(mod4)$.
∴　$15b^2 = x^2 + 2x + 7 \equiv 0,3(mod4)$, it means $2(mod4) \equiv 0,3(mod4)$, it is self-contradiction, this shows that (3) has no integer points as well.
Second step:$2 | a$.
∵ $2 | a$, let $a = 2e, e \in Z$, and $x - 2 = 15a^2$.
∴ $x = 60e^2 + 2$.
Go a step further $x^2 + 2x + 7 = (x+1)^2 + 6 = (60e^2 + 3)^2 + 6 = 15b^2$, it is $(12e^2 + 1)^2 + 96e^4 = b^2$, it is equivalent to:
$$(b + 12e^2 + 1)(b - 12e^2 - 1) = 96e^4 \qquad (4)$$
∵ $2 | a$, and $x - 2 = 15a^2$.
∴ $2 | 15a^2 + 2$.
∴ $2 | x$.
∵ $x^2 + 2x + 7 = 15b^2$.
∴ $2 \nmid x^2 + 2x + 7$.
∴ $2 \nmid 15b^2$.
∴ $2 \nmid b$.
Taken together, $2 | [b - (12e^2 + 1)]$.
∴ $\gcd(b + 12e^2 + 1, b - 12e^2 - 1) = \gcd(24e^2 + 2, b - 12e^2 - 1)$
$$= 2\gcd(12e^2 + 1, b - 12e^2 - 1)$$
$$= 2\gcd(12e^2 + 1, b).$$
Let $d = \gcd(12e^2 + 1, b)$, Then $d | b, d | 12e^2 + 1$, so, $d | (b + 12e^2 + 1)$, Therefore, $d | 96e^4$.
∵ $\gcd(12e^2 + 1, 96e^4) = 1$.
∴ $d = 1$. it means $\gcd(b + 12e^2 + 1, b - 12e^2 - 1) = 2$.

ICEESE 2017

IOP Publishing

IOP Conf. Series: Earth and Environmental Science **128** (2018) 012108   doi:10.1088/1755-1315/128/1/012108

Furthermore $96 = 3 \times 2^5$, equation (4) can be divided into:

$$\begin{cases} b + 12e^2 + 1 = 2cf^4 \\ b - 12e^2 - 1 = \frac{48}{c}g^4, \ gcd(f,g) = 1, gcd\left(c, \frac{24}{c}\right) = 1, c = 1,3,8,24. \\ e = fg \end{cases} \quad (5)$$

From the first two formulas (5), we will get:

$$12e^2 + 1 = cf^4 - \frac{24}{c}g^4 \quad (6)$$

Making an equivalent of the modulus 4 on (6), we will get:

$$1 \equiv cf^4 - \frac{24}{c}g^4 \ (mod4) \quad (7)$$

Making an equivalent of the modulus 3 on (6), we will get:

$$1 \equiv cf^4 - \frac{24}{c}g^4 \ (mod3) \quad (8)$$

When $c = 24$, (7) is equivalent to:

$$1 \equiv -g^4 \ (mod4) \quad (9)$$

$\because g^2 \equiv 0,1 \ (mod4)$.

$\therefore -g^4 \equiv 0,3 \ (mod4)$.

(9) means $1 \equiv 0,3 \ (mod4)$, it is self-contradiction, this shows that (3) has no integer points.
When $c = 3$, (7) is equivalent to:

$$1 \equiv 3g^4 \ (mod4) \quad (10)$$

$\because f^2 \equiv 0,1 \ (mod4)$.

$\therefore 3f^4 \equiv 0,3 \ (mod4)$.

(10) means $1 \equiv 0,3 \ (mod4)$, it is self-contradiction, this shows that (3) has no integer points.
When $c = 2$, (8) is equivalent to:

$$1 \equiv 2f^4 \ (mod3) \quad (11)$$

Because the Legendre symbol value is $\left(\frac{1}{3}\right) = 1$, while the Legendre symbol value is $\left(\frac{2f^4}{3}\right) = \left(\frac{2}{3}\right) = -1$, it is self-contradiction, this shows that (3) has no integer points.

When $c = 1$, the (6) is equivalent to $12e^2 + 1 = f^4 - 24g^4$, from $e = fg$ and $12f^2g^2 + 1 = f^4 - 24g^4$, we will get:

$$(f^2 - 6g^2)^2 - 60g^4 = 1 \quad (12)$$

Let $s = f^2 - 6g^2, s \in Z$, (12) is equivalent to:

$$s^2 - 60g^4 = 1 \quad (13)$$

Let $t = 2g^2, t \in N$, (13) is equivalent to:

$$s^2 - 15t^2 = 1 \quad (14)$$

From the key Lemma we will get equation (13) and it will have a set of positive integer solutions (s, g) at most, and if it is so, equation (14) will have a positive integer solution $(s, t) = (s, 2g^2)$.

Let $(s_1, t_1) = (4,1)$ to be the basic solution of Pell equation (14), then all the positive integer solutions of equation (14) can be expressed as: $s_n + t_n\sqrt{15} = (4 + \sqrt{15})^n, n \in Z^+$.

It can be seen that the positive integer solution of (12) is satisfied:

$$(f^2 - 6g^2) + 2g^2\sqrt{15} = (4 + \sqrt{15})^n, n \in Z^+. \quad (15)$$

From (15) we will get:

$$2g^2 = \begin{cases} \displaystyle\sum_{i=0}^{n-1} \binom{n}{2i} \times 4^{2i} \times 15^{\frac{n-2i-1}{2}}, 2 \nmid n \\ \displaystyle\sum_{i=1}^{\frac{n}{2}+1} \binom{n}{2i-1} \times 4^{2i-1} \times 15^{\frac{n-2i}{2}}, 2 \mid n \end{cases}.$$

4

ICEESE 2017　　　　　　　　　　　　　　　　　　　　　　　　　　　　IOP Publishing

IOP Conf. Series: Earth and Environmental Science **128** (2018) 012108　　doi:10.1088/1755-1315/128/1/012108

it means:

$$2g^2 = \begin{cases} 15^{\frac{n-1}{2}} + \sum_{i=1}^{n-1} \binom{n}{2i} \times 4^{2i} \times 15^{\frac{n-2i-1}{2}}, & 2 \nmid n \\ \sum_{i=1}^{\frac{n}{2}+1} \binom{n}{2i-1} \times 4^{2i-1} \times 15^{\frac{n-2i}{2}}, & 2 \mid n \end{cases} \qquad .(16)$$

When $2 \nmid n$, $15^{\frac{n-1}{2}} + \sum_{i=1}^{n-1} \binom{n}{2i} \times 4^{2i} \times 15^{\frac{n-2i-1}{2}}$ is an odd number, while $2g^2$ is an even number, it is self-contradiction. Therefore, it means $2 \mid n$, we know that n in equation (15) should be $n = 2$ from the Lemma.

When $n = 2$, we will get $(f^2 - 6g^2) + 2g^2\sqrt{15} = (4 + \sqrt{15})^n = 31 + 8\sqrt{15}$, therefore $f^2 - 6g^2 = 31, 2g^2 = 8$, then $f^2 = 55$, Obviously there is no integer solution, it is self-contradiction, this shows that (3) has no integer points.

In conclusion, elliptic curve $y^2 = x^3 + 3x - 14$ has only one integer point $(x, y) = (2,0)$.

References

[1]　D. Zagier. Lager Integral Point on Elliptic Curves [J]. Math Comp, 1987,48:425-436.

[2]　Zhu H L, Chen J H. Integral point on $y^2 = x^3 + 27x - 62$ [J]. J Math Study, 2009, 42(2): 117-125.

[3]　Wu H M . Points on the Elliptic Curves $y^2 = x^3 - 27x - 62$ [J] . J Acta Mathematica Curve Sinica,2010,53(1):205-208.

[4]　Li Y Z, Cui B J. Points on the Elliptic $\quad y^2 = x^3 - 21x - 90$ [J]. Journal of Yanan University (Natural Science Edition), 2015, 34(3):14-15.

[5]　Guo J. The Integral Points on the Elliptic Curve $y^2 = x^3 + 27x + 62$ [J]. Journal of Chongqing Normal University (Natural Science), 2016,33(5): 50- 53.

[6]　Guo J. The Positive Integral Points on the Elliptic Curve $y^2 = x^3 - 21x + 90$ [J]. Mathematics in Practice and Theory,2017,47(8):288-291.

[7]　Togbé A.,Voutier P.M.,and Walsh P.G.,Solving a family of Thue equations with an application to the equation $x^2 - dy^2 = 1$.Acta. Arith.,2005,120(1) :39-58.

IOP Conference Series: Materials Science and Engineering

On the Diophantine equations $x^2 - 47y^2 = 1$ and $y^2 - Pz^2 = 49$

To cite this article: Jianhong Zhao and Lixing Yang 2018 *IOP Conf. Ser.: Mater. Sci. Eng.* **382** 052037

View the article online for updates and enhancements.

Related content

- On the Diophantine Equation Ax 2 -- KX Y + Y2 + Lx = 0
  J D Urrutia, J M E Arañas, J A C L Lara et al.

- Nonlinear Diophantine equation 11x +13y = z2
  A Sugandha, A Tripena, A Prabowo et al.

- AN ASYMPTOTIC FORMULA FOR THE NUMBER OF SOLUTIONS OF A DIOPHANTINE EQUATION
  M I Israilov

IOP ebooks™

Bringing you innovative digital publishing with leading voices to create your essential collection of books in STEM research.

Start exploring the collection - download the first chapter of every title for free.

AMIMA 2018

IOP Publishing

IOP Conf. Series: Materials Science and Engineering **382** (2018) 052037   doi:10.1088/1757-899X/382/5/052037

On the Diophantine equations $x^2 - 47y^2 = 1$ and $y^2 - Pz^2 = 49$

Jianhong Zhao, Lixing Yang*

Department of Teachers and Education, Lijiang Teachers College,Lijiang, Yunnan 674199, China

*Corresponding author's E-mail: 375734990@qq.com

Abstract. The integer solution of diophantine equations $x^2 - D_1y^2 = m, (D_1 \in Z^+, m \in Z)$ and $y^2 - D_2z^2 = n, (D_2 \in Z^+, n \in Z)$ is a matter of great concern. Researchers study for different m,n and D_1, D_2, and obtained some correlation results as follows.
When $m = 1, n = 1$, the diophantine equations turns into $x^2 - D_1y^2 = 1$ and $y^2 - D_2z^2 = 1$. At present, there are only a few conclusions on it, see Ref [1] and [2]. When $m = 1, n = 4$, the integer solution see Ref [3] - [9]. When $m = 1, n = 16$, the previous conclusions see Ref [10].
When $m = 1, n = 49$, the diophantine equations turns into $x^2 - D_1y^2 = 1$ and $y^2 - D_2z^2 = 49$. In this case, $D_1 = 47, D_2$ can be expressed as $2^t p_1{}^{a_1} p_2{}^{a_2} p_3{}^{a_3} p_4{}^{a_4}$ where $a_i = 0$ or 1 for $1 \leq i \leq 4$, and $t \in Z^+, p_s(1 \leq s \leq 4)$ are different odd primes. Up to now, there is no relevant result on the integer solution of $x^2 - 47y^2 = 1$ and $y^2 - Pz^2 = 49$, this paper mainly discusses the integer solution of it.

1. Introduction

The integer solution of diophantine equations

$$x^2 - D_1y^2 = m, (D_1 \in Z^+, m \in Z) \text{ and } y^2 - D_2z^2 = n, (D_2 \in Z^+, n \in Z) \quad (1)$$

is a matter of great concern. Researchers study for different m, n and D_1, D_2, and obtain some correlation results as follows.

When $m = 1, n = 1$, diophantine equations (1) turns into:

$$x^2 - D_1y^2 = 1 \text{ and } y^2 - D_2z^2 = 1 \quad (2)$$

At present, there are only a few conclusions on (2), see Ref [1] and [2].

When $m = 1, n = 4$, diophantine equations (1) turns into:

$$x^2 - D_1y^2 = 1 \text{ and } y^2 - D_2z^2 = 4 \quad (3)$$

For even numbers D_1, D_2, the integer solution of (3), see Ref [3] - [9].

When $m = 1, n = 16$, diophantine equations (1) turns into:

$$x^2 - D_1y^2 = 1 \text{ and } y^2 - D_2z^2 = 16 \quad (4)$$

The previous conclusions on (4), see Ref [10].

When $m = 1, n = 49$, the diophantine equations turns into:

$$x^2 - D_1y^2 = 1 \text{ and } y^2 - D_2z^2 = 49 \quad (5)$$

In this case, $D_1 = 47, D_2$ can be expressed as $2^t p_1{}^{a_1} p_2{}^{a_2} p_3{}^{a_3} p_4{}^{a_4}$ where $a_i = 0$ or 1 for $1 \leq i \leq 4$, and $t \in Z^+, p_s(1 \leq s \leq 4)$ are different odd primes. Up to now, there is no relevant result on the

1

AMIMA 2018

IOP Publishing

IOP Conf. Series: Materials Science and Engineering **382** (2018) 052037   doi:10.1088/1757-899X/382/5/052037

integer solution of $x^2 - 47y^2 = 1$ and $y^2 - Pz^2 = 49$, this paper mainly discusses the integer solution of it.

2. Critical lemma

Lemma 1[11]   Let p be an odd prime number, there is no integer solution of the diophantine equation $x^4 - py^2 = 1$ except $p = 5, x = 3, y = 4$ and $p = 29, x = 99, y = 1820$.

Lemma 2[12]   There is 1 sets of solutions of the diophantine equation $ax^4 - by^2 = 1$ at most when a is a square number which is greater than 1.

Lemma 3[13]   Let D be a square-free positive integer, then the equation $x^2 - Dy^4 = 1$ has two sets of positive integer solutions (x, y) at most. Furthermore, the necessary and sufficient condition of it is $D = 1785$ or $D = 28560$, or $2x_0$ and y_0 are square numbers where (x_0, y_0) is the basic solution of $x^2 - Dy^4 = 1$.

Lemma 4   Suppose that all the integer solution on Pell equation $x^2 - 7y^2 = 1$ could be $(x_n, y_n), n \in Z$, for the arbitrary $n \in Z$, it has the following properties on (x_n, y_n):

(I) x_n is a square number if and only if $n = 0$.

(II)$\frac{x_n}{48}$ is a square number if and only if $n = \pm 1$.

(III) $\frac{y_n}{7}$ is a square number if and only if $n = 0$ or $n = 1$.

Proof: (I) Let $x_n = a^2$, we will get $a^4 - 47y^2 = 1$, from Lemma 1 we can get there are only 2 integer solution $(a, y) = (\pm 1, 0)$ on $a^4 - 47y^2 = 1$, so $x_n = 1, n = 0$. On the contrary, it also holds.

(II) Let $\frac{x_n}{48} = a^2$, we will get $2304a^4 - 47y^2 = 1$, from Lemma 2 we can get there are only 4 integer solution $(a, y) = (\pm 1, \pm 7)$ on $2304a^4 - 47y^2 = 1$, so $x_n = 48, n = \pm 1$. On the contrary, it also holds.

(III) Let $\frac{y_n}{7} = b^2$, we will get $x^2 - 2303b^4 = 1$, from Lemma 3 we can get there are only 6 integer solution $(x, b) = (\pm 1, 0), (\pm 48, \pm 1)$ on $x^2 - 2303b^4 = 1$, so $y_n = 0$ or $y_n = 7. n = 0$ or $n = 1$. On the contrary, it also holds.

3. Proof of main theorem

By using elementary method such as congruence, the integer solution of the diophantine equations on $x^2 - 47y^2 = 1$ and $y^2 - Pz^2 = 9$ can be obtained.

3.1. Theorem

Let $p_s (1 \le s \le 4)$ are diverse odd primes, $P = 2^k p_1^{a_1} \cdots p_s^{a_s} (a_i = 0 \text{ or } 1, 1 \le i \le 4, k \in Z^+)$, then the diophantine equations

$$x^2 - 47y^2 = 1 \text{ and } y^2 - Pz^2 = 49 \qquad (6)$$

(i) has common solution $(x, y, z) = (\pm 48, \pm 7, 0)$ and nontrivial solution $(x, y, z) = (\pm 442224, \pm 64505, \pm 672)$ when $P = 2 \times 17 \times 271$.

(ii) has common solution $(x, y, z) = (\pm 48, \pm 7, 0)$ and nontrivial solution $(x, y, z) = (\pm 442224, \pm 64505, \pm 336)$ when $P = 2^3 \times 17 \times 271$.

(iii) has common solution $(x, y, z) = (\pm 48, \pm 7, 0)$ and nontrivial solution $(x, y, z) = (\pm 442224, \pm 64505, \pm 168)$ when $P = 2^5 \times 17 \times 271$.

(iv) has common solution $(x, y, z) = (\pm 48, \pm 7, 0)$ and nontrivial solution $(x, y, z) = (\pm 442224, \pm 64505, \pm 84)$ when $P = 2^7 \times 17 \times 271$.

(v) has common solution $(x, y, z) = (\pm 48, \pm 7, 0)$ and nontrivial solution $(x, y, z) = (\pm 442224, \pm 64505, \pm 42)$ when $P = 2^9 \times 17 \times 271$.

(vi) has common solution $(x, y, z) = (\pm 48, \pm 7, 0)$ and nontrivial solution $(x, y, z) = (\pm 442224, \pm 64505, \pm 21)$ when $P = 2^{11} \times 17 \times 271$.

(vii) has only nontrivial solution $(x, y, z) = (\pm 48, \pm 7, 0)$ when $P \ne 2^\alpha \times 17 \times 127 (\alpha = 1, 3, 5, 7, 9, 11)$.

2

AMIMA 2018
IOP Publishing
IOP Conf. Series: Materials Science and Engineering **382** (2018) 052037 doi:10.1088/1757-899X/382/5/052037

3.2. Proof of main theorem

3.2.1. Primary analysis.

Let (x_1, y_1) be the basic solution of the Pell equation $x^2 - 47y^2 = 1$, then $(x_1, y_1) = (48,7)$. It means that all solution of the Pell equation $x^2 - 47y^2 = 1$ is:

$$x_n + y_n\sqrt{P} = (48 + 7\sqrt{47})^n, n \in Z.$$

It is easily shown that

(i) $y_n{}^2 - 49 = y_{n+1}y_{n-1}$;

(ii) $y_{2n} = 2x_n y_n$;

(iii) $y_{2n+1} \equiv 1 (mod2)$;

(iv) $x_{2n} \equiv 1(mod2), x_{2n+1} \equiv 48(mod96)$;

(v) $\gcd(x_n, y_n) = 1, \gcd(x_{n+1}, y_{n+1}) = 1, \gcd(x_n, x_{n+1}) = 1, \gcd(y_n, y_{n+1}) = 7$;

(vi) $\gcd(x_{2n}, y_{2n+1}) = \gcd(x_{2n+2}, y_{2n+1}) = 1, \gcd(x_{2n+1}, y_{2n}) = \gcd(x_{2n+1}, y_{2n+2}) = 48$.

(vii) $y_{2n+2} = 96y_{n+1} - y_n, y_0 = 0, y_1 = 7; x_{n+2} = 96x_{n+1} - x_n, x_0 = 1, x_1 = 48$.

Suppose that $(x, y, z) = (x_n, y_n, z_n), n \in Z$ is the integer solution of the diophantine equation (6), from(I), we can get:

$$y_n{}^2 - 49 = y_{n+1}y_{n-1} \tag{7}$$

from $y^2 - Pz^2 = 49$ of (6), we can get:

$$Pz^2 = y_{n+1}y_{n-1} \tag{8}$$

As a result the equation (8) will be:

Case 1 n is an positive odd number.

Case 2 n is an positive even number.

3.2.2. Discusion on Case 1

Let $n = 2m - 1, m \in Z$,(8) is equivalent to:

$$Pz^2 = y_{2(m-1)}y_{2m} \tag{9}$$

from (II),(9) is equivalent to:

$$Pz^2 = 4x_{m-1}y_{m-1}x_m y_m \tag{10}$$

1. m is an positive odd number.

From (V), we can get $\gcd(x_{m-1}, y_{m-1}) = \gcd(x_m, y_m) = 1$, $\gcd(x_m, x_{m-1}) = 1$, $\gcd(x_{m-1}, y_m) = 1, \gcd(y_m, y_{m-1}) = 7$, it means $\gcd\left(\frac{y_m}{7}, \frac{y_{m-1}}{7}\right) = 1$. From (VI), we can get $\gcd(x_m, y_{m-1}) = 48$, it means $\gcd\left(\frac{x_m}{48}, \frac{y_{m-1}}{48}\right) = 1$.

Therefore, $x_{m-1}, \frac{y_{m-1}}{336}, \frac{x_m}{48}, \frac{y_m}{7}$ are pairwise coprime.

1.1 k is an positive odd number.

Let $k = 2l - 1$,(10) is equivalent to:

$$Pz^2 = 8x_{4(l-1)}x_{4l-3}x_{2(l-1)}y_{2(l-1)}y_{4l-3} \tag{11}$$

From (II),(11) is equivalent to:

$$Pz^2 = 16x_{4(l-1)}x_{4l-3}x_{2(l-1)}x_{l-1}y_{l-1}y_{4l-3} \tag{12}$$

From (V), we can get $\gcd(x_{l-1}, y_{l-1}) = 1$, it means $x_{4(l-1)}, x_{l-1}, \frac{y_{l-1}}{336}, x_{2(l-1)}, \frac{x_{4l-3}}{48}, \frac{y_{4l-3}}{7}$ are pairwise coprime when l is an odd number, and $x_{4(l-1)}, \frac{x_{l-1}}{48}, \frac{y_{l-1}}{7}, x_{2(l-1)}, \frac{x_{4l-3}}{48}, \frac{y_{4l-3}}{7}$ are pairwise coprime when l is an even number.

From (III), we can get $y_{4l-3} \equiv 1(mod2)$, it means $2 \nmid y_{4l-3}$, so y_{4l-3} is an odd number. From (IV), we can get $x_{4(l-1)}, x_{2(l-1)}, \frac{y_{4l-3}}{48}$ are odd numbers. x_{l-1} is an odd number when l is an odd number and $\frac{x_{l-1}}{48}$ is an odd number when l is an even number. Therefore, $x_{4(l-1)}, x_{l-1}, x_{2(l-1)}, \frac{x_{4l-3}}{48}, \frac{y_{4l-3}}{7}$ are odd

3

AMIMA 2018

IOP Publishing

IOP Conf. Series: Materials Science and Engineering **382** (2018) 052037   doi:10.1088/1757-899X/382/5/052037

numbers when l is an odd number and $x_{4(l-1)}, \frac{x_{l-1}}{48}, x_{2(l-1)}, \frac{x_{4l-3}}{48}, \frac{y_{4l-3}}{7}$ are odd numbers when l is an even number.

From Lemma 4, we can get $x_{4(l-1)}, x_{2(l-1)}, x_{l-1}, \frac{x_{4l-3}}{48}, \frac{y_{4l-3}}{7}$ are square numbers if and only if $l = 1$, $\frac{x_{l-1}}{48}$ is a square number if and only if $l = 2$ or $l = 0$.

So, $x_{4(l-1)}, x_{2(l-1)}, x_{l-1}, \frac{x_{4l-3}}{48}, \frac{y_{4l-3}}{7}$ are non-square numbers when odd number $l \neq 1$, and it has 5 diverse odd primes. Therefore, (12) is impossible, which means (6) have no integer solution.

$x_{4(l-1)}, x_{2(l-1)}, \frac{x_{l-1}}{48}, \frac{x_{4l-3}}{48}, \frac{y_{4l-3}}{7}$ are non-square numbers when even number $l \neq 0, 2$, and it has 5 diverse odd primes, which is contradict with $P = 2^t p_1{}^{a_1} p_2{}^{a_2} p_3{}^{a_3} p_4{}^{a_4}$, Therefore, (12) is impossible, which means (6) have no integer solution.

When $l = 1$, (11) is equivalent to: $z^2 = 8x_0{}^2 x_1 y_0 y_1 = 0$, so $z = 0$, it means that diophantine equation (6) has and only has common solution $(x, y, z) = (\pm 48, \pm 7, 0)$.

When $l = 0, 2$, (12) is equivalent to: $Pz^2 = 16x_{-4}x_{-3}x_{-2}x_{-1}y_{-1}y_{-3} = 16x_4 x_3 x_2 x_1 y_1 y_3$, From (IV), we can get x_4, x_2 are odd numbers, from (III), we can get y_1, y_3 are odd numbers, it means $Pz^2 = 2^8 \times 3 \times x_4 x_3 x_2 y_1 y_3$, Therefore, the right part of (12) has 5 diverse odd primes at least, which is contradict with $P = 2^t p_1{}^{a_1} p_2{}^{a_2} p_3{}^{a_3} p_4{}^{a_4}$, Therefore, (12) is impossible, which means (6) have no integer solution.

1.2 k is an positive even number.

From (III), we can get y_{k-1}, y_{2k-1} are odd numbers, it means $\frac{y_{k-1}}{7}, \frac{y_{2k-1}}{7}$ are odd numbers too, From (IV), we can get $x_{2k-1}, \frac{x_{k-1}}{48}, \frac{x_{2k-1}}{48}$ are odd numbers.

From Lemma 4, we can get $x_{2(k-1)}, \frac{y_{2k-1}}{7}$ are square numbers if and only if $k = 1$. $\frac{x_{k-1}}{48}$ is a square number if and only if $k = 0$ or $k = 2$. $\frac{x_{2k-1}}{48}$ is a square number if and only if $k = 0$ or $k = 1$. $\frac{y_{k-1}}{7}$ is a square number if and only if $k = 1$ or $k = 2$. So, $x_{2k-1}, \frac{x_{k-1}}{48}, \frac{x_{2k-1}}{48}, \frac{y_{k-1}}{7}, \frac{y_{2k-1}}{7}$ are non-square numbers when even number $k \neq 0, 2$, and it has 5 diverse odd primes. Therefore, (12) is impossible, which means (6) have no integer solution.

When $k = 0$, (11) is equivalent to: $Pz^2 = 8x_2 y_1{}^2 x_1{}^2 = 2^{11} \times 3^2 \times 7^2 \times 17 \times 271$, so $z = 21, P = 2^{11} \times 17 \times 271$ or $z = 42, P = 2^9 \times 17 \times 271$ or $z = 84, P = 2^7 \times 17 \times 271$ or $z = 168, P = 2^5 \times 17 \times 271$ or $z = 336, P = 2^3 \times 17 \times 271$ or $z = 672, P = 2 \times 17 \times 271$, From (6), we can get:

(6) has common solution $(x, y, z) = (\pm 48, \pm 7, 0)$ when $P = 2^\alpha \times 17 \times 127 (\alpha = 1, 3, 5, 7, 9, 11)$ and has nontrivial solution $(x, y, z) = (\pm 442224, \pm 64505, \pm 672)$ when $P = 2 \times 17 \times 271$. $(x, y, z) = (\pm 442224, \pm 64505, \pm 336)$ when $P = 2^3 \times 17 \times 271$, $(x, y, z) = (\pm 442224, \pm 64505, \pm 168)$ when $P = 2^5 \times 17 \times 271$., $(x, y, z) = (\pm 442224, \pm 64505, \pm 84)$ when $P = 2^7 \times 17 \times 271$, $(x, y, z) = (\pm 442224, \pm 64505, \pm 42)$ when $P = 2^9 \times 17 \times 271$, $(x, y, z) = (\pm 442224, \pm 64505, \pm 21)$ when $P = 2^{11} \times 17 \times 271$.

When $k = 2$, (11) is equivalent to: $Pz^2 = 8x_2 x_1{}^2 y_1 y_3 y = 2^{11} \times 7^2 \times 53^2 \times 17 \times 19 \times 97 \times 271$, it has 5 diverse odd primes, which is contradict with $P = 2^t p_1{}^{a_1} p_2{}^{a_2} p_3{}^{a_3} p_4{}^{a_4}$, Therefore, (12) is impossible, which means (6) have no integer solution.

2. m is an positive even number.

Imitating the previous proof of 1 we can get the diophantine equation (7) only has common solution $(x, y, z) = (\pm 48, \pm 7, 0)$.

3.2.3. Discusion on Case 2

From (III), we can get $y_{n-1} \equiv y_{n+1} \equiv 1 \pmod 2$, it means y_{n-1}, y_{n+1} are odd numbers. Therefore, the left part of (8) is an even number, when it's right is an odd number, it is self-contradiction. Therefore, diophantine equation (6) have no integer solution.

To sum up, the theorem is proved.

4

附件三 相关研究论文

AMIMA 2018

IOP Publishing

IOP Conf. Series: Materials Science and Engineering **382** (2018) 052037 doi:10.1088/1757-899X/382/5/052037

4. Conclusion

The integer solution of diophantine equations $x^2 - D_1 y^2 = m, (D_1 \in Z^+, m \in Z)$ and $y^2 - D_2 z^2 = n, (D_2 \in Z^+, n \in Z)$ is a matter of great concern.

By using elementary number theory methods, we solved the common solution and nontrivial solution on the diophantine equation when $m = 1$, $n = 49$, $D_1 = 47$, D_2 can be expressed as $2^t p_1{}^{a_1} p_2{}^{a_2} p_3{}^{a_3} p_4{}^{a_4}$ where $a_i = 0$ or 1 for $1 \le i \le 4$, and $t \in Z^+$, $p_s (1 \le s \le 4)$ are different odd primes.

Acknowledgment

Supported by Scientific Research Project fund of Education Department of Yunnan Province: 2018JS608.

References

[1] Ljunggren W. Litt om Simultane Pellske Ligninger [J]. Norsk Mat. Tidsskr.,1941,23:132-138.

[2] Pan Jia-yu, Zhang Yu-ping, Zou Rong. The Pell Equations $x^2 - ay^2 = 1$ and $y^2 - Dz^2 = 1$ [J]. Chinese Quarterly Journal of Mathematics,1999,14(1):73-77.

[3] Guan Xun-gui. On the integer solution of Pell's equation $x^2 - 2y^2 = 1$ and $y^2 - Dz^2 = 4$ [J]. Journal of Huazhong Normal University (Natural Sciences):2012,46(3):267-269+278.

[4] Hu Yong-zhong, Han Qing. On the integer solution of the simultaneous equations $x^2 - 2y^2 = 1$ and $y^2 - Dz^2 = 4$ [J]. Journal of Hua Zhong Normal University (Natural Sciences):2002,36(1):17-19.

[5] Du Xian-cun, Guan Xun-gui, Yang Hui-zhang, On the system of Diophantine equations $x^2 - 6y^2 = 1$ and $y^2 - Dz^2 = 4$ [J] . Journal of Hua Zhong Normal University (Natural Sciences):2014.48(3):5-8.

[6] Zhao jian-hong, Wan Fei. On the Common Solution of Pell Equations $x^2 - 3y^2 = 1$ and $y^2 - 2^n y^2 = 16$ [J]. Journal of Hubei University for Nationalities (Natural Science Edition), 2016,2(6):146-148.

[7] He La-rong, Zhang Shu-jing, Yuan Jin. Diophantine Equation $x^2 - 6y^2 = 1$, $y^2 - Dz^2 = 4$ [J]. Journal of Yunnan University of Nationalities (Natural Sciences Edition),2012,21(1):57-58.

[8] Ran Yan-ping. Integer Solution of the Simultaneous Diophantine Equations $x^2 - 10y^2 = 1$ and $y^2 - Dz^2 = 4$ [J]. Journal of Yanan University (Natural Science Edition),2012,31(1):8-10.

[9] Guo Jing, Du Xian-cun. On the System of Pell Equations $x^2 - 30y^2 = 1$ and $y^2 - Dz^2 = 4$[J]. Mathematics in Practice and Theory.2015,45(1):309-314.

[10] Wan Fei, Du Xian-cun. On the Simultaneous Diophantine Equations $x^2 - 5y^2 = 1$ and $y^2 - Dz^2 = 16$ [J]. Journal of Chongqing Normal University (Natural Science),2016,33(4):107-111.

[11] Sun Qi, Yuan Ping-zhi, ON THE DIOPHANTINE EQUATION $x^4 - Dy^2 = 1$ [J]. Journal Of Sichuan University (Natural Science Edition),1997:34(3):265-267.

[12] Le Mao-hua. A Family of Binary Quartic Diophantine Equations [J]. Journal of Yunnan Normal University (Natural Sciences Edition),2010,30(1):12-17.

[13] Walsh G. A note on a theorem of Ljunggren and the Diophantine equations $x^2 - kxy^2 + y^4 = 1$ or 4[J]. Arch. Math.1999,73(2):504-513.

[14] Chen Yong-gao. Pell Equation $x^2 - 2y^2 = 1$ and $y^2 - Dz^2 = 4$ [J]. Acta Scientia rum Naturahum Universitatis Pekinesis,1994,30(3):298-302.

[15] Wang Guan-min, Li Bing-rong. The Integer Solution of The Pell's Equations $x^2 - 6y^2 = 1$ and $y^2 - Dz^2 = 4$ [J]. Journal of ZhangZhou Teachers College，2002,15(4):9-14.

第 39 卷第 8 期
Vol. 39　No. 8

西南大学学报（自然科学版）
Journal of Southwest University (Natural Science Edition)

2017 年 8 月
Aug.　2017

DOI：10.13718/j.cnki.xdzk.2017.08.009

关于 Pell 方程组 $x^2-s^2(s^2-1)y^2=1$ 与 $y^2-Dz^2=4$ 的解①

赵建红¹，　　杜先存²

1. 丽江师范高等专科学校 数学与计算机科学系，云南 丽江 674199；2. 红河学院 教师教育学院，云南 蒙自 661100

摘要：设 $D=p_1\cdots p_j(1\leqslant j\leqslant 3)$，$p_1,\cdots,p_j(1\leqslant j\leqslant 3)$ 是互异的奇素数. 利用初等方法讨论了 Pell 方程组 $x^2-s^2(s^2-1)y^2=1(s\in \mathbf{Z}^+,\ s\geqslant 2)$ 与 $y^2-Dz^2=4$ 的解的情况.

关 键 词：Pell 方程；基本解；整数解；奇素数；递归序列

中图分类号：O156　　　**文献标志码：A**　　　**文章编号**：1673-9868(2017)08-0065-08

近年来，Pell 方程 $x^2-D_1y^2=k(D_1\in \mathbf{Z}^+,\ k\in \mathbf{Z})$ 与 $y^2-D_2z^2=m(D_2\in \mathbf{Z}^+,\ m\in \mathbf{Z})$ 的公解问题一直备受人们的关注. 当 $k=1$，$m=4$，D_1 为偶数，D_2 为奇数时，已有如下结果：

（i）文献[1−5]对 $D_1=2$ 的情况做了一些研究；

（ii）文献[6−7]对 $D_1=6$ 的情况做了一些研究.

本文主要讨论当 $D_1=s^2(s^2-1)$，$(s\in \mathbf{Z}^+,\ s\geqslant 2)$ 时及 D_2 为奇数时的情况，即证明了以下定理：

定理 1　若 $s\in \mathbf{Z}^+$，$s\geqslant 2$，$D=p_1\cdots p_j(1\leqslant j\leqslant 3)$，其中 $p_1,\cdots,p_j(1\leqslant j\leqslant 3)$ 是互异的奇素数，则 Pell 方程组

$$\begin{cases}x^2-s^2(s^2-1)y^2=1\\ y^2-Dz^2=4\end{cases}\tag{1}$$

除 D 为 $16s^4-16s^2+3$ 中不超过 3 个次数为奇次的素因子之积外均只有平凡解

$$(x,y,z)=(\pm(2s^2-1)\pm2,0)$$

1　关键性引理

引理 1[8]　不定方程 $x^4-Dy^2=1$ 除当 $D=1\,785$，$4\times1\,785$，$16\times1\,785$ 时分别有两组正整数解 $(x,y)=(13,4)$，$(239,1\,352)$；$(x,y)=(13,2)$，$(239,676)$；$(x,y)=(13,1)$，$(239,338)$ 外，最多只有一组正整数解 (x_1,y_1)，且满足 $x_1^2=x_0$ 或 $2x_0^2-1$，这里 $\varepsilon=x_0+y_0\sqrt{D}$ 是 Pell 方程 $x^2-Dy^2=1$ 的基本解.

引理 2[9]　当 $a>1$ 且 a 是平方数时，方程 $ax^4-by^2=1$ 至多有一组正整数解.

引理 3[10]　若 D 是一个非平方的正整数，方程 $x^2-Dy^4=1$ 至多有两组正整数解，而且方程恰有两组正整数解的充要条件是 $D=1\,785$ 或 $D=28\,560$ 或 $2x_1$ 和 y_1 都是平方数，这里 (x_1,y_1) 是方程 $x^2-Dy^2=1$

① 收稿日期：2016-07-14

　基金项目：云南省科技厅应用基础研究计划青年项目（2013FD060）；云南省科技厅应用基础研究计划青年项目（Y0120160010）；红河学院校级教学改革项目（JJJG151010）.

　作者简介：赵建红（1981-），男，云南巍山人，副教授，主要从事初等数论和课程与教学论研究.

　通信作者：杜先存，副教授.

的基本解.

引理 4[11] 当 $A \in \mathbf{N}, A > 1, B \in \mathbf{N}$ 且 AB 不是平方数时,方程 $Ax^2 - By^4 = 1$ 至多只有 1 组正整数解.

引理 5 设 Pell 方程 $x^2 - s^2(s^2-1)y^2 = 1$ 的基本解为 (x_1, y_1),其全部整数解为 (x_n, y_n),$n \in \mathbf{Z}$,则对任意 $n \in \mathbf{Z}$,x_n,y_n 具有如下性质:

(i) 若 $2s^2 - 1$ 及 $8s^2(s^2-1) + 1$ 不为平方数,那么 x_n 为平方数当且仅当 $n = 0$;若 $2s^2 - 1$ 为平方数,那么 x_n 为平方数当且仅当 $n = 1$ 或 $n = 0$;若 $8s^2(s^2-1) + 1$ 为平方数,那么 x_n 为平方数当且仅当 $n = 2$ 或 $n = 0$;

(ii) $\dfrac{x_n}{2s^2 - 1}$ 为平方数当且仅当 $n = 1$;

(iii) $\dfrac{y_n}{2}$ 为平方数当且仅当 $n = 0$ 或 $n = 1$.

证 设 (x_1, y_1) 是 Pell 方程 $x^2 - s^2(s^2-1)y^2 = 1$ 的基本解,(x_n, y_n) $(n \in \mathbf{Z})$ 是 Pell 方程 $x^2 - s^2(s^2-1)y^2 = 1$ 的整数解.

(i) 若 $x_n = a^2$,将其代入原方程得

$$a^4 - s^2(s^2-1)y^2 = 1 \tag{2}$$

又 Pell 方程 $x^2 - s^2(s^2-1)y^2 = 1$ 的基本解为 $(x_1, y_1) = (2s^2-1, 2)$,则

$$x_1 = 2s^2 - 1 \qquad 2x_1^2 - 1 = 8s^2(s^2-1) + 1$$

根据引理 1 得,方程(2)至多有一组正整数解. 当 $2s^2 - 1$ 及 $8s^2(s^2-1) + 1$ 均不为平方数时,根据引理 1 得,方程(2)仅有平凡解 $(a, y) = (\pm 1, 0)$,故 $x_n = 1$,从而 $n = 0$;当 $2s^2 - 1$ 为平方数时,根据引理 1 得,方程(2)有整数解 $(a, y) = (\sqrt{2s^2-1}, 2)$ 及平凡解 $(a, y) = (\pm 1, 0)$,那么 $x_n = 2s^2 - 1$ 或 $x_n = 1$,从而 $n = 1$ 或 $n = 0$;$8s^2(s^2-1) + 1$ 为平方数时,根据引理 1 得,方程(2)有整数解 $(a, y) = (\sqrt{8s^2(s^2-1)+1}, 4(2s^2-1))$ 及平凡解 $(a, y) = (\pm 1, 0)$,那么 $x_n = 8s^2(s^2-1) + 1$ 或 $x_n = 1$,从而 $n = 2$ 或 $n = 0$. 反之,显然.

(ii) 若 $\dfrac{x_n}{2s^2-1} = a^2$,则 $x_n = (2s^2-1)a^2$,代入原方程得

$$(2s^2-1)^2 a^4 - s^2(s^2-1)y^2 = 1 \tag{3}$$

根据引理 2 得,方程(3)仅有整数解 $(a, y) = (\pm 1, \pm 2)$,此时 $x_n = 2s^2 - 1$,从而 $n = 1$. 反之,显然.

(iii) 若 $\dfrac{y_n}{2} = b^2$,则 $y_n = 2b^2$,代入原方程得

$$x^2 - 4s^2(s^2-1)b^4 = 1 \tag{4}$$

根据引理 3 得,方程(4)有整数解 $(x, b) = (\pm(2s^2-1), \pm 1)$ 及平凡解 $(x, b) = (\pm 1, 0)$,此时 $y_n = 2$ 或 0,从而 $n = 1$ 或 $n = 0$. 反之,显然.

2 定理 1 的证明

证 设 (x_1, y_1) 为 Pell 方程 $x^2 - s^2(s^2-1)y^2 = 1$($s \in \mathbf{Z}^+$,$s \geqslant 2$)的基本解,则有 $(x_1, y_1) = (2s^2-1, 2)$,故 Pell 方程 $x^2 - s^2(s^2-1)y^2 = 1$($s \in \mathbf{Z}^+$,$s \geqslant 2$)的全部正整数解为:

$$x_n + y_n \sqrt{s^2(s^2-1)} = (x_1 + \sqrt{s^2(s^2-1)}\, y_1)^n = (2s^2 - 1 + 2\sqrt{s^2(s^2-1)})^n, \quad n \in \mathbf{Z}^+$$

容易验证以下性质成立:

性质 a $y_n^2 - 4 = y_{n-1}y_{n+1}$;

性质 b $y_{2n} = 2x_n y_n$;

性质 c $x_{2n} = 2x_n^2 - 1$;

性质 d　$\gcd(x_n, y_n)=1$, $\gcd(x_n, x_{n+1})=1$, $\gcd(y_n, y_{n+1})=2$;

性质 e　$\gcd(x_{2n}, y_{2n+1})=\gcd(x_{2n+2}, y_{2n+1})=1$, $\gcd(x_{2n+1}, y_{2n})=\gcd(x_{2n+1}, y_{2n+2})=2s^2-1$;

性质 f　$x_n\equiv1(\bmod 2)$, $x_n\equiv\pm1(\bmod s^2)$, $x_n\equiv1(\bmod(s^2-1))$, $x_{2n+1}\equiv0(\bmod(2s^2-1))$, $x_{2n}\equiv\pm1(\bmod(2s^2-1))$;

性质 g　$y_{2n+1}\equiv2(\bmod 4)$, $y_{2n}\equiv0(\bmod 4)$, $y_{2n+1}\equiv\pm2(\bmod(2s^2-1))$, $y_{2n}\equiv0(\bmod(2s^2-1))$.

情形 1　n 为正偶数，由(1)式得
$$s^2(s^2-1)Dz^2=x_n^2-(2s^2-1)^2=(x_n+2s^2-1)(x_n-2s^2+1) \tag{5}$$

令 $n=2m(m\in\mathbf{Z}^+)$，则(5)式成为
$$s^2(s^2-1)Dz^2=(x_{2m}+2s^2-1)(x_{2m}-2s^2+1) \tag{6}$$

由性质 c 得，(6)式可化为
$$s^2(s^2-1)Dz^2=(2x_m^2+2s^2-2)(2x_m^2-2s^2)=4(x_m^2+s^2-1)(x_m^2-s^2)$$

即
$$s^2(s^2-1)Dz^2=4(x_m^2+s^2-1)(x_m^2-s^2) \tag{7}$$

由性质 f 知
$$x_m^2\equiv0,1(\bmod 2s^2-1)$$

则
$$x_m^2-s^2\equiv\pm s^2(\bmod 2s^2-1)$$

则
$$\gcd(x_m^2+s^2-1, x_m^2-s^2)=\gcd(2s^2-1, \pm s^2)=\gcd(1, s^2)=1$$

由性质 f 知，
$$x_m^2\equiv1(\bmod s^2)\qquad x_m^2\equiv1(\bmod(s^2-1))$$

所以(7)式可分解为下面 2 种情形：

情形 Ⅰ　$4(x_m^2+s^2-1)=D_1s^2z_1^2$, $x_m^2-s^2=D_2(s^2-1)z_2^2$, $D=D_1D_2$, $z=z_1z_2$

情形 Ⅱ　$x_m^2+s^2-1=D_1s^2z_1^2$, $4(x_m^2-s^2)=D_2(s^2-1)z_2^2$, $D=D_1D_2$, $z=z_1z_2$

其中 $\gcd(D_1, D_2)=1$, $\gcd(z_1, z_2)=1$.

先讨论情形 Ⅰ：

由
$$4(x_m^2+s^2-1)=D_1s^2z_1^2 \text{ 及 } x_m^2-s^2(s^2-1)y_m^2=1$$

得
$$D_1z_1^2=4[x_m^2-(s^2-1)^2y_m^2]=4[x_m+(s^2-1)y_m][x_m-(s^2-1)y_m]$$

即
$$D_1z_1^2=4[x_m+(s^2-1)y_m][x_m-(s^2-1)y_m] \tag{8}$$

又
$$\gcd(x_m+(s^2-1)y_m, x_m-(s^2-1)y_m)=\gcd(2(s^2-1)y_m, x_m-(s^2-1)y_m)=$$
$$\gcd(s^2-1, x_m)=1$$

则(8)式可分解为下式：
$$x_m+(s^2-1)y_m=D_3z_3^2, x_m-(s^2-1)y_m=D_4z_4^2, D_1=D_3D_4, z_1=2z_3z_4 \tag{9}$$

其中
$$\gcd(D_3, D_4)=1\qquad\gcd(z_3, z_4)=1$$

由
$$x_m^2-s^2=D_2(s^2-1)z_2^2 \text{ 及 } x_m^2-s^2(s^2-1)y_m^2=1$$

得

$$D_2 z_2^2 = s^4 y_m^2 - x_m^2 = (s^2 y_m + x_m)(s^2 y_m - x_m)$$

即

$$D_2 z_2^2 = (s^2 y_m + x_m)(s^2 y_m - x_m) \tag{10}$$

又

$$\gcd(s^2 y_m + x_m, s^2 y_m - x_m) = \gcd(2 x_m, s^2 y_m - x_m) = \gcd(x_m, s^2) = \gcd(1, s^2) = 1$$

则(10)可分解为下式:

$$s^2 y_m + x_m = D_5 z_5^2, \quad s^2 y_m - x_m = D_6 z_6^2, \quad D_2 = D_5 D_6, \quad z_2 = z_5 z_6 \tag{11}$$

其中

$$\gcd(D_5, D_6) = 1 \qquad \gcd(z_5, z_6) = 1$$

由(9),(11)式得 $D = D_1 D_2 = D_3 D_4 D_5 D_6$,且 D_3, D_4, D_5, D_6 两两互素,而 $D = p_1 \cdots p_s$ 且 $p_1, \cdots, p_s (1 \leqslant s \leqslant 3)$ 是互异的奇素数,则 D_3, D_4, D_5, D_6 中至少有一个为 1.

当 $D_3 = 1$ 时,(9)式的第一式为

$$x_m + (s^2 - 1) y_m = z_3^2$$

则

$$x_m = z_3^2 - (s^2 - 1) y_m$$

代入 $x_m^2 - s^2(s^2 - 1) y_m^2 = 1$,整理得

$$s^2 z_3^4 - (s^2 - 1)(y_m + z_3^2)^2 = 1$$

由引理 2 得

$$s^2 z_3^4 - (s^2 - 1)(y_m + z_3^2)^2 = 1$$

仅有一组正整数解 $(z_3, y_m + z_3^2) = (1, 1)$,则有

$$(z_3, y_m) = (1, 0)$$

此时方程(1)无正整数解.

当 $D_4 = 1$ 时,(9)式的第二式为

$$x_m - (s^2 - 1) y_m = z_4^2$$

则有

$$x_m = (s^2 - 1) y_m + z_4^2$$

代入

$$x_m^2 - s^2(s^2 - 1) y_m^2 = 1$$

整理得

$$s^2 z_4^4 - (s^2 - 1)(y_m - z_4^2)^2 = 1$$

由引理 2 得

$$s^2 z_4^4 - (s^2 - 1)(y_m - z_4^2)^2 = 1$$

仅有一组正整数解

$$(z_4, |y_m - z_4^2|) = (1, 1)$$

则有 $(z_4, y_m) = (1, 0)$ 或 $(1, 2)$,从而 $m = 1$,则 $n = 2$,此时有

$$D z^2 = y_2^2 - 4 = (8 s^2 - 4)^2 - 4 = 2^2 \times (16 s^4 - 16 s^2 + 3)$$

则若 $16 s^4 - 16 s^2 + 3$ 中次数为奇次的素因子的个数不超过 3 个,则此时方程(1)有正整数解;若 $16 s^4 - 16 s^2 + 3$ 中次数为奇次的素因子的个数超过 3 个,则此时方程(1)仅有平凡解.

当 $D_5 = 1$ 时,(11)式的第一式为 $s^2 y_m + x_m = z_5^2$,则有

$$x_m = z_5^2 - s^2 y_m$$

代入

$$x_m^2 - s^2(s^2 - 1) y_m^2 = 1$$

整理得

$$s^2(y_m-z_5^2)^2-(s^2-1)z_5^4=1$$

由引理 4 得

$$s^2(y_m-z_5^2)^2-(s^2-1)z_5^4=1$$

仅有 1 组正整数解

$$(\,|\,y_m-z_5{}^2\,|\,,\,z_5)=(1,\,1)$$

则有 $(z_5,\,y_m)=(1,\,0)$ 或 $(1,\,2)$，从而 $m=1$，则 $n=2$，$m=2$，此时方程(1)的解的情况与 $D_4=1$ 的情形相同.

当 $D_6=1$ 时，(11)式的第二式为 $s^2y_m-x_m=z_6^2$，则有

$$x_m=s^2y_m-z_6^2$$

代入

$$x_m^2-s^2(s^2-1)y_m^2=1$$

整理得

$$s^2(y_m-z_6^2)^2-(s^2-1)z_6^4=1$$

由引理 4 得

$$s^2(y_m-z_6^2)^2-(s^2-1)z_6^4=1$$

仅有 1 组正整数解

$$(\,|\,y_m-z_6{}^2\,|\,,\,z_6)=(1,\,1)$$

则有 $(z_6,\,y_m)=(1,\,0)$ 或 $(1,\,2)$，从而 $m=1$，则 $n=2$，$m=2$，此时方程(1)的解的情况与 $D_4=1$ 的情形相同.

综上知情形 Ⅰ 不成立.

下面讨论情形 Ⅱ：

由

$$(x_m^2+s^2-1)=D_1s^2z_1^2 \text{ 及 } x_m^2-s^2(s^2-1)y_m^2=1$$

得

$$D_1z_1^2=x_m^2-(s^2-1)^2y_m^2=[x_m+(s^2-1)y_m][x_m-(s^2-1)y_m]$$

即

$$D_1z_1^2=[x_m+(s^2-1)y_m][x_m-(s^2-1)y_m] \tag{12}$$

又

$$\gcd(x_m+(s^2-1)y_m \qquad x_m-(s^2-1)y_m)=1$$

则(12)式可分解为下式：

$$x_m+(s^2-1)y_m=D_7z_7^2,\ x_m-(s^2-1)y_m=D_8z_8^2,\ D_1=D_7D_8,\ z_1=z_7z_8 \tag{13}$$

其中

$$\gcd(D_7,\,D_8)=1 \qquad \gcd(z_7,\,z_8)=1$$

由

$$4(x_m^2-s^2)=D_2(s^2-1)z_2^2 \text{ 及 } x_m^2-s^2(s^2-1)y_m^2=1$$

得

$$D_2z_2^2=4(s^4y_m^2-x_m^2)=4(s^2y_m+x_m)(s^2y_m-x_m)$$

即

$$D_2z_2^2=4(s^2y_m+x_m)(s^2y_m-x_m) \tag{14}$$

又

$$\gcd(s^2y_m+x_m,\,s^2y_m-x_m)=1$$

则(14)式可分解为下式：

$$s^2 y_m + x_m = D_9 z_9^2, \quad s^2 y_m - x_m = D_{10} z_{10}^2, \quad D_2 = D_9 D_{10}, \quad z_2 = 2 z_9 z_{10} \tag{15}$$

其中

$$\gcd(D_9, D_{10}) = 1 \qquad \gcd(z_9, z_{10}) = 1$$

由(13),(15)式得 $D = D_1 D_2 = D_7 D_8 D_9 D_{10}$ 且 D_7, D_8, D_9, D_{10} 两两互素，而 $D = p_1 \cdots p_s$，且 $p_1, \cdots,$ $p_s (1 \leqslant s \leqslant 3)$ 是互异的奇素数，则 D_7, D_8, D_9, D_{10} 中至少有一个为 1.

$D_7 = 1$ 时仿 $D_3 = 1$ 的证明可知此时情形 Ⅱ 不成立；$D_8 = 1$ 时仿 $D_4 = 1$ 的证明可知此时情形 Ⅱ 不成立；$D_9 = 1$ 时仿 $D_5 = 1$ 的证明可知此时情形 Ⅱ 不成立；$D_{10} = 1$ 时仿 $D_6 = 1$ 的证明可知此时情形 Ⅱ 不成立. 综上可知情形 Ⅱ 不成立.

情形 2　n 为正奇数，令 $n = 2m - 1$，$m \in \mathbf{Z}^+$. 则由性质 a 得

$$Dz^2 = y_{2m-1}^2 - 4 = y_{2(m-1)} y_{2m} \tag{16}$$

由性质 b 知，(16)式可化为

$$Dz^2 = 4 x_{m-1} y_{m-1} x_m y_m \tag{17}$$

由性质 d 知

$$\gcd(x_{m-1}, y_{m-1}) = \gcd(x_m, y_m) = 1 \qquad \gcd(x_m, x_{m-1}) = 1 \qquad \gcd(y_m, y_{m-1}) = 2$$

则

$$\gcd\left(\frac{y_m}{2}, \frac{y_{m-1}}{2}\right) = 1$$

m 为正偶数时，由性质 e 知

$$\gcd(x_m, y_{m-1}) = 1 \qquad \gcd(x_{m-1}, y_m) = 2 s^2 - 1$$

则

$$\gcd\left(\frac{x_{m-1}}{2 s^2 - 1}, \frac{y_m}{2 s^2 - 1}\right) = 1$$

则有 $x_m, \dfrac{y_{m-1}}{2}, \dfrac{x_{m-1}}{2 s^2 - 1}, \dfrac{y_m}{2(2 s^2 - 1)}$ 两两互素.

由性质 f 知 $x_m, \dfrac{x_{m-1}}{2 s^2 - 1}$ 均为奇数，由性质 g 知 m 为偶数时，$\dfrac{y_{m-1}}{2}$ 为奇数；由引理 5(ⅰ)知仅当 $m = 0$ 时，x_m 为平方数；由引理 5(ⅱ)知仅当 $m = 2$ 时，$\dfrac{x_{m-1}}{2 s^2 - 1}$ 为平方数；由引理 5(ⅲ)知仅当 $m = 1$ 或 $m = 2$ 时，$\dfrac{y_{m-1}}{2}$ 为平方数. 故有 $x_m, \dfrac{y_{m-1}}{2}, \dfrac{x_{m-1}}{2 s^2 - 1}, \dfrac{y_m}{2(2 s^2 - 1)}$ 两两互素，因此 $m > 2$ 为正偶数时，$x_m, \dfrac{y_{m-1}}{2}, \dfrac{x_{m-1}}{2 s^2 - 1}$ 均为奇数且均不为平方数，故 $x_m, \dfrac{y_{m-1}}{2}, \dfrac{x_{m-1}}{2 s^2 - 1}$ 至少为 D 提供 3 个互异的奇素数.

$m > 2$ 为正偶数时，因为 $D = p_1 \cdots p_s (1 \leqslant s \leqslant 3)$，$p_1, \cdots, p_s (1 \leqslant s \leqslant 3)$ 是互异的奇素数，所以 $\dfrac{y_m}{2(2 s^2 - 1)}$ 为平方数. 令 $m = 2k$，$k \in \mathbf{Z}^+$，且 $k \geqslant 2$，则有

$$\frac{y_m}{2(2 s^2 - 1)} = \frac{y_{2k}}{2(2 s^2 - 1)} = \frac{2 x_k y_k}{2(2 s^2 - 1)} = \frac{x_k y_k}{2 s^2 - 1}$$

即

$$\frac{y_m}{2(2 s^2 - 1)} = \frac{x_k y_k}{2 s^2 - 1} \tag{18}$$

若 $k \geqslant 2$ 为正偶数，则由性质 b 知

$$\frac{y_m}{2(2 s^2 - 1)} = x_k \cdot \frac{y_k}{2 s^2 - 1}$$

由性质 d 知

$$\gcd\left(x_k,\frac{y_k}{2s^2-1}\right)=1$$

则 $\dfrac{y_m}{2(2s^2-1)}$ 为平方数需 x_k，$\dfrac{y_k}{2s^2-1}$ 同时为平方数，由引理 5 的 (i) 知 $k\geqslant 2$ 时 x_k 不为平方数，则

$\dfrac{y_m}{2(2s^2-1)}$ 不为平方数，故此时方程(1)无正整数解.

若 $k\geqslant 3$ 为正奇数，由性质 b 知

$$\frac{y_m}{2(2s^2-1)}=y_k\cdot\frac{x_k}{2s^2-1}$$

由性质 d 知

$$\gcd\left(y_k,\frac{x_k}{2s^2-1}\right)=1$$

则 $\dfrac{y_m}{2(2s^2-1)}$ 为平方数，需 y_k，$\dfrac{x_k}{2s^2-1}$ 同时为平方数，由引理 6(i) 知 $k\geqslant 3$ 时，x_k 不为平方数，则

$\dfrac{y_m}{2(2s^2-1)}$ 不为平方数，故此时方程(1)无正整数解.

$m=0$ 时，(17) 式为

$$Dz^2=4x_{-1}y_{-1}x_0y_0=0$$

从而得方程(1)的平凡解

$$(x,y,z)=(\pm(2s^2-1),\pm 2,0)$$

$m=2$ 时，(17) 式为

$$Dz^2=4x_1y_1x_2y_2=4\times(2s^2-1)\times 2\times(8s^4-8s^2+1)\times 4(2s^2-1)=$$
$$2^5(2s^2-1)^2(8s^4-8s^2+1)$$

即

$$Dz^2=2^5(2s^2-1)^2(8s^4-8s^2+1) \tag{19}$$

又 D 为奇数，故(19)式左边 2 的次数为偶数次，而(19)式右边 2 的次数为 5 次，矛盾，故此时方程(1)无正整数解.

m 为正奇数时，仿情形 1 的证明可知此时方程(1)亦无正整数解.

3　相关推论

推论 1　Pell 方程组 $x^2-12y^2=1$ 与 $y^2-195z^2=4$ 有解 $(x,y,z)=(\pm 97,\pm 28,\pm 2)$，$(\pm 7;\pm 2,0)$；

推论 2　Pell 方程组 $x^2-72y^2=1$ 与 $y^2-1155z^2=4$ 有解 $(x,y,z)=(\pm 577,\pm 68,\pm 2)$，$(\pm 17;\pm 2,0)$；

推论 3　Pell 方程组 $x^2-240y^2=1$ 与 $y^2-427z^2=4$ 有解 $(x,y,z)=(\pm 1921,\pm 124,\pm 6)$，$(\pm 31;\pm 2,0)$.

参考文献：

[1] 陈建华. 关于 Pell 方程 $x^2-2y^2=1$ 与 $y^2-Dz^2=4$ 的公解 [J]. 武汉大学学报(理学版)，1990，(1)：8—12.

[2] 刘玉记. 关于 Pell 方程 $x^2-2y^2=1$ 与 $y^2-Dz^2=4$ 的公解 [J]. 内蒙古民族师范学院学报：自然科学版，1994，9(2)：9—11.

[3] 曹珍富. 关于 Pell 方程 $x^2-2y^2=1$ 与 $y^2-Dz^2=4$ 的公解 [J]. 科学通报，1986，31(6)：476.

72　　　　　　　西南大学学报(自然科学版)　　　http://xbbjh.swu.edu.cn　　　第 39 卷

[4]　曾登高. 也说 Pell 方程 $x^2 - 2y^2 = 1$ 与 $y^2 - Dz^2 = 4$ 的公解 [J]. 数学的实践与认识, 1995(1): 81−84.

[5]　陈永高. Pell 方程组 $x^2 - 2y^2 = 1$ 与 $y^2 - Dz^2 = 4$ 的公解 [J]. 北京大学学报(自然科学版), 1994, 30(3): 298−302.

[6]　苏小燕. 关于 Pell 方程 $x^2 - 6y^2 = 1$ 与 $y^2 - Dz^2 = 4$ 的公解 [J]. 漳州师范学院学报(自然科学版), 2000, 13(3): 35−38.

[7]　冉银霞, 冉延平. 不定方程组 $x^2 - 6y^2 = 1$, $y^2 - Dz^2 = 4$ [J]. 延安大学学报(自然科学版), 2008, 27(4): 19−21.

[8]　孙　琦, 袁平之. 关于不定方程 $x^4 - Dy^2 = 1$ 的一个注记 [J]. 四川大学学报(自然科学版), 1997, 34(3): 265−267.

[9]　乐茂华. 一类二元四次 Diophantine 方程 [J]. 云南师范大学学报(自然科学版), 2010, 30(1): 12−17.

[10]　WALSH G. A Note on a Theorem of Ljunggren and the Diophantine Equations $x^2 - kxy^2 + y^4 = 1$ or 4 [J]. ArchMath, 1999, 73(2): 504−513.

[11]　LJUNGGREN W. EinSatzüber Die Diophantische Gleichung $Ax^2 - By^4 = C(C = 1, 2, 4)$ [J]. Tolfte Skand Mat Lund, 1953, 8(2): 188−194.

On the Solutions of System of Pell Equations
$x^2 - s^2(s^2-1)y^2 = 1$ and $y^2 - Dz^2 = 4$

ZHAO Jian-hong[1],　　DU Xian-cun[2]

1. Department of Mathematics and Computer Science, Lijiang Teachers College, Lijiang Yunnan 674199, China;

2. College of Teachers Education, Honghe University, Mengzi Yunnan 661100, China

Abstract: Let D be not a perfect square positive integer which has at most three distinct prime factors. The integer solutions of the system of Pell equations in title are discussed with the help of the elementary method.

Key words: Pell equation; fundamental solution; integer solution; odd prime; recursive sequence

責任编辑　张　枸

第 43 卷　第 4 期　　　　　　西南师范大学学报（自然科学版）　　　　　　2018 年 4 月
Vol. 43　No. 4　　Journal of Southwest China Normal University (Natural Science Edition)　　Apr. 2018

DOI：10. 13718/j. cnki. xsxb. 2018. 04. 003

椭圆曲线 $y^2 = x^3 - 17x + 114$ 的正整数点[①]

赵建红[1]，　　杜先存[2]

1. 丽江师范高等专科学校 教师教育学院，云南 丽江 674199；

2. 红河学院 教师教育学院，云南 蒙自 661199

摘要：利用唯一分解定理、同余的性质、Legendre 符号的性质、奇偶数的性质、Pell 方程的解的性质等初等方法证明了椭圆曲线 $y^2 = x^3 - 17x + 114$ 无正整数点.

关　键　词：椭圆曲线；正整数点；Pell 方程；同余；Legendre 符号

中图分类号：O156. 1　　　　**文献标志码**：A　　　　**文章编号**：1000-5471(2018)04-0011-04

　　椭圆曲线的正整数点和整数点是数论和算术代数几何学中很重要的问题，在密码学等领域有着广泛的应用. 关于椭圆曲线

$$y^2 = x^3 + ax + b \qquad a, b \in \mathbf{Z} \tag{1}$$

的整数点问题，目前已经有如下结果：文献[1]研究了 $a = -27$，$b = 62$ 时椭圆曲线 $y^2 = x^3 - 27x + 62$ 的整数点的问题，此问题对于研究椭圆曲线的算术性质有非常重要的意义；文献[2—3]得出了该椭圆曲线的全部整数点；文献[4]得出了 $a = 27$，$b = 62$ 时椭圆曲线 $y^2 = x^3 + 27x + 62$ 的整数点；文献[5]得出了 $a = -21$，$b = -90$ 时椭圆曲线 $y^2 = x^3 - 21x - 90$ 的全部整数点；文献[6]得出了 $a = -21$，$b = 90$ 时椭圆曲线 $y^2 = x^3 - 21x + 90$ 的全部整数点. 而对于 $a = -17$，$b = 114$ 时椭圆曲线 $y^2 = x^3 - 17x + 114$ 的整数点问题，目前还没有相关结论.

　　定理　椭圆曲线

$$y^2 = x^3 - 17x + 114 \tag{2}$$

无正整数点.

　　证　设 (x, y) 是椭圆曲线(2)的正整数点，则

$$y^2 = (x + 6)(x^2 - 6x + 19) \tag{3}$$

因为

$$\gcd(x + 6, x^2 - 6x + 19) = \gcd(x + 6, 91) = 1, 7, 13, 91$$

故(3)式可分解为以下 4 种情况：

　　情形 Ⅰ　$x + 6 = u^2$，$x^2 - 6x + 19 = v^2$，$y = uv$，$\gcd(u, v) = 1(u, v \in \mathbf{Z}_+)$；

　　情形 Ⅱ　$x + 6 = 7u^2$，$x^2 - 6x + 19 = 7v^2$，$y = 7uv$，$\gcd(u, v) = 1(u, v \in \mathbf{Z}_+)$；

　　情形 Ⅲ　$x + 6 = 13u^2$，$x^2 - 6x + 19 = 13v^2$，$y = 13uv$，$\gcd(u, v) = 1(u, v \in \mathbf{Z}_+)$；

　　情形 Ⅳ　$x + 6 = 91u^2$，$x^2 - 6x + 19 = 91v^2$，$y = 91uv$，$\gcd(u, v) = 1(u, v \in \mathbf{Z}_+)$.

　　情形 Ⅰ　因为 $u^2 \equiv 0, 1 \pmod 4$，故 $x = u^2 - 6 \equiv 2, 3 \pmod 4$，因此

① 收稿日期：2017-04-24

　　基金项目：云南省科技厅应用基础研究计划青年项目(2017FD166)；江西省教育厅科学技术研究项目(GJJ160782)；江西科技师范大学重点课题(2015XJZD002)；红河学院中青年学术骨干培养资助项目(2015GG0207)；红河学院教改课题(JJJG151010).

　　作者简介：赵建红(1981-)，男，云南巍山人，副教授，主要从事初等数论及数学教学论的研究.

　　通信作者：杜先存，副教授.

$$x^2 - 6x + 19 \equiv 2,3 (\bmod 4)$$

而 $v^2 \equiv 0,1 (\bmod 4)$，故有

$$2,3 (\bmod 4) \equiv x^2 - 6x + 19 = v^2 \equiv 0,1 (\bmod 4)$$

矛盾，故情形 Ⅰ 不成立.

情形 Ⅱ　因为 $u^2 \equiv 0,1,4 (\bmod 8)$，故

$$x = 7u^2 - 6 \equiv 1,2,6 (\bmod 8)$$

因此

$$x^2 - 6x + 19 \equiv 3,6 (\bmod 8)$$

而 $v^2 \equiv 0,1,4 (\bmod 8)$，则 $7v^2 \equiv 0,4,7 (\bmod 8)$，故有

$$3,6 (\bmod 8) \equiv x^2 - 6x + 19 = 7v^2 \equiv 0,4,7 (\bmod 8)$$

矛盾，故情形 Ⅱ 不成立.

情形 Ⅲ　因为 $u^2 \equiv 0,1 (\bmod 4)$，故 $x = 13u^2 - 6 \equiv 2,3 (\bmod 4)$，因此

$$x^2 - 6x + 19 \equiv 2,3 (\bmod 4)$$

而 $v^2 \equiv 0,1 (\bmod 4)$，则 $13v^2 \equiv 0,1 (\bmod 4)$，故有

$$2,3 (\bmod 4) \equiv x^2 - 6x + 19 = 13v^2 \equiv 0,1 (\bmod 4)$$

矛盾，故情形 Ⅲ 不成立.

情形 Ⅳ　若 $2 \nmid u$，则有 $u^2 \equiv 1 (\bmod 8)$，故有 $x = 91u^2 - 6 \equiv 5 (\bmod 8)$，因此

$$x^2 - 6x + 19 \equiv 6 (\bmod 8)$$

而 $v^2 \equiv 0,1,4 (\bmod 8)$，则 $91v^2 \equiv 0,3,4 (\bmod 8)$，故有

$$6 (\bmod 8) \equiv x^2 - 6x + 19 = 91v^2 \equiv 0,3,4 (\bmod 8)$$

矛盾，因此 $2 \nmid u$ 不成立，所以 $2 \mid u$. 令 $u = 2f (f \in \mathbf{Z}_+)$，则 $x + 6 = 91u^2$ 为 $x + 6 = 364f^2$，将 $x + 6 = 364f^2$ 代入 $x^2 - 6x + 19 = 91v^2$，得 $160f^4 + (36f^2 - 1)^2 = v^2$，即

$$(v + 36f^2 - 1)(v - 36f^2 + 1) = 160f^4 \tag{4}$$

易知

$$\gcd(v + 36f^2 - 1, v - 36f^2 + 1) = \gcd(72f^2 - 2, v - 36f^2 + 1) =$$
$$\gcd(2(36f^2 - 1), v - (36f^2 - 1))$$

因为 $2 \mid u$，$\gcd(u, v) = 1$，所以 $2 \nmid v$，故 $2 \mid [v - (36f^2 - 1)]$. 易知

$$\gcd(2(36f^2 - 1), v - (36f^2 - 1)) = 2\gcd(36f^2 - 1, v - (36f^2 - 1)) = 2\gcd(36f^2 - 1, v)$$

设 $\gcd(36f^2 - 1, v) = d$，则 $d \mid v$，$d \mid (36f^2 - 1)$，故由(4)式知 $d \mid 160f^4$. 因为：

$$\gcd(36f^2 - 1, 2) = 1 \qquad \gcd(36f^2 - 1, f) = 1$$

故

$$\gcd(36f^2 - 1, 160f^4) = \gcd(36f^2 - 1, 2^5 \times 5) =$$
$$\gcd(36f^2 - 1, 5) = 1,5$$

因为 $160 = 2^5 \times 5$，故当

$$\gcd(a, b) = 1 \qquad \gcd\left(h, \frac{40}{h}\right) = 1 \qquad h = 1,5,2^3,2^3 \times 5$$

时，(4)式可分解为：

$$v + 36f^2 - 1 = 2ha^4 \qquad v - 36f^2 + 1 = \frac{80}{h}b^4 \qquad f = ab \tag{5}$$

由(5)式得

$$36f^2 - 1 = ha^4 - \frac{40}{h}b^4 \tag{6}$$

对(6)式两边取 mod 5，得

$$f^2 + \frac{40}{h}b^4 - ha^4 \equiv 1 (\bmod 5) \tag{7}$$

若 $\gcd(36f^2 - 1, v) = 5$，则当 $h = 1, 2^3$ 时，因为 $\gcd(36f^2 - 1, v) = 5$，故 $5 \mid a$. 又因为 $f = ab$，所以 $5 \mid f$，因此(7)式为 $0 \equiv 1 \pmod 5$ 显然不成立，故 $h = 1, 2^3$ 时(7)式不成立，则(4)式不成立.

当 $h = 5, 2^3 \times 5$ 时，因为 $\gcd(36f^2 - 1, v) = 5$，故 $5 \mid b$. 又因为 $f = ab$，所以 $5 \mid f$，因此(7)式为 $0 \equiv 1 \pmod 5$ 显然不成立，故 $h = 5, 2^3 \times 5$ 时(7)式不成立，则(4)式不成立.

综上所述，有 $\gcd(36f^2 - 1, v) = 5$ 不成立，故 $\gcd(36f^2 - 1, v) = 1$，则

$$\gcd(v + 36f^2 - 1, v - 36f^2 + 1) = 2$$

对(6)式两边取 $\bmod 4$，得

$$-1 \equiv ha^4 - \frac{40}{h}b^4 \pmod 4 \tag{8}$$

当 $h = 1, 5$ 时，(8)式为

$$-1 \equiv a^4 \pmod 4 \tag{9}$$

因为 $a^2 \equiv 0, 1 \pmod 4$，故 $a^4 \equiv 0, 1 \pmod 4$，则(9)式为 $3 \equiv a^4 \equiv 0, 1 \pmod 4$，矛盾. 故(9)式不成立. 因此 $g = 1, 5$ 时(4)式不成立，即情形 Ⅳ 不成立.

当 $h = 2^3$ 时，(6)式为 $36f^2 - 1 = 8a^4 - 5b^4$，将(5)式中的 $f = ab$ 代入，得

$$36a^2b^2 - 1 = 8a^4 - 5b^4 \tag{10}$$

由(10)式得

$$72a^2b^2 - 2 = 16a^4 - 10b^4$$

配方得

$$(4a^2 - 9b^2)^2 - 91b^4 = -2 \tag{11}$$

对(11)式两边取 $\bmod 7$，得

$$(4a^2 - 9b^2)^2 \equiv -2 \pmod 7 \tag{12}$$

因为 Legendre 符号值 $\left(\dfrac{-2}{7}\right) = -1$，故(12)式不成立，因此方程(11)无整数解，所以 $h = 2^3$ 时(4)式不成立，即情形 Ⅳ 不成立.

当 $h = 2^3 \times 5$ 时，(6)式为

$$36f^2 - 1 = 40a^4 - b^4$$

将(5)式中的 $f = ab$ 代入得 $36a^2b^2 - 1 = 40a^4 - b^4$，配方得

$$(18a^2 + b^2)^2 - 364a^4 = 1 \tag{13}$$

令 $s = 18a^2 + b^2 (s \in \mathbf{Z}_+)$，此时(13)式为

$$s^2 - 364a^4 = 1 \tag{14}$$

令 $r = 2a (r \in \mathbf{Z}_+)$，则(14)式为

$$s^2 - 91r^2 = 1 \tag{15}$$

因为 $s^2 - 364a^2 = 1$ 的基本解为 $(1\,574, 165)$，故由文献[8]的定理1得，方程(14)至多有1组正整数解 (s, a)，且若方程(14)有正整数解 (s, a)，则方程(15)有正整数解 $(s, r) = (s, 2a^2)$. 因为 Pell 方程(15)的基本解为 $(1\,574, 165)$，则方程(15)的全部正整数解可表示为

$$s_n + r_n\sqrt{91} = (1\,574 + 165\sqrt{91})^n \qquad n \in \mathbf{Z}_+$$

由此可知方程(13)的全部正整数解满足

$$(18a^2 + b^2) + a^2\sqrt{364} = (6a^2 + b^2) + 2a^2\sqrt{91} = (1\,574 + 165\sqrt{91})^n \tag{16}$$

其中 $n \in \mathbf{Z}_+$. 则有：

$$18a^2 + b^2 = s_n \qquad 2a^2 = r_n \qquad n \in \mathbf{Z}_+$$

容易验证

$$r_{n+2} = 3\,148r_{n+1} - r_n \qquad r_0 = 0,\ r_1 = 165 \tag{17}$$

对递归序列(17)取 $\bmod 2$，得周期为 2 的剩余类数列 $0, 1, 0, 1, \cdots$，且当 $n \equiv 1 \pmod 2$ 时，有 $r_n \equiv 1 \pmod 2$；$n \equiv 0 \pmod 2$ 时，有 $r_n \equiv 0 \pmod 2$. 因为 $2a^2 = r_n$，所以 r_n 为偶数，故(17)式成立需 $n \equiv 0 \pmod 2$.

由文献[8]的定理 1 知(14)式成立需满足 $n=2$ 或 $2 \nmid n$，因此 $n=2$. 由(16)式得

$$(18a^2+b^2)+a^2\sqrt{364}=(18a^2+b^2)+2a^2\sqrt{91}=$$
$$(1\,574+165\sqrt{91})^2=4\,954\,951+519\,420\sqrt{91}$$

因此有 $18a^2+b^2=4\,954\,951$，$2a^2=519\,420$，则 $a^2=259\,710$，显然无整数解，故方程(13)无整数解，因此 $h=2^3\times5$ 时(4)式不成立，即情形 Ⅳ 不成立.

综上所述，定理得证.

参考文献：

[1]　ZAGIER D. Lager Integral Point on Elliptic Curves [J]. Math Comp, 1987, 48：425−436.

[2]　ZHU H L, CHEN J H. Integral Point on $y^2=x^3+27x-62$ [J]. J Math Study, 2009, 42(2)：117−125.

[3]　吴华明. 椭圆曲线 $y^2=x^3+27x-62$ 的整数点 [J]. 数学学报(中文版), 2010, 53(1)：205−208.

[4]　过　静. 椭圆曲线 $y^2=x^3+27x+62$ 的整数点 [J]. 重庆师范大学学报(自然科学版), 2016, 33(5)：50−53.

[5]　李亚卓, 崔保军. 关于椭圆曲线 $y^2=x^3-21x-90$ 的正整数点 [J]. 延安大学学报(自然科学版), 2015, 34(3)：14−15.

[6]　过　静. 关于椭圆曲线 $y^2=x^3-21x+90$ 的整数点 [J]. 数学的实践与认识, 2017, 67(8)：288−291.

On Positive Integral Points on Elliptic Curve $y^2=x^3-17x+114$

ZHAO Jian-hong[1], 　　DU Xian-cun[2]

1. Department of Teachers Education, Lijiang Teachers College, Lijiang Yunnan 674199, China;

2. College of Teachers Education, Honghe University, Mengzi Yunnan 661199, China

Abstract： In this paper, we are to prove the elliptic curve $y^2=x^3-17x+114$ has no positive integer points by the unique decomposition theorem, some properties of congruence, Legendre symbols, odd and even number, solution of the Pell equation.

Key words： elliptic curve; positive integer point; Pell equation; congruence; Legendre symbols

责任编辑　廖　坤　崔玉洁

第 48 卷第 1 期
2018 年 1 月

数学的实践与认识
MATHEMATICS IN PRACTICE AND THEORY

Vol.48, No. 1
Jan., 2018

关于不定方程 $x^2 - 51y^2 = 1$ 与 $y^2 - Dz^2 = 49$ 的公解

赵建红

(丽江师范高等专科学校 教师教育学院, 云南 丽江 674199)

摘 要: 设 $p_s(1 \leq s \leq 4)$ 是互异的奇素数, $D = 2^t p_1^{a_1} p_2^{a_2} p_3^{a_3} p_4^{a_4}$ ($a_i = 0$ 或 1, $1 \leq i \leq 4, t \in Z^+$) 时, 不定方程 $x^2 - 51y^2 = 1$ 与 $y^2 - Dz^2 = 49$ 仅当 $D = 2^t \times 4999$ ($t = 1, 3, 5$) 时有非平凡公解 $(x, y, z) = (\pm 50, \pm 7, 0)$.

关键词: 整数解; 公解; 不定方程; Pell 方程; 同余; 递归序列

Diophantine 方程
$$x^2 - D_1 y^2 = k \ (D_1 \in Z^+, k \in Z) \text{ 与 } y^2 - D_2 z^2 = m \ (D_2 \in Z^+, m \in Z) \tag{1}$$
的公解问题一直受到人们的关注. $k = 1, m = 1$ 时方程 (1) 成为:
$$x^2 - D_1 y^2 = 1 \text{ 与 } y^2 - D_2 z^2 = 1. \tag{2}$$
关于方程 (2) 的公解的情况, 目前结论还比较少, 详见文 [1-2];

$k = 1, m = 4$ 时方程 (1) 成为:
$$x^2 - D_1 y^2 = 1 \text{ 与 } y^2 - D_2 z^2 = 4. \tag{3}$$
D_1, D_2 均为偶数时, 方程 (3) 的公解的情况已有结果详见文 [3-9];

$k = 1, m = 16$ 时方程 (1) 成为:
$$x^2 - D_1 y^2 = 1 \text{ 与 } y^2 - D_2 z^2 = 16. \tag{4}$$
关于方程 (4) 的公解的情况已有结果详见文 [10];

$k = 1, m = 49$ 时方程 (1) 成为:
$$x^2 - D_1 y^2 = 1 \text{ 与 } y^2 - D_2 z^2 = 49. \tag{5}$$
方程 (5) 的公解情况目前无相关结果. 本文主要讨论 $D_1 = 51, D_2 = 2^t p_1^{a_1} p_2^{a_2} p_3^{a_3} p_4^{a_4}$ 时方程 (5) 的公解的情况.

1 关键性引理

引理 1[11] 若 D 是一个非平方的正整数, 则方程 $x^2 - Dy^4 = 1$ 至多有两组正整数解 (x, y), 而且方程恰有两组正整数解的充要条件是 $D = 1785$ 或 $D = 28560$ 或 $2x_0$ 和 y_0 都是平方数, 这里 (x_0, y_0) 是方程 $x^2 - Dy^4 = 1$ 的基本解.

引理 2[12] 设 D 是一个非平方的正整数, 则方程 $x^2 - Dy^4 = 1$ 至多有 2 组正整数解 (x, y). 如果方程 $x^2 - Dy^4 = 1$ 恰有两组正整数解, 则当 $D = 2^{4s} \times 1785$, 其中 $s \in \{0, 1\}$ 时, $(x_1, y_1) = (169, 2^{1-s})$ 且 $(x_2, y_2) = (6525617281, 2^{1-s} \times 6214)$; 当 $D \neq 2^{4s} \times 1785$ 时,

收稿日期: 2017-06-12
资助项目: 云南省科技厅应用基础研究计划青年项目 (2017FD166)

$(x_1, y_1) = (u_1, \sqrt{v_1})$ 且 $(x_2, y_2) = (u_2, \sqrt{v_2})$, 这里 (u_n, v_n) 是 Pell 方程 $U^2 - DV^2 = 1$ 的正整数解.

引理 3[13]　不定方程 $x^2 - Dy^4 = 1$ 除开当 $D = 1785, 4 \times 1785, 16 \times 1785$ 分别有两组正整数解 $(x, y) = (13, 4), (239, 1352)$; $(x, y) = (13, 2), (239, 676)$; $(x, y) = (13, 1), (239, 338)$ 外, 最多只有一组正整数解 (x_1, y_1), 且满足 $x_1 = x_0$ 或 $x_1 = 2x_0^2 - 1$, 这里 $\varepsilon = x_0 + y_0\sqrt{D}$ 为 Pell 方程 $x^2 - Dy^2 = 1$ 的基本解.

引理 4[14]　当 $a > 1$ 且 a 是平方数, 方程 $ax^4 - by^2 = 1$ 至多有一组正整数解.

引理 5　设 (x_n, y_n), $n \in Z$ 为 Pell 方程 $x^2 - 51y^2 = 1$ 的所有解, 则对任意 x_n, y_n 具有性质:

1) $\dfrac{y_n}{7}$ 为平方数, 当且仅当 $n = 0$ 或 $n = 1$ 或 $n = 2$;

2) x_n 为平方数, 当且仅当 $n = 0$; $\dfrac{x_n}{50}$ 为平方数, 当且仅当 $n = 1$ 或 $n = -1$.

证明　设 $(x_n, y_n)(n \in Z)$ 是 Pell 方程 $x^2 - 51y^2 = 1$ 的整数解.

1) 则 $y_n = 7b^2$, 代入原方程得 $x^2 - 2499b^4 = 1$, 因 $u^2 - 2499v^2 = 1$ 的基本解是 $u_1 = 50, v_1 = 1$, 因此 $2u_1 = 100$, 所以 $2u_1 = 100, v_1 = 1$ 均为平方数, 所以据引理 1 知 $x^2 - 2499y^4 = 1$ 有 2 组正整数解. 又 $u_2 = 4999, v_2 = 100$, 故据引理 2 知 $x^2 - 2499y^4 = 1$ 仅有正整数解 $(x, y) = (50, 1), (4999, 10)$, 故 $x^2 - 2499b^4 = 1$ 仅有整数解 $(x, b) = (\pm 1, 0), (\pm 50, \pm 1), (\pm 4999, \pm 10)$, 则方程 $x^2 - 51y^2 = 1$ 仅有整数解 $(x, y) = (\pm 1, 0), (\pm 50, \pm 7)$, $(\pm 4999, \pm 700)$, 此时 $y_n = 0$ 或 $y_n = 7$ 或 $y_n = 700$, 从而 $n = 0$ 或 $n = 1$ 或 $n = 2$, 反之, 显然.

2) 设 $(x_n, y_n)(n \in Z)$ 是 Pell 方程 $x^2 - 51y^2 = 1$ 的正整数解, 若 $x_n = a^2$, 代入得 Pell 方程 $a^4 - 51y^2 = 1$. 因为方程 $u^2 - 2499v^2 = 1$ 的基本解是 $u_1 = 50, v_1 = 1$, 而 $u_1 = 50, 2u_1^2 - 1 = 4999$, 故 $u_1, 2u_1^2 - 1$ 均不为平方数, 所以据引理 3 知, 方程 $a^4 - 51y^2 = 1$ 仅有平凡解 $(a, y) = (\pm 1, 0)$, 即 $x^4 - 51y^2 = 1$ 仅有平凡解 $(x, y) = (\pm 1, 0)$, 此时 $x_n = 1$, 从而 $n = 0$, 反之, 显然;

若 $\dfrac{x_n}{50} = a^2$, 则 $x_n = 50a^2$, 代入原方程得 $2500a^4 - 51y^2 = 1$, 据引理 4 知, $2500a^4 - 51y^2 = 1$ 至多只有一个正整数解, 而 $(a, y) = (\pm 1, \pm 7)$ 为 $2500a^4 - 51y^2 = 1$ 仅有整数解, 故 $x^2 - 51y^2 = 1$ 仅有整数解 $(a, y) = (\pm 50, \pm 7)$, 此时 $x_n = 50$, 从而 $n = 1$ 或 $n = -1$, 反之, 显然.

2　定理及证明

定理　若 $p_s(1 \leq s \leq 4)$ 是互异的奇素数, 则当 $D = 2^t p_1^{a_1} p_2^{a_2} p_3^{a_4} p_4^{a_4}$($a_i = 0$ 或 1, $1 \leq i \leq 4$, $t \in Z^+$) 时, 不定方程

$$x^2 - 51y^2 = 1 \text{ 与 } y^2 - Dz^2 = 49. \tag{6}$$

1) $D = 2 \times 4999$ 时, 有非平凡公解 $(x, y, z) = (\pm 499850, \pm 69993, \pm 700)$ 和平凡公解 $(x, y, z) = (\pm 50, \pm 7, 0)$;

2) $D = 2^2 \times 4999$ 时, 有非平凡公解 $(x, y, z) = (\pm 499850, \pm 69993, \pm 350)$ 和平凡公解 $(x, y, z) = (\pm 50, \pm 7, 0)$;

3) $D = 2^3 \times 4999$ 时, 有非平凡公解 $(x, y, z) = (\pm 499850, \pm 69993, \pm 175)$ 和平凡公解 $(x, y, z) = (\pm 50, \pm 7, 0)$;

4) $D \neq 2^\alpha \times 4999$($\alpha = 1, 3, 5$) 时, 方程 (6) 只有平凡公解 $(x, y, z) = (\pm 50, \pm 7, 0)$.

证明 设 (x_1, y_1) 为 Pell 方程 $x^2 - 51y^2 = 1$ 的基本解, 则有 $(x_0, y_0) = (50, 7)$, 故 Pell 方程 $x^2 - 51y^2 = 1$ 的全部整数解为: $x_n + y_n\sqrt{D} = (50 + 5\sqrt{51})^n, n \in Z$.

容易验证以下性质成立:

1) $y_n^2 - 49 = y_{n+1}y_{n-1}$;

2) $y_{2n} = 2x_n y_n$;

3) $\gcd(x_{2n}, y_{2n+1}) = \gcd(x_{2n+2}, y_{2n+1}) = 1, \gcd(x_{2n}, y_{2n}) = \gcd(x_{2n+1}, y_{2n+2}) = 50$;

4) $x_{2n} \equiv 1(\mathrm{mod}2)$, $x_{2n+1} \equiv 0(\mathrm{mod}50)$, $x_{2n+1} \equiv 50(\mathrm{mod}100)$, $y_{2n+1} \equiv 1(\mathrm{mod}2)$, $y_{2n} \equiv 0(\mathrm{mod}50)$, $y_n \equiv 0(\mathrm{mod}7)$;

5) $\gcd(x_n, y_n) = \gcd(x_{n-1}, y_{n-1}) = 1, \gcd(x_n, x_{n+1}) = 1, \gcd(y_n, y_{n+1}) = 7$.

设 $(x, y, z) = (x_n, y_n, z), n \in Z$ 是 (6) 的整数解, 据引理 (1) 有 $Dz^2 = y_n^2 - 49 = y_{n-1}y_{n+1}$, 即

$$Dz^2 = y_{n-1}y_{n+1}. \tag{7}$$

情形 1 n 为奇数, 令 $n = 2m - 1, m \in Z$, 此时 (7) 式成为:

$$Dz^2 = y_{2m-1}^2 - 49 = y_{2(m-1)}y_{2m}. \tag{8}$$

据 (2) 知, (8) 式可化为:

$$Dz^2 = 4x_{m-1}y_{m-1}x_m y_m. \tag{9}$$

情形 1.1 m 为偶数, 据 3) 知, $\gcd(x_m, y_{m-1}) = 1, \gcd(x_{m-1}, y_{m-1}) = \gcd(x_m, y_m) = 1$, $\gcd(y_m, y_{m-1}) = 7$, 则 $\gcd(\frac{y_m}{7}, \frac{y_{m-1}}{7}) = 1$; 再据 3) 知, $\gcd(x_m, y_{m-1}) = 1, \gcd(x_{m-1}, y_m) = 50$, 则 $\gcd(\frac{x_{m-1}}{50}, \frac{y_m}{50}) = 1$; 所以 $\frac{x_{m-1}}{50}, \frac{y_{m-1}}{7}, x_m, \frac{y_m}{350}$ 两两互素.

令 $m = 2k, k \in Z^+$, 由 (2) 式得, (9) 式成为:

$$Dz^2 = 4x_{2k-1}y_{2k-1}x_{2k}y_{2k}. \tag{10}$$

由 (2) 式得, (10) 式可化为:

$$Dz^2 = 8x_{2k-1}y_{2k-1}x_{2k}x_k y_k. \tag{11}$$

据 3) 知, $\gcd(x_{m-1}, y_{m-1}) = 1$, 又据 4) 知, 当 k 是奇数时, $\frac{x_{2k-1}}{50}, \frac{y_{2k-1}}{7}, x_{2k}, \frac{x_k}{50}, \frac{y_k}{7}$ 两两互素; 当 k 是偶数时, $\frac{x_{2k-1}}{50}, \frac{y_{2k-1}}{7}, x_{2k}, x_k, \frac{y_k}{350}$ 两两互素.

情形 1.1.1 k 为奇数, 据 5) 知, $\frac{y_{2k-1}}{7}, x_{2k}, \frac{y_k}{7}$ 均为奇数. 据 5) 知 $50|x_k, 50|x_{2k-1}$, 故 $\frac{x_k}{50}$, $\frac{x_{2k-1}}{50}$ 均为奇数, 故 k 是奇数时 $\frac{y_{2k-1}}{7}, x_{2k}, \frac{y_k}{7}, \frac{x_k}{50}, \frac{x_{2k-1}}{50}$ 均为奇数.

据引理 5 知, 仅当 $k = 0$ 或 1 时 $\frac{x_{2k-1}}{50}$ 是平方数, 仅当 $k = 0$ 时 x_{2k} 是平方数, 仅当 $k = -1$ 或 1 时, $\frac{x_k}{50}$ 是平方数, 仅当 $k = 0$ 或 1 或 2 时, $\frac{y_k}{7}$ 是平方数, 仅当 $k = 1$ 时 $\frac{y_{2k-1}}{7}$ 是平方数.

故 $k = \pm 1$ 是奇数时, $\frac{x_{2k-1}}{50}, \frac{y_{2k-1}}{7}, x_{2k}, \frac{x_k}{50}, \frac{y_k}{7}$ 都不是平方数, 此时 (11) 式为:

$$Dz^2 = 2^5 \times 5^4 \times 7^2 x_{2k} \cdot \frac{x_{2k-1}}{50} \cdot \frac{x_k}{50} \cdot \frac{y_k}{7} \cdot \frac{y_{2k-1}}{7}. \tag{12}$$

(12) 式右端为 $2^5 \times 5^4 \times 7^2$ 与至少 5 个奇素数的乘积, 故 $z = 700$, D 是 2 与至少 5 个奇素数的乘积; 或 $z = 350$, D 是 4 与至少 5 个奇素数的乘积; 或 $z = 175$, D 是 8 与至少 5 个奇素数的乘积; 与题设 "$D = 2^t p_1^{a_1} p_2^{a_2} p_3^{a_3} p_4^{a_4}$ ($a_i = 0$ 或 $1, 1 \leq i \leq 4, t \in Z^+$)" 矛盾, 故 (12) 式不成立, (6) 无公解.

$k = 1$ 时, (11) 式为 $Dz^2 = 8x_1y_1x_2x_1y_1 = x_1^2y_1^2x_2 = 2^5 \times 5^4 \times 7^2 \times 4999$, 故 $z = 700$, $D = 2 \times 4999$, 或 $z = 350$, $D = 2^2 \times 499$, 或 $z = 175$, $D = 2^3 \times 4999$, 则方程 (6) 有非平凡公解

$(x, y, z) = (\pm 499850, \pm 69993, \pm 700), (\pm 499850, \pm 69993, \pm 350), (\pm 499850, \pm 69993, \pm 175)$ 和平凡公解 $(x, y, z) = (\pm 50, \pm 7, 0)$.

$k = -1$ 时, (11) 式为 $Dz^2 = 8x_{-1}y_{-1}x_{-2}x_{-1}y_{-1} = 8x_1y_1x_2x_1y_1 = 8x_1^2y_1^2x_2$, 此时方程 (6) 的解的情况与 $k = 1$ 的情况相同.

情形 1.1.2　k 为偶数, 令 $k = 2l, l \in Z$, 此时 (11) 式成为:

$$Dz^2 = 8x_{4l-1}y_{4l-1}x_{4l}x_{2l}y_{2l}. \tag{13}$$

据 2) 知, (13) 式可化为:

$$Dz^2 = 16x_{4l-1}y_{4l-1}x_{4l}x_{2l}x_ly_l. \tag{14}$$

据 3) 知, $\gcd(x_{l-1}, y_{l-1}) = 1$, 又据 4) 知, 当 l 为奇数时, $x_{2l}, x_{4l}, \frac{x_l}{50}, \frac{x_{4l-1}}{50}, \frac{y_l}{7}, \frac{y_{4l-1}}{7}$ 两两互素; 当 l 为偶数时, $x_l, x_{2l}, x_{4l}, \frac{x_{4l-1}}{50}, \frac{y_l}{350}, \frac{y_{4l-1}}{7}$ 两两互素.

据 4) 知 $x_{4l}, x_{2l}, \frac{y_{4l-1}}{7}$ 为奇数, $50|x_{4k-1}$, 故 $\frac{x_{4l-1}}{50}$ 为奇数; l 为偶数时 $x_l, x_{4l}, x_{2l}, x_l, \frac{x_{4l-1}}{50}, \frac{y_{4l-1}}{7}$ 都是奇数; l 为奇数时 $50|x_k$, 故 $\frac{x_l}{50}, x_{4l}, x_{2l}, \frac{x_{4l-1}}{50}, \frac{x_l}{50}, \frac{y_l}{7}, \frac{y_{4l-1}}{7}$ 都是奇数;

据引理 5 知, 仅当 $l = 0$ 时 $x_l, x_{2l}, x_{4l}, \frac{x_{4l-1}}{50}$ 为平方数, 仅当 $l = 0$ 或 1 或 2 时 $\frac{y_l}{7}$ 为平方数, 仅当 $l = -1$ 或 1 时 $\frac{x_l}{50}$ 为平方数, 对 $\forall l \in Z$, $\frac{y_{4l-1}}{7}$ 都不是平方数.

故 $l \neq \pm 1$ 为奇数时, $x_{4l}, x_{2l}, \frac{x_l}{50}, \frac{x_{4l-1}}{50}, \frac{y_l}{7}, \frac{y_{4l-1}}{7}$ 均不为平方数, 故 $l \neq 0, 2$ 为偶数时, $x_l, x_{2l}, x_{4l}, \frac{x_{4l-1}}{50}, \frac{y_{4l-1}}{7}$ 均不为平方数. 此时 (14) 式成为:

$$Dz^2 = 2^6 \times 5^4 \times 7^2 x_{2l} \cdot x_{4l} \cdot \frac{x_l}{50} \cdot \frac{x_{4l-1}}{50} \cdot \frac{y_l}{7} \cdot \frac{y_{4l-1}}{7}. \tag{15}$$

(15) 式右边至少含 6 个互异奇素数, 与题设矛盾, (14) 式不成立, 故 $l \neq \pm 1$ 为奇数时 (6) 无公解.

故 $l \neq 0, 2$ 为偶数时, $\frac{x_{4l-1}}{50}, x_{4l}, x_{2l}, x_l, \frac{y_{4l-1}}{7}$ 均不为平方数, 此时 (14) 式成为:

$$Dz^2 = 2^6 \times 5^4 \times 7^2 x_l \cdot x_{2l} \cdot x_{4l} \cdot \frac{x_{4l-1}}{50} \cdot \frac{y_{4l-1}}{7} \cdot \frac{y_l}{350}. \tag{16}$$

(16) 式右边至少含 5 个互异奇素数, 与题设矛盾, (15) 式不成立, 故 $l \neq 0, 2$ 为偶数时 (6) 无公解.

$l = 1$ 时, (16) 式为 $Dz^2 = 2^6 \times 5^4 \times 7^2 x_2 \cdot x_4 \cdot \frac{x_1}{50} \cdot \frac{x_3}{50} \cdot \frac{y_1}{7} \cdot \frac{y_3}{7} = 2^6 \times 5^4 \times 7^2 x_2 \cdot x_4 \cdot \frac{x_3}{50} \cdot \frac{y_3}{7} = 2^6 \times 5^4 \times 7^2 \times 4999 \times 49980001 \times 9997 \times 9999$, 即

$$Dz^2 = 2^6 \times 3^2 \times 5^4 \times 7^2 \times 11 \times 13 \times 101 \times 769 \times 4999 \times 49980001. \tag{17}$$

(17) 式右边至少含 6 个互异奇素数, 与题设矛盾, (17) 式不成立, 故 $l = 1$ 时 (6) 无公解.

$l = -1$ 时, (16) 式为 $Dz^2 = 2^6 \times 5^4 \times 7^2 x_{-2} \cdot x_{-4} \cdot \frac{x_{-1}}{50} \cdot \frac{x_{-3}}{50} \cdot \frac{y_{-1}}{7} \cdot \frac{y_{-3}}{7} = 2^6 \times 5^4 \times 7^2 x_2 \cdot x_4 \cdot \frac{x_1}{50} \cdot \frac{x_3}{50} \cdot \frac{y_1}{7} \cdot \frac{y_3}{7}$, 此时方程 (6) 的解的情况与 $l = 1$ 的情况相同.

$l = 0$ 时, (16) 式为 $Dz^2 = 2^6 \times 5^4 \times 7^2 x_0^3 \cdot \frac{x_{-1}}{50} \cdot \frac{y_{-1}}{7} \cdot \frac{y_0}{350} = 0$, 则 $z = 0$, 此时给出方程 (6) 的平凡公解 $(x, y, z) = (\pm 50, \pm 7, 0)$.

$l = 2$ 时, (16) 式为 $Dz^2 = 2^6 x_2 \cdot x_4 \cdot x_6 \cdot \frac{x_7}{50} \cdot \frac{y_7}{7} \cdot \frac{y_2}{350} = 2^7 \times 5^4 \times 7^2 x_2 \cdot x_4 \cdot x_6 \cdot \frac{x_7}{50} \cdot \frac{y_7}{7}$, 即

$$Dz^2 = 2^7 \times 3^2 \times 5^4 \times 7^2 \times x_2 \cdot x_4 \cdot x_6 \cdot \frac{x_7}{50} \cdot \frac{y_7}{7}. \tag{18}$$

因为 $x_2, x_4, x_6, \frac{x_7}{50}, \frac{y_7}{7}$ 均为奇数, 故 (18) 式右边至少含 5 个不同的奇素数, 与题设矛盾, (18) 式不成立, 所以 $l = 2$ 时方程 (6) 无公解.

情形 1.2　m 为奇数

仿情形 1.1 的证明可得方程 (6) 仅有公解 $(x, y, z) = (\pm 50, \pm 7, 0)$.

情形 2 n 为偶数, 据 (4) 知, $y_{n-1} \equiv y_{n+1} \equiv 1 \pmod 2$, 则 $2 \nmid y_{n-1}$, $2 \nmid y_{n+1}$, 所以 (7) 式为 $Dz^2 = y_{n+1}y_{n-1}$, D 的左边为偶数, D 的右边为奇数, 矛盾.

综上所述, 定理得证.

参考文献

[1] Ljunggren W. Litt om Simultane Pellske Ligninger [J]. Norsk Mat Tidsskr, 1941, 23: 132-138.

[2] Pan Jiayu, Zhang Yuping, Zou Rong. The Pell Equations $x^2 - ay^2 = 1$ and $y^2 - Dz^2 = 1$ [J]. Chinese Quarterly Journal of Mathematics, 1999, 14(1): 73-77.

[3] 管训贵. 关于 Pell 方程 $x^2 - 2y^2 = 1$ 与 $y^2 - Dz^2 = 4$ 的公解 [J]. 华中师范大学学报: 自然科学版: 2012, 46(3): 267-269+278.

[4] 胡永忠, 韩清. 也谈不定方程组 $x^2 - 2y^2 = 1$ 与 $y^2 - Dz^2 = 4$[J]. 华中师范大学学报: 自然科学版: 2002, 36(1): 17-19.

[5] 杜先存, 管训贵, 杨慧章. 关于不定方程组 $x^2 - 6y^2 = 1$ 与 $y^2 - Dz^2 = 4$ 的公解 [J]. 华中师范大学学报: 自然科学版, 2014, 48(3): 5-8.

[6] 赵建红, 万飞. 关于 Pell 方程 $x^2 - 3y^2 = 1$ 与 $y^2 - 2^n y^2 = 16$ 的公解 [J]. 湖北民族学院学报: 自然科学版, 2016, 2(6): 146-148.

[7] 贺腊荣, 张淑静, 袁进. 关于不定方程组 $x^2 - 6y^2 = 1$, $y^2 - Dz^2 = 4$, [J]. 云南民族大学学报: 自然科学版, 2012, 21(1): 57-58.

[8] 冉延平. 不定方程组 $x^2 - 10y^2 = 1$, $y^2 - Dz^2 = 4$ [J]. 延安大学学报: 自然科学版, 2012, 31(1): 8-10.

[9] 过静, 杜先存. 关于 Pell 方程 $x^2 - 30y^2 = 1$ 与 $y^2 - Dz^2 = 4$ 的公解 [J]. 数学的实践与认识, 2015, 45(1): 309-314

[10] 万飞, 杜先存. 关于不定方程组 $x^2 - 30y^2 = 1$ 与 $y^2 - Dz^2 = 4$ 的公解 [J]. 重庆师范大学学报: 自然科学版, 2016, 33(4): 107-111

[11] Walsh G. A note on a theorem of Ljunggren and the diophantine equations $x^2 - kxy^2 + y^4 = 1$ or 4 [J]. Arch Math 1999, 73(2): 504-513.

[12] Togbé A, Voutier P M, and Walsh P G. Solving a family of Thue equations with an application to the equation $x^2 - Dy^4 = 1$. Acta Arith, 2005, 120(1): 39-58.

[13] 孙琦, 袁平之. 关于不定方程 $x^4 - Dy^2 = 1$ 的一个注记 [J]. 四川大学学报 (自然科学版), 1997, 34(3): 13-16.

[14] 乐茂华. 一类二元四次 Diophantine 方程 [J]. 云南师范大学学报 (自然科学版), 2010, 30(1): 12-17.

On the Simultaneous Indefinite Equations $x^2 - 51y^2 = 1$ and $y^2 - Dz^2 = 49$

ZHAO Jian-hong

(Department of Teachers and Education, LiJiang Teachers College, Lijiang 674199, China)

Abstract: Let $D = 2^t p_1^{a_1} p_2^{a_2} p_3^{a_4} p_4^{a_4}$ ($a_i = 0$ or $1, 1 \leq i \leq 4, t \in Z^+$), where are distinct odd primes, the simultaneous Diophantine equations in the title has a positive integer solution only when $D = 2^t \times 4999$ ($t = 1, 3, 5$).

Keywords: integer solution; common solution; indefinite equation; Pell equation; congruence; recursive sequence

第 48 卷第 3 期　　　　　　数学的实践与认识　　　　　　Vol.48, No. 3

2018 年 2 月　　　MATHEMATICS IN PRACTICE AND THEORY　　　Feb., 2018

关于不定方程 $x^2 - 23y^2 = 1$ 与 $y^2 - Dz^2 = 25$ 的公解

赵建红

(丽江师范高等专科学校 教师教育学院, 云南 丽江 674199)

摘　要: 设 $p_s(1 \leq s \leq 4)$ 是互异的奇素数, $D = 2^t p_1^{a_1} p_2^{a_2} p_3^{a_3} p_4^{a_4}$ ($a_i = 0$ 或 1, $1 \leq i \leq 4, t \in Z^+$) 时, 不定方程 $x^2 - 23y^2 = 1$ 与 $y^2 - Dz^2 = 25$ 仅当 $D = 2^t \times 1151$ ($t = 1, 3, 5, 7, 9$) 时有正整数解.

关键词: 整数解; 公解; 不定方程; 同余; 递归序列

近年来, 不定方程

$$x^2 - D_1 y^2 = m \ (D_1 \in Z^+, m \in Z) \ \text{与} \ y^2 - D_2 z^2 = n \ (D_2 \in Z^+, n \in Z) \tag{1}$$

的求解问题一直受到人们的关注. $m = 1, n = 1$ 时方程组 (1) 成为:

$$x^2 - D_1 y^2 = 1 \ \text{与} \ y^2 - D_2 z^2 = 1. \tag{2}$$

关于方程组 (2) 的解的情况, 目前主要集中在解数范围和解数估计上, 对于方程组 (2) 的具体解, 结论还比较少, 主要结论详见文 [1–3];

$m = 1, n = 4$ 时方程 (1) 成为:

$$x^2 - D_1 y^2 = 1 \ \text{与} \ y^2 - D_2 z^2 = 4. \tag{3}$$

关于方程组 (3) 的解的情况, 当 D_2 为奇数时, 主要结论见文 [4–6], 当 D_2 为偶数时, 主要结论见文 [7–11];

$m = 1, n = 25$ 时方程 (1) 成为:

$$x^2 - D_1 y^2 = 1 \ \text{与} \ y^2 - D_2 z^2 = 25. \tag{4}$$

关于方程 (4) 的解的情况目前无相关结果. 本文主要讨论 $m = 1, n = 25, D_1 = 23$ 时方程 (4) 的公解的情况.

1 关键性引理

引理 1[12]　设 p 是一个奇素数, 则丢番图方程 $x^4 - py^2 = 1$ 除开 $p = 5, x = 3, y = 4$ 和 $p = 29, x = 99, y = 1820$ 外, 无其他正整数解.

引理 2[13]　当 $a > 1$ 且 a 是一个平方数时, 方程 $ax^4 - by^2 = 1$ 至多有一组正整数解.

引理 3[14]　若 D 是一个非平方的正整数, 则方程 $x^2 - Dy^4 = 1$ 至多有两组正整数解 (x, y). 而且方程恰有两组正整数解的充要条件是 $D = 1785$ 或 $D = 28560$, 或 $2x_0$ 和 y_0 都是平方数, 这里 (x_0, y_0) 是方程 $x^2 - Dy^2 = 1$ 的基本解.

收稿日期: 2017-07-11

资助项目: 云南省科技厅应用基础研究计划青年项目 (2017FD166); 云南省科技厅应用基础研究计划青年项目 (2013FD060)

引理 4 设 (x_n, y_n), $n \in Z$ 为 Pell 方程 $x^2 - 23y^2 = 1$ 的所有解, 则任意 x_n, y_n 有如下性质:

i)x_n 为平方数, 当且仅当 $n = 0$; $\frac{x_n}{24}$ 为平方数, 当且仅当 $n = 1$ 或 $n = -1$;

ii)$\frac{y_n}{5}$ 为平方数, 当且仅当 $n = 0$ 或 $n = 1$.

证明 设 $(x_n, y_n)(n \in Z)$ 是 Pell 方程 $x^2 - 23y^2 = 1$ 的整数解.

i) 令 $x_n = a^2$, 代入 $x^2 - 23y^2 = 1$, 得 $a^4 - 23y^2 = 1$, 据引理 1 知: $a^4 - 23y^2 = 1$ 仅有平凡解 $(a, y) = (\pm 1, 0)$, 故 $x_n = 1$, 从而 $n = 0$; 令 $\frac{x_n}{24} = a^2$, 代入 $x^2 - 23y^2 = 1$, 得 $576a^4 - 23y^2 = 1$, 据引理 2 知: $576a^4 - 23y^2 = 1$ 仅有整数解 $(a, y) = (\pm 1, \pm 5)$, 故 $x_n = 24$, 从而 $n = 1$ 或 $n = -1$.

ii) 令 $\frac{y_n}{5} = b^2$, 代入 $x^2 - 23y^2 = 1$ 得 $x^2 - 575b^4 = 1$, 据引理 3 知: $x^2 - 575b^4 = 1$ 仅有整数解 $(x, b) = (\pm 24, \pm 1), (\pm 1, 0)$, 故 $\frac{y_n}{5} = 1$ 或 0, 从而 $n = 1$ 或 $n = 0$.

2 定理及证明

定理 若 $p_s(1 \le s \le 4)$ 是互异的奇素数, 则当 $D = 2^t p_1^{a_1} p_2^{a_2} p_3^{a_3} p_4^{a_4}(a_i = 0$ 或 $1, 1 \le i \le 4$, $t \in Z^+)$ 时, 不定方程

$$x^2 - 23y^2 = 1 \text{ 与 } y^2 - Dz^2 = 25. \tag{5}$$

i)$D = 2 \times 1151$ 时, 方程组 (5) 有非平凡公解 $(x, y, z) = (\pm 55224, \pm 11515, \pm 240)$ 和平凡公解 $(x, y, z) = (\pm 24, \pm 5, 0)$;

ii)$D = 2^3 \times 1151$ 时, 方程组 (5) 有非平凡公解 $(x, y, z) = (\pm 55224, \pm 11515, \pm 120)$ 和平凡公解 $(x, y, z) = (\pm 24, \pm 5, 0)$;

iii)$D = 2^5 \times 1151$ 时, 方程组 (5) 有非平凡公解 $(x, y, z) = (\pm 55224, \pm 11515, \pm 60)$ 和平凡公解 $(x, y, z) = (\pm 24, \pm 5, 0)$;

iv)$D = 2^7 \times 1151$ 时, 方程组 (5) 有非平凡公解 $(x, y, z) = (\pm 55224, \pm 11515, \pm 30)$ 和平凡公解 $(x, y, z) = (\pm 24, \pm 5, 0)$;

v)$D = 2^9 \times 1151$ 时, 方程组 (5) 有非平凡公解 $(x, y, z) = (\pm 55224, \pm 11515, \pm 15)$ 和平凡公解 $(x, y, z) = (\pm 24, \pm 5, 0)$;

vi)$D \neq 2^\alpha \times 1151(\alpha = 1, 3, 5, 7, 9)$ 时, 方程组 (5) 只有平凡解 $(x, y, z) = (\pm 24, \pm 5, 0)$.

证明 设 (x_1, y_1) 为 Pell 方程 $x^2 - 23y^2 = 1$ 的基本解, 则有 $(x_1, y_1) = (24, 5)$, 故 Pell 方程 $x^2 - 5y^2 = 1$ 的全部整数解为: $x_n + y_n\sqrt{D} = (24 + 5\sqrt{23})^n, n \in Z$.

容易验证以下性质成立:

I) $y_{2n+1} \equiv 1(\bmod 2), y_{2n} \equiv 0(\bmod 2), y_{2n} \equiv 0(\bmod 24), y_{2n+1} \equiv \pm 5(\bmod 24), y_n \equiv 0(\bmod 5)$;

II) $x_{2n+1} \equiv 0(\bmod 2), x_{2n} \equiv 1(\bmod 2), x_{2n} \equiv \pm 1(\bmod 24), x_{2n+1} \equiv 0(\bmod 24), x_{2n+1} \equiv 24(\bmod 48), x_{2n} \equiv 1(\bmod 5), x_{2n+1} \equiv -1(\bmod 5)$;

III) $\gcd(x_n, y_n) = 1, \gcd(x_n, x_{n+1}) = 1, \gcd(y_n, y_{n+1}) = 5$.

IV) $\gcd(x_{2n}, y_{2n+1}) = \gcd(x_{2n+2}, y_{2n+1}) = 1, \gcd(x_{2n+1}, y_{2n}) = \gcd(x_{2n+1}, y_{2n+2}) = 24$;

V) $x_{n+2} = 48x_{n+1} - x_n, x_0 = 1, x_1 = 24, y_{n+2} = 48y_{n+1} - y_n, y_0 = 0, y_1 = 5$;

设 $(x, y, z) = (x_n, y_n, z_n)$, $n \in Z$ 为不定方程 (5) 的整数解, 则对任意的 $n \in Z$, 有 $y_n^2 - 25 = y_n^2 - 25(x_n^2 - 23y_n^2) = 576y_n^2 - 25x_n^2 = (24y_n + 5x_n)(24y_n - 5x_n) = y_{n-1}y_{n+1}$, 即:

$$y_n^2 - 25 = y_{n-1}y_{n+1}. \tag{6}$$

由 (5) 式的 $Dz^2 = y_n^2 - 25$, 得:

$$Dz^2 = y_{n-1}y_{n+1}. \tag{7}$$

情形 1 n 为偶数

由 I) 的 $y_{2n+1} \equiv 1 \pmod{2}$ 知, $y_{n-1} \equiv y_{n+1} \equiv 1 \pmod{2}$, 于是 (7) 式右边为奇数. 又因为 $D = 2^t p_1^{a_1} p_2^{a_2} p_3^{a_3} p_4^{a_4} (a_i = 0$ 或 $1, 1 \le i \le 4, t \in Z^+)$, 故 (7) 式左边为偶数, 显然矛盾, 所以此时无整数解, 故方程 (5) 无公解.

情形 2 n 为奇数

设 $n = 2m - 1, m \in Z$, 此时 (7) 式成为:

$$Dz^2 = y_{2(m-1)}y_{2m}. \tag{8}$$

由 $y_{2m} = 2x_m y_m, y_{2(m-1)} = 2x_{m-1}y_{m-1}$ 知, (8) 式可化为:

$$Dz^2 = 4x_{m-1}y_{m-1}x_m y_m. \tag{9}$$

情形 2.1 m 为偶数

由 III) 知, $\gcd(x_{m-1}, y_{m-1}) = 1$, $\gcd(x_m, x_{m-1}) = 1$, $\gcd(y_m, y_{m-1}) = 5$, 故 $\gcd(\frac{y_m}{5}, \frac{y_{m-1}}{5}) = 1$; 再据 IV) 知, $\gcd(x_m, y_{m-1}) = 1$, $\gcd(x_{m-1}, y_m) = 24$, 故 $\gcd(\frac{x_{m-1}}{24}, \frac{y_m}{24}) = 1$. 所以 $\frac{x_{m-1}}{24}$, $x_m, \frac{y_{m-1}}{5}, \frac{y_m}{120}$ 两两互素.

设 $m = 2k, k \in Z$, 此时 (9) 式成为:

$$Dz^2 = 4x_{2k-1}y_{2k-1}x_{2k}y_{2k}. \tag{10}$$

由 $y_{2k} = 2x_k y_k$ 知, (10) 式可化为:

$$Dz^2 = 8x_{2k-1}y_{2k-1}x_{2k}x_k y_k. \tag{11}$$

据 III) 知, $\gcd(x_k, y_k) = 1$, 由 I) 知, $5|y_k, k$ 为偶数时 $24|y_k$, 由 II) 知 k 为奇数时 $24|x_k$. 则 k 为偶数时, $x_{2k}, \frac{y_k}{120}, \frac{x_{2k-1}}{24}, \frac{y_{2k-1}}{5}, x_k$ 两两互素; k 为奇数时, $x_{2k}, \frac{x_k}{24}, \frac{y_k}{5}, \frac{x_{2k-1}}{24}, \frac{y_{2k-1}}{5}$ 两两互素.

情形 2.1.1 k 为偶数

设 $k = 2l, l \in Z$, 此时 (11) 式成为:

$$Dz^2 = 8x_{4l}y_{2l}x_{2l}x_{4l-1}y_{4l-1}. \tag{12}$$

由 $y_{2l} = 2x_l y_l$ 知, (12) 式可化为:

$$Dz^2 = 16x_{4l-1}y_{4l-1}x_{4l}x_l y_l x_{2l}. \tag{13}$$

据 (III) 知, $\gcd(x_l, y_l) = 1$, 由 (I) 知, $5|y_l, l$ 为偶数时 $24|y_l$, 由 (II) 知 l 为奇数时 $24|x_l$. 则 l 为偶数时, $x_{4l}, x_l, \frac{y_l}{120}, x_{2l}, \frac{x_{4l-1}}{24}, \frac{y_{4l-1}}{5}$ 两两互素; l 为奇数时, $x_{4l}, \frac{x_l}{24}, x_{2l}, \frac{y_l}{5}, \frac{y_{4l-1}}{5}, \frac{x_{4l-1}}{24}$ 两两互素.

由 (I) 知, $\frac{y_{4l-1}}{5}$ 为奇数, l 为奇数时 $\frac{y_l}{5}$ 为奇数; 由 (II) 知, x_{2l}, x_{4l} 为奇数, l 为偶数时 x_l 为奇数;

由 (II) 知, $x_{4l-1} \equiv 24 \pmod{48}$, 故 l 为奇数时 $x_l \equiv 24 \pmod{48}$, $\frac{x_{4l-1}}{24}$ 为奇数, 而此时 $\frac{x_l}{24}$ 为奇数. 故 l 为偶数时, $x_{4l}, x_l, x_{2l}, \frac{x_{4l-1}}{24}, \frac{y_{4l-1}}{5}$ 均为奇数; l 为奇数时, $x_{4l}, \frac{x_l}{24}, x_{2l}, \frac{x_{4l-1}}{24}, \frac{y_{4l-1}}{5}$ 均为奇数.

由引理 4 的 (I) 知, 仅当 $l = 0$ 时, $x_{4l}, x_{2l}, x_l, \frac{x_{4l-1}}{24}$ 为平方数; 仅当 $l = 1$ 或 $l = -1$ 时, $\frac{x_l}{24}$ 为平方数. 又由引理 4 的 (II) 知, $\frac{y_l}{5}$ 为平方数当且仅当 $l = 1$ 或 $l = 0$; 对于任意的 $l \in Z$, $\frac{y_{4l-1}}{5}$ 不为平方数. 所以 $l \ne 0$ 为偶数时 $x_{4l}, x_l, x_{2l}, \frac{x_{4l-1}}{24}, \frac{y_{4l-1}}{5}$ 都不是平方数, 此时他们至少含有 5 个不同的奇素数, 从而 (13) 式不成立, 所以 $l \ne 0$ 为偶数时方程 (5) 无公解. 当 $l \ne \pm 1$ 为奇

数时, $x_{4l}, \frac{x_l}{24}, x_{2l}, \frac{x_{4l-1}}{24}, \frac{y_{4l-1}}{5}$ 均不为奇数, 此时他们至少含有 5 个不同的奇素数, 从而 (13) 式不成立, 所以 $l \neq \pm 1$ 为偶数时方程 (5) 无公解.

$l = 0$ 时, 由 (V) 知 (13) 式为 $Dz^2 = 16x_0^3 y_0 x_{-1} y_{-1}$, 则有 $z = 0$, 此时方程 (5) 仅有平凡解 $(x, y, z) = (\pm 9, \pm 4, 0)$.

$l = 1$ 时,(13) 式为

$$Dz^2 = 480^2 \cdot x_2 \cdot x_4 \cdot \frac{x_3}{24} \cdot \frac{y_3}{5}. \tag{14}$$

又由 (V) 知 $x_2 = 1151, \frac{x_3}{24} = 3 \times 13 \times 59, \frac{y_3}{5} = 2303 = 7^2 \times 47, x_4 = 2649601$, 故 (14) 式右边至少含有 6 个不同的素数, 从而 (14) 式不成立, 所以 $l = 1$ 时方程 (5) 无公解.

$l = -1$ 时,(13) 式为

$$Dz^2 = 16 \cdot 24^2 \cdot 5^2 \cdot x_2 \cdot x_4 \cdot \frac{x_5}{24} \cdot \frac{y_5}{5}. \tag{15}$$

又由 (V) 知 $x_2 = 1151, \frac{x_5}{24} = 5296901, \frac{y_5}{5} = 5301505 = 5 \times 11 \times 41 \times 2351, x_4 = 2649601$, 故 (15) 式右边至少含有 6 个不同的素数, 从而 (15) 式不成立, 所以 $l = -1$ 时方程 (5) 无公解.

情形 2.1.2 k 为奇数

令 $k = 2l - 1, l \in Z$, 此时 (11) 式可化为:

$$Dz^2 = 8x_{4l-3} y_{4l-3} x_{2(2l-1)} x_{2l-1} y_{2l-1}. \tag{16}$$

故 $\frac{x_{4l-3}}{24}, \frac{y_{4l-3}}{24}, x_{2(2l-1)}, \frac{x_{2l-1}}{24}, \frac{y_{2l-1}}{5}$ 两两互素. 由 (I) 知 $\frac{y_{4l-3}}{5}, \frac{y_{2l-1}}{5}$ 为奇数; 由 (II) 知 $x_{2(2l-1)}$ 为奇数, $x_{4l-3} \equiv 24 (\bmod 48), x_{2l-1} \equiv 24 (\bmod 48)$, 故 $\frac{x_{2l-1}}{24}, \frac{x_{4l-3}}{24}$ 均为奇数.

故 $\frac{x_{4l-3}}{24}, \frac{y_{4l-3}}{5}, x_{2(2l-1)}, \frac{x_{2l-1}}{24}, \frac{y_{2l-1}}{5}$ 两两互素且均为奇数.

由引理 4 的 (I) 知, 仅当 $l = 0$ 或 1 时 $\frac{x_{2l-1}}{24}$ 是平方数, 仅当 $l = 1$ 时 $\frac{x_{4l-3}}{24}$ 是平方数; 对于任意的 $l \in Z, x_{2(2l-1)}, \frac{y_{4l-1}}{5}$ 不是平方数. 又由引理 4 的 (II) 知, 仅当 $l = 1$ 时, $\frac{y_{2l-1}}{5}$ 和 $\frac{y_{4l-3}}{5}$ 是平方数. 所以 $l \neq 0, 1$ 时, $\frac{x_{4l-3}}{24}, \frac{y_{4l-3}}{5}, x_{2(2l-1)}, \frac{x_{2l-1}}{24}, \frac{y_{2l-1}}{5}$ 都不是平方数, 此时他们至少含有 5 个不同的奇素数, 从而 (16) 不成立, 所以 $l \neq 0, 1$ 时方程无公解.

$l = 0$ 时, (16) 式为

$$Dz^2 = 2^9 \cdot 3^2 \cdot 5^2 \cdot x_2 \cdot \frac{x_3}{24} \cdot \frac{y_3}{5}. \tag{17}$$

又由 (V) 知, $x_2 = 1151, \frac{x_3}{24} = 2301 = 3 \times 13 \times 59, \frac{y_3}{5} = 2303 = 7^2 \times 47$, 故 (17) 式右边含有 5 个不同的素数, 从而 (17) 式不成立, 所以 $l = 0$ 时方程 (5) 无公解.

$l = 1$ 时, (16) 式成为 $Dz^2 = 8 \cdot x_1 \cdot y_1 \cdot x_2 \cdot x_1 \cdot y_1 = 2^9 \times 3^2 \times 5^2 \times 1151$. 于是 $z = 240, D = 2 \times 1151$ 或 $z = 120, D = 2^3 \times 1151$ 或 $z = 60, D = 2^7 \times 1151$ 或 $z = 30, D = 2^9 \times 1151$. 所以当 $D = 2 \times 1151$ 时, 方程组 (5) 有非平凡公解 $(x, y, z) = (\pm 55224, \pm 11515, \pm 240)$; 当 $D = 2^3 \times 1151$ 时, 方程组 (5) 有非平凡公解 $(x, y, z) = (\pm 55224, \pm 11515, \pm 120)$; 当 $D = 2^5 \times 1151$ 时, 方程组 (5) 有非平凡公解 $(x, y, z) = (\pm 55224, \pm 11515, \pm 60)$; 当 $D = 2^7 \times 1151$ 时, 方程组 (5) 有非平凡公解 $(x, y, z) = (\pm 55224, \pm 11515, \pm 30)$; 当 $D = 2^9 \times 1151$ 时, 方程组 (5) 有非平凡公解 $(x, y, z) = (\pm 55224, \pm 11515, \pm 15)$;

情形 2.2 m 为偶数

仿情形 2.1 的证明可得方程 (5) 仅有公解 $(x, y, z) = (\pm 24, \pm 5, 0)$.

综上所述, 定理得证.

参考文献

[1] Ljunggren W. Litt om Simultane Pellske Ligninger [J]. Norsk Mat Tidsskr, 1941, 23: 132-138.

[2] Pan Jiayu, Zhang Yuping, Zou Rong. The Pell Equations $x^2 - ay^2 = 1$ and $y^2 - Dz^2 = 1$ [J]. Chinese Quarterly Journal of Mathematics, 1999, 14(1): 73-77.

[3] 乐茂华. 关于联立 Pell 方程组 $x^2 - 4D_1y^2 = 1$ 与 $y^2 - D_2z^2 = 4$ [J]. 佛山科学奇数学院学报: 自然科学版, 2004, 22(2): 1-3+9.

[4] 陈永高. Pell 方程组 $x^2 - 2y^2 = 1$ 与 $y^2 - Dz^2 = 4$ 的公解 [J]. 北京大学学报, 1994, 30(3): 298-302.

[5] 任小枝. Pell 方程 $x^2 - 2y^2 = 1$ 与 $y^2 - Dz^2 = 4$ 的公解 [J]. 南京大学学报: 自然科学版, 2014, 37(2): 1-4.

[6] 过静, 杜先存. 关于不定方程 $x^2 - 12y^2 = 1$ 与 $y^2 - Dz^2 = 4$ 的解 [J]. 数学的实践与认识, 2015, 45(9): 289-293.

[7] 胡永忠, 韩清. 也谈不定方程组 $x^2 - 2y^2 = 1$ 与 $y^2 - Dz^2 = 4$[J]. 华中师范大学学报: 自然科学版, 2002, 36(1): 17-19.

[8] 管训贵. 关于 Pell 方程 $x^2 - 2y^2 = 1$ 与 $y^2 - Dz^2 = 4$ 的公解 [J]. 华中师范大学学报: 自然科学版, 2012, 46(3): 267-269+278.

[9] 杜先存, 李玉龙. 关于 Pell 方程 $x^2 - 6y^2 = 1$ 与 $y^2 - Dz^2 = 4$ 的公解 [J]. 安徽大学学报: 自然科学版, 2015, 39(6): 19-22.

[10] 杜先存, 管训贵, 杨慧章. 关于不定方程组 $x^2 - 6y^2 = 1$ 与 $y^2 - Dz^2 = 4$ 的公解 [J]. 华中师范大学学报: 自然科学版, 2014, 48(3): 5-8.

[11] 过静, 杜先存. 关于 Pell 方程 $x^2 - 30y^2 = 1$ 与 $y^2 - Dz^2 = 4$ 的公解 [J]. 数学的实践与认识, 2015, 45(1): 309-314

[12] 柯召, 孙琦. 谈谈不定方程 [M]. 哈尔滨: 哈尔滨工业大学出版社,2011:64.

[13] 孙琦, 袁平之. 关于不定方程 $x^4 - Dy^2 = 1$ 的一个注记 [J]. 四川大学学报 (自然科学版), 1997, 34(3): 13-16.

[14] 乐茂华. 一类二元四次 Diophantine 方程 [J]. 云南师范大学学报 (自然科学版), 2010, 30(1): 12-17.

[15] Walsh G.A note on a theorem of Ljunggren and the diophantine equations $x^2 - kxy^2 + y^4 = 1$ or 4 [J]. Arch Math, 1999, 73(2): 504-513.

On the Simultaneous Indefinite Equations $x^2 - 23y^2 = 1$ and $y^2 - Dz^2 = 25$

ZHAO Jian-hong

(Department of Teachers and Education, LiJiang Teachers College, LiJiang 674199, China)

Abstract: Let $D = 2^t p_1^{a_1} p_2^{a_2} p_3^{a_3} p_4^{a_4}$ ($a_i = 0$ or $1, 1 \le i \le 4$, $t \in Z^+$), where $p_s (1 \le s \le 4)$ are distinct odd primes, the simultaneous Diophantine equations in the title has a positive integer solution only when $D = 2^t \times 1151$ ($t = 1, 3, 5, 7, 9$).

Keywords: integer solution; common solution; indefinite equation; Pell equation; congruence; recursive sequence

第 39 卷第 6 期　　　　　　　西南大学学报（自然科学版）　　　　　　　2017 年 6 月

Vol. 39　No. 6　　　Journal of Southwest University (Natural Science Edition)　　　Jun.　2017

DOI：10. 13718/j. cnki. xdzk. 2017. 06. 011

椭圆曲线 $y^2=(x+2)(x^2-2x+p)$ 的整数点[①]

杜先存[1]，　　赵建红[2]，　　万　飞[1]

1. 红河学院 教师教育学院，云南 蒙自 661199；2. 丽江师范高等专科学校 数学与计算机科学系，云南 丽江 674199

摘要：利用 Legendre 符号、同余式、Pell 方程的解的性质等初等方法证明了：当 $p=36s^2-5(s\in Z_+,2\nmid s)$，而 $6s^2-1,12s^2+1$ 均为素数时，椭圆曲线 $y^2=(x+2)(x^2-2x+p)$ 仅有整数点为 $(x,y)=(-2,0)$.

关　键　词：椭圆曲线；整数点；Pell 方程；Legendre 符号；同余

中图分类号：O156. 2　　　　**文献标志码**：A　　　　**文章编号**：1673-9868(2017)06-0069-05

3 次不定方程是一类基本而又重要的方程，目前关于 3 次不定方程的结论已经比较多[1-2]．而椭圆曲线 $y^2=(x+a)(x^2-ax+p)(a,p\in Z)$ 是 3 次不定方程中的一类特殊的方程．椭圆曲线的整数点问题是数论和算术代数几何学中基本而又重要的问题，其结果有着广泛的应用．近年来，寻找椭圆曲线的整数点问题引起了人们的兴趣．关于椭圆曲线的整数点问题，目前已有一些结论．当 $a=-2$ 时，椭圆曲线的结论见文献[3-7]；当 $a=2$ 时，椭圆曲线的结果仅限于 $p=31$ 的情况，见文献[8]．本文将对 $a=2,p=36s^2-5(s\in Z_+,2\nmid s)$ 时椭圆曲线的整数点问题进行研究．

定理 1　设 $p=36s^2-5(s\in Z_+,2\nmid s)$，而 $6s^2-1,12s^2+1$ 均为素数，则椭圆曲线

$$y^2=(x+2)(x^2-2x+p) \tag{1}$$

仅有整数点为 $(x,y)=(-2,0)$.

证　设 (x,y) 是椭圆曲线(1)的整数点．因为 $12s^2+1$ 为素数，所以

$$\gcd(x+2,x^2-2x+p)=\gcd(x+2,p+8)=$$
$$\gcd(x+2,3(12s^2+1))=$$
$$1,3,12s^2+1,3(12s^2+1)$$

设 $t=12s^2+1$，故椭圆曲线(1)可分解为以下 4 种可能的情形：

情形 Ⅰ　$x+2=u^2,x^2-2x+p=v^2,y=uv,\gcd(u,v)=1(u,v\in Z)$；

情形 Ⅱ　$x+2=3u^2,x^2-2x+p=3v^2,y=3uv,\gcd(u,v)=1(u,v\in Z)$；

情形 Ⅲ　$x+2=tu^2,x^2-2x+p=tv^2,y=tuv,\gcd(u,v)=1(u,v\in Z)$；

情形 Ⅳ　$x+2=3tu^2,x^2-2x+p=3tv^2,y=3tuv,\gcd(u,v)=1(u,v\in Z)$.

下面分别讨论这 4 种情形下椭圆曲线(1)的整数点的情况．

情形 Ⅰ　因为 $p=36s^2-5$，故 $p\equiv-1(\bmod 4)$．又因 $x=u^2-2\equiv2,3(\bmod 4)$，因此 $x^2-2x+p\equiv$

---

①　收稿日期：2016-05-24

　　基金项目：云南省科技厅应用基础研究计划青年项目(2013FD060)；云南省教育厅科研基金(2014Y462)；红河学院校级课题
　　　　　　　(XJ15Y22)；红河学院中青年学术骨干培养资助项目(2015GG0207)；喀什大学校级课题((14)2507)；江西省教育厅科学技术
　　　　　　　研究项目(GJJ160782).

　　作者简介：杜先存(1981-)，女，云南凤庆人，副教授，主要从事初等数论的研究.

　　通信作者：万　飞，副教授.

$2,3 (\bmod 4)$，故有

$$2,3(\bmod 4) \equiv x^2 - 2x + p = v^2 \equiv 0,1(\bmod 4)$$

显然矛盾，故该情形下椭圆曲线(1)无整数点.

情形 Ⅱ 因为 $p = 36s^2 - 5$，$2 \nmid s$，故 $p \equiv -1(\bmod 8)$. 又因 $x = 3u^2 - 2 \equiv 1,2,6(\bmod 8)$，则 $x^2 - 2x + p \equiv 6,7(\bmod 8)$，故有

$$6,7(\bmod 8) \equiv x^2 - 2x + p = 3v^2 \equiv 0,3,4(\bmod 8)$$

显然矛盾，故该情形下椭圆曲线(1)无整数点.

情形 Ⅲ 因为 $p = 36s^2 - 5$，故 $p \equiv -1(\bmod 4)$. 又因 $x = 13u^2 - 2 \equiv 2,3(\bmod 4)$，因此 $x^2 - 2x + p \equiv 2,3(\bmod 4)$. 因 $t = 12s^2 + 1 \equiv 1(\bmod 4)$，故 $tv^2 \equiv 0,1(\bmod 4)$，故有

$$2,3(\bmod 4) \equiv x^2 - 2x + p = tv^2 \equiv 0,1(\bmod 4)$$

显然矛盾，故该情形下椭圆曲线(1)无整数点.

情形 Ⅳ 当 $2 \nmid u$ 时，有 $u^2 \equiv 1(\bmod 4)$. 因为 $t = 12s^2 + 1 \equiv 1(\bmod 4)$，因此 $x = 3tu^2 - 2 \equiv 1(\bmod 4)$. 又因 $p = 36s^2 - 5$，故 $p \equiv -1(\bmod 4)$，因此 $x^2 - 2x + p \equiv 2(\bmod 4)$. 而 $3tv^2 \equiv 0,3(\bmod 4)$，故有

$$2(\bmod 4) \equiv x^2 - 2x + p = 3tv^2 \equiv 0,3(\bmod 4)$$

矛盾，因此 $2 \nmid u$ 不成立，所以 $2 \mid u$. 令 $u = 2m(m \in \mathbf{Z})$，则 $x + 2 = 3tu^2$ 为 $x + 2 = 12tm^2$，代入 $x^2 - 2x + p = 3tv^2$ 得 $(12tm^2 - 3)^2 + p - 1 = 3tv^2$. 又因为 $p = 36s^2 - 5$，$t = 12s^2 + 1$，所以 $p = 3t - 8$，代入 $(12tm^2 - 3)^2 + p - 1 = 3tv^2$，配方得 $(12m^2 - 1)^2 + 48(t - 3)m^4 = v^2$，即

$$(v + 12m^2 - 1)(v - 12m^2 + 1) = 48(t - 3)m^4 \tag{2}$$

因为 $2 \mid u$，故由 $x + 2 = 3tu^2$ 知 $2 \mid x$，则 $x^2 - 2x + p = 3tv^2$ 得 $2 \nmid v$，故 $2 \mid [v - (12m^2 - 1)]$. 因此

$$\gcd(v + 12m^2 - 1, v - 12m^2 + 1) = 2\gcd(12m^2 - 1, v)$$

设 $\gcd(12m^2 - 1, v) = d$，则由(2)式知 $d \mid 48(t - 3)m^4$. 又因 $6s^2 - 1$ 为素数，则

$$\gcd(12m^2 - 1, 48(t - 3)m^4) = \gcd(12m^2 - 1, 4(12s^2 - 2)m^2) =$$
$$\gcd(6s^2 - 1, 2m^2 - s^2) =$$
$$1, 6s^2 - 1$$

若 $\gcd(12m^2 - 1, 48(t - 3)m^4) = 6s^2 - 1$，则有 $12m^2 - 1 \equiv 0(\bmod (6s^2 - 1))$，即 $12m^2 \equiv 1(\bmod (6s^2 - 1))$. 因 Legendre 符号值 $\left(\dfrac{12m^2}{6s^2 - 1}\right) = \left(\dfrac{3}{6s^2 - 1}\right) = \left(\dfrac{6s^2 - 1}{3}\right) = -1$，故 $12m^2 \equiv 1(\bmod (6s^2 + 1))$ 不成立，则 $\gcd(12m^2 - 1, 48(t - 3)m^2) = 1$，因此 $d = 1$，即 $\gcd(12m^2 - 1, v) = 1$. 所以

$$\gcd(v + 12m^2 - 1, v - 12m^2 + 1) = 2$$

因为 $48(t - 3) = 96(6s^2 - 1)$，而 $6s^2 - 1$ 为素数，故(2)式可分解为：

$$\begin{cases} v + 12m^2 - 1 = 2ra^4 \\ v - 12m^2 + 1 = \dfrac{48(6s^2 - 1)}{r}b^4 \end{cases} \tag{3}$$

其中 $m = ab$，$\gcd(a, b) = 1$，$\gcd\left(r, \dfrac{24(6s^2 - 1)}{r}\right) = 1$，$a, b \in \mathbf{Z}$，且

$$r = 1, 2^3, 3, 2^3 \times 3 \times (6s^2 - 1), 2^3 \times (6s^2 - 1), 6s^2 - 1, 2^3 \times 3, 3 \times (6s^2 - 1)$$

由(3)式得

$$12m^2 - 1 = ra^4 - \dfrac{24(6s^2 - 1)}{r}b^4 \tag{4}$$

对(4)式两边取模 4，得

$$-1 \equiv ra^4 - \dfrac{24(6s^2 - 1)}{r}b^4(\bmod 4) \tag{5}$$

对(4)式两边取模 3，得

$$-1 \equiv ra^4 - \frac{24(6s^2-1)}{r}b^4 \pmod{3} \tag{6}$$

当 $r=6s^2-1$ 时，(5)式为

$$-1 \equiv ra^4 \pmod{4} \tag{7}$$

因为 $2 \nmid s$，所以 $ra^4=(6s^2-1)a^4 \equiv 0,1 \pmod 4$，则(7)式为 $-1 \equiv ra^4 \equiv 0,1 \pmod 4$，显然矛盾，故(7)式不成立，因此 $r=6s^2-1$ 时(3)式不成立，即情形 Ⅳ 不成立.

当 $r=2^3,2^3 \times (6s^2-1)$ 时，(5)式为

$$1 \equiv \frac{24(6s^2-1)}{r}b^4 \pmod{4} \tag{8}$$

因为 $2 \nmid s$，所以 $r=2^3$ 时，有

$$\frac{24(6s^2-1)}{r}b^4 = 3(6s^2-1)b^4 \equiv 0,3 \pmod 4$$

而 $r=2^3(6s^2-1)$ 时，有

$$\frac{24(6s^2-1)}{r}b^4 = 3b^4 \equiv 0,3 \pmod 4$$

故当 $r=2^3,2^3 \times (6s^2-1)$ 时，(8)式为 $1 \equiv 0,3 \pmod 4$，显然矛盾，故(8)式不成立，因此当 $r=2^3,2^3 \times (6s^2-1)$ 时(3)式不成立，即情形 Ⅳ 不成立.

当 $r=1$ 时，(6)式为

$$-1 \equiv ra^4 \pmod{3} \tag{9}$$

因为 $r=1$ 时，Legendre 符号值 $\left(\dfrac{ra^4}{3}\right)=\left(\dfrac{a^4}{3}\right)=1$，而 Legendre 符号值 $\left(\dfrac{-1}{3}\right)=-1$，故(9)式不成立，因此当 $r=1$ 时(3)式不成立，即情形 Ⅳ 不成立.

当 $r=3(6s^2-1),2^3 \times 3$ 时，(6)式为

$$1 \equiv \frac{24(6s^2-1)}{r}b^4 \pmod{3} \tag{10}$$

因为 Legendre 符号值 $\left(\dfrac{1}{3}\right)=1$，而 $r=3(6s^2-1)$ 时 Legendre 符号值 $\left(\dfrac{\frac{24(6s^2-1)}{r}b^4}{3}\right)=\left(\dfrac{8b^4}{3}\right)=-1$，故(10)式不成立；$r=2^3 \times 3$ 时，Legendre 符号值 $\left(\dfrac{\frac{24(6s^2-1)}{r}b^4}{3}\right)=\left(\dfrac{6s^2-1}{3}\right)=\left(\dfrac{-1}{3}\right)=-1$，故 $r=2^3 \times 3$ 时(10)式不成立. 因此 $r=3(6s^2-1),2^3 \times 3$ 时(3)式不成立，即情形 Ⅳ 不成立.

当 $r=2^3 \times 3 \times (6s^2-1)$ 时，(4)式为 $12m^2-1=2^3 \times 3 \times (6s^2-1)a^4-b^4$. 将(3)式的 $m=ab$ 代入并配方，得

$$(6a^2+b^2)^2-12(12s^2+1)a^4=1 \tag{11}$$

令 $w=6a^2+b^2$，则(11)式为

$$w^2-12(12s^2+1)a^4=1 \tag{12}$$

由文献[9]的定理 1 得，方程(12)至多有 1 组正整数解 (w,a). 又由文献[7]的引理 6 得，Pell 方程

$$w^2-12(12s^2+1)a^2=1$$

的基本解为 $(w_1,a_1)=(24s^2+1,2s)$，则方程(12)的全部正整数解可表示为

$$w_n+a_n\sqrt{12(12s^2+1)} = w_n+2a_n\sqrt{3(12s^2+1)} = ((24s^2+1)+4s\sqrt{3(12s^2+1)})^n \qquad n \in \mathbf{Z}_+$$

由此可知方程(11)的正整数解满足

$$(6a^2+b^2)+2a^2\sqrt{3(12s^2+1)} = ((24s^2+1)+4s\sqrt{3(12s^2+1)})^n \qquad n \in \mathbf{Z}_+ \tag{13}$$

由(13)式有

$$2a^2 = \begin{cases} 4sn(24s^2+1)^{n-1} + \sum_{i=1}^{\frac{n-1}{2}} \binom{n}{2i+1} \times (24s^2+1)^{n-2i-1} \times (4s)^{2i+1} \times 3^i \times (12s^2+1)^i & 2\nmid n \\ \sum_{i=1}^{\frac{n}{2}} \binom{n}{2i-1} \times (24s^2+1)^{n-2i+1} \times (4s)^{2i-1} \times 3^{i-1} \times (12s^2+1)^{i-1} & 2\mid n \end{cases} \tag{14}$$

则由文献[7]的定理 1 得知 $n=2, 2\nmid n$.

若 $2\nmid n$, 因为 $2\nmid s$, 故 $4ns \equiv 4 \pmod 8$, $24s^2+1 \equiv 1 \pmod 8$. (14) 式两边取模 8, 得 $a^2 \equiv 2 \pmod 8$, 显然不成立, 故方程(12)无正整数解, 因此方程(11)仅有平凡解 $(w,a)=(1,0)$. 由 $w=6a^2+b^2=1$, 得 $a=0, b=\pm 1$, 此时得出椭圆曲线(1)有整数点 $(x,y)=(-2,0)$, 故 $r=2^3 \times 3 \times (6s^2-1)$ 时椭圆曲线 (1) 有整数点 $(x,y)=(-2,0)$.

若 $n=2$, 由(13)式得
$$6a^2+b^2 = (24s^2+1)^2+(4s)^2[3(12s^2+1)] =$$
$$576s^4+48s^2+1+576s^4+48s^2 = 1\,152s^4+96s^2+1$$
即
$$6a^2+b^2 = 1\,152s^4+96s^2+1 \tag{15}$$
由(14)式得
$$2a^2 = \binom{2}{1} \times (24s^2+1) \times 4s = 8s(24s^2+1)$$
即
$$a^2 = 4s(24s^2+1) \tag{16}$$
由(16)式的 $a^2=4s(24s^2+1)$ 知 $2\mid a$. 令 $a=2t, t\in \mathbf{Z}$, 则(16)式变为
$$t^2 = s(24s^2+1) \tag{17}$$
又因 $\gcd(s, 24s^2+1)=1$, 故(17)式可分解为:
$$s=c^2 \qquad 24s^2+1=e^2 \qquad t=ce \qquad \gcd(c,e)=1 \tag{18}$$
将(18)式的 $s=c^2$ 代入 $24s^2+1=e^2$, 得 $24c^4+1=e^2$, 即
$$e^2-24c^4=1 \tag{19}$$
因为方程(19)有正整数解 $(e,c)=(5,1)$, 故由文献[9]的定理 1 得方程(19)仅有正整数解 $(e,c)=(5,1)$. 于是 $s=c^2=1, 1\,152s^4+96s^2+1=1\,249$, 所以 $a^2=4s(24s^2+1)=100$, 则 $a=10$. 代入(16) 式, 得 $6a^2+b^2=600+b^2=1\,249$, 则有 $b^2=1\,249-600=649$, 显然无解, 此时椭圆曲线(1)没有整数点.

当 $r=3$ 时, (4)式为 $12m^2-1=3a^4-8(6s^2-1)b^4$, 将(3)式的 $m=ab$ 代入并配方, 得
$$4(12s^2+1)b^4-3(a^2-2b^2)^2=1 \tag{20}$$
令 $f=2b^2, t=a^2-2b^2, r,f\in \mathbf{N}$, 则(20)式为
$$(12s^2+1)f^2-3t^2=1 \tag{21}$$
又因 $(1,2s)$ 为方程(21)的基本解, 则方程(21)的一切整数解可表示为
$$f\sqrt{12s^2+1}+t\sqrt{3} = \pm(\sqrt{12s^2+1}+2s\sqrt{3})^{2n+1} \qquad n\in \mathbf{Z}$$
由此方程(20)的一切正整数解 (b^2, a^2-2b^2) 满足
$$2b^2\sqrt{12s^2+1}+|a^2-2b^2|\sqrt{3} = (\sqrt{12s^2+1}+2s\sqrt{3})^{2n+1} \qquad n\in \mathbf{N} \tag{22}$$
由(22)式, 得
$$2b^2 = \sum_{i=0}^{n} \binom{2n+1}{2i} \times \sqrt{12s^2+1}^{\,2(n-i)} \times (2s)^{2i} \times \sqrt{3}^{\,2i} =$$

$$(12s^2+1)+\sum_{i=1}^{n}\binom{2n+1}{2i}\times(12s^2+1)^{n-i}\times s^{2i}\times 3^i\times 4^i$$

即

$$2b^2=(12s^2+1)+\sum_{i=1}^{n}\binom{2n+1}{2i}\times(12s^2+1)^{n-i}\times s^{2i}\times 3^i\times 4^i \tag{23}$$

因为 $\sum\limits_{i=1}^{n}\binom{2n+1}{2i}\times(12s^2+1)^{n-i}\times s^{2i}\times 3^i\times 4^i$ 为偶数，则(23)式左边为偶数，右边为奇数，矛盾，所以方程(23)无整数解，故 $r=3$ 时(3)式不成立，即情形 IV 不成立.

综上所述，定理 1 得证.

参考文献：

[1] 万 飞，杜先存. 关于不定方程 $x^3\pm1=6pqDy^2$ 的整数解 [J]. 西南师范大学学报（自然科学版），2016，41(12)：16—19.

[2] 呼家源，李小雪. Diophantine 方程 $x^3+8=py^2$ 有本原正整数解的必要条件 [J]. 西南大学学报（自然科学版），2017，39(2)：50—54.

[3] ZAGIER D. Lager Integral Point on Elliptic Curves [J]. Math Comp, 1987, 48：425—436.

[4] ZHU H L, CHEN J H. Integral Point on $y^2=x^3+27x-62$ [J]. J Math Study, 2009, 42(2)：117—125.

[5] 吴华明. 椭圆曲线 $y^2=x^3+27x-62$ 的整数点 [J]. 数学学报（中文版），2010，53(1)：205—208.

[6] 贺艳峰. 数论函数的均值分布及整点问题的研究 [D]. 西安：西北大学，2010：20—25.

[7] 管训贵. 椭圆曲线 $y^2=x^3+(p-4)x-2p$ 的整数点 [J]. 数学进展，2014，43(4)：521—526.

[8] 过 静. 椭圆曲线 $y^2=x^3+27x+62$ 的整数点 [J]. 重庆师范大学学报（自然科学版），2016，33(5)：50—53.

[9] TOGBÉ A, VOUTIER P M, WALSH P G. Solving a Family of Thue Equations with an Application to the Equation $x^2-Dy^4=1$ [J]. Acta Arith, 2005, 120(1)：39—58.

The Integral Points on the Elliptic Curve
$y^2=(x+2)(x^2-2x+p)$

DU Xian-cun[1], 　ZHAO Jian-hong[2], 　WAN Fei[1]

1. College of Teacher Education, Honghe University, Mengzi Yunnan 661199, China;

2. Department of Mathematics and Computer Science, Lijiang Teachers College, Lijiang Yunnan 674199, China

Abstract：Let $p=36s^2-5(s\in Z_+, 2\nmid s)$, where is a positive odd number satisfying that $6s^2-1$ and $12s^2+1$ are primes. It is proved in this paper with the help of the Legendre symbol, congruence and some properties of the solutions to the Pell equation that the elliptic curve $y^2=(x+2)(x^2-2x+p)$ has only integer point $(x, y)=(-2, 0)$.

Key words：elliptic curve; integer point; Pell equation; Legendre symbol; congruence

责任编辑 廖 坤

第 47 卷 第 1 期
2018 年 1 月

内蒙古师范大学学报（自然科学汉文版）
Journal of Inner Mongolia Normal University（Natural Science Edition）

Vol. 47 No. 1
Jan. 2018

关于 D. H. Lehmer 数及其相关问题

赵建红[1]，李小雪[2]

(1. 丽江师范高等专科学校 教师教育学院，云南 丽江 674199；

2. 西安航空学院 理学院，陕西 西安 710077)

摘　要：设 p 是奇素数. 对任一整数 a 且 $1 \leqslant a \leqslant p-1$，显然存在唯一的整数 $0 \leqslant b \leqslant p-1$，使得 $ab \equiv 1 \bmod p$. 设 $N(p)$ 表示同余方程 $ab \equiv 1 \bmod p$ 满足 $1 \leqslant a, b \leqslant p-1$，且 a 和 b 具有相反的奇偶性的所有整数 a 的集合，$S(p)$ 表示满足 $a+b \equiv 1 \bmod p$ 的所有 $a, b \in N(p)$ 的解的个数. 利用解析方法以及 Gauss 和的性质，研究了 D. H. Lehmer 数的相关问题，证明了存在两个整数 $a, b \in N(p)$，使得 $a+b \equiv 1 \bmod p$，并得到了关于 $S(p)$ 的一个较强的渐近公式.

关键词：Lehmer 问题；渐近公式；解析方法；Gauss 和

中图分类号：O 156.4　　**文献标志码**：A　　**文章编号**：1001-8735(2018)01-0012-04

doi：10.3969/j.issn.1001-8735.2018.01.003

设 $q \geqslant 3$ 是一个奇素数，c 是一个给定的整数，且 $(c, q) = 1$. 对任一整数 a 且 $1 \leqslant a \leqslant q-1$，显然存在唯一的整数 $0 \leqslant b \leqslant q-1$ 使得 $ab \equiv c \bmod q$. 设 $M(c, q)$ 表示满足 $ab \equiv c \bmod q$ 且 a 和 b 具有相反奇偶性的解的个数. 在文献[1]中，D. H. Lehmer 教授建议研究 $M(1, p)$ 或者至少说一些它的非平凡性质，其中 p 为素数. 我们知道，当 $p \equiv \pm 1 \bmod 4$ 时，$M(1, p) \equiv 2$ 或 $0 \bmod 4$. 张文鹏[2-3]研究了 $M(1, q)$ 的渐近性质，并得到了较强的渐近公式

$$M(1, q) \equiv \frac{1}{2}\varphi(q) + O\left(q^{\frac{1}{2}} d(q) \ln q\right),$$

其中 $d(n)$ 表示 Dirichlet 除数函数.

设 $R(a, p) = M(a, p) - \dfrac{p-1}{2}$. 张文鹏[4]研究了 $R(a, p)$ 的平方均值，并证明了渐近公式

$$\sum_{a=1}^{p-1} R^2(a, p) = \frac{3}{4}p^2 + O\left(p \exp\left(\frac{3\ln p}{\ln \ln p}\right)\right),$$

其中 $\exp(y) = e^y$. 我们设 $N(p)$ 表示同余方程 $ab \equiv 1 \bmod p$ 满足 $1 \leqslant a, b \leqslant p-1$，且 a 和 b 具有相反的奇偶性的所有整数 a 的集合. 那么，由文献[2-3]，有

$$|N(p)| = M(1, p) = \frac{1}{2}p + O(p^{\frac{1}{2}} \ln p). \tag{1}$$

其中 $|N(p)|$ 表示集合 $N(p)$ 中所有元素的个数. 为了方便，称 $N(p)$ 中的数为 D. H. Lehmer 数. 在本文中，我们考虑以下两个问题：

A. 是否存在两个 D. H. Lehmer 数 a 和 b，使得 $a+b \equiv 1 \bmod p$？如果存在，设 $S(p)$ 表示满足 $a+b \equiv 1 \bmod p$ 的所有解 $a, b \in N(p)$ 的个数. 那么，$S(p)$ 的渐近性如何？

B. 是否存在整数 $1 < a < p-1$，使得 a 和 $a+1$ 是 D. H. Lehmer 数？如果存在，设 $H(p)$ 表示满足 $1 < a < p-1$ 且 a 和 $a+1$ 都是 D. H. Lehmer 数的整数 a 的个数. 那么，$H(p)$ 的渐近性如何？

关于这两个问题，至今未见有相关研究. 在某些条件下，它们的本质是对 D. H. Lehmer 问题的进一步推广和延伸. 本文主要利用解析方法以及 Gauss 和的性质研究这两个问题，并证明以下两个定理：

收稿日期：2017-07-04

基金项目：国家自然科学基金资助项目(11371291)

作者简介：赵建红(1981—)，男，云南大理人，丽江师范高等专科学校副教授，主要从事初等数论的教学与研究.

定理 1　设 $p > 3$ 是任意素数，则有渐近公式 $S(p) = \dfrac{1}{4}p + O(p^{\frac{1}{2}} \ln^2 p)$.

定理 2　设 $p > 3$ 是任意素数，则有渐近公式 $H(p) = \dfrac{1}{4}p + O(p^{\frac{1}{2}} \ln^2 p)$.

对于一般的奇数 $q > 3$，是否存在 $S(q)$ 和 $H(q)$ 的一个较强的渐近公式，这是两个公开的问题.

1　若干引理

下文中需要用到同余以及三角和的性质，文献[5]对这些性质进行了详细的描述，这里不再赘述.

引理 1　设 p 是一个奇素数，那么对任意整数 m 和 n，有估计

$$S(m,n;p) = \sum_{a=1}^{p-1} e\left(\frac{ma + n\overline{a}}{p}\right) \leqslant p^{\frac{1}{2}}(m,n,p)^{\frac{1}{2}},$$

其中 $e(y) = e^{2\pi i y}$，$a\overline{a} \equiv 1 \bmod p$，$(m,n,p)$ 表示 m,n 和 p 的最大公约数.

证明　参阅文献[6-8].

引理 2　设 p 是一个奇素数，那么有估计 $\sum_{a=1}^{p-2} (-1)^{\overline{a}+\overline{a+1}} = O(p^{\frac{1}{2}} \ln^2 p)$.

证明　首先证明估计

$$\sum_{a=1}^{p-1} (-1)^{a+\overline{a}} = O(p^{\frac{1}{2}} \ln^2 p). \tag{2}$$

事实上，利用三角恒等式

$$\sum_{a=1}^{p} e\left(\frac{ma}{p}\right) = \begin{cases} p, & \text{如果}(m,p) = p, \\ 0, & \text{如果}(m,p) = 1. \end{cases} \tag{3}$$

和引理 1，有

$$\sum_{a=1}^{p-1} (-1)^{a+\overline{a}} = \frac{1}{p^2} \sum_{\substack{a=1 \\ ab \equiv 1 \bmod p}}^{p-1} \sum_{b=1}^{p-1} \sum_{c=1}^{p-1} (-1)^c \sum_{d=1}^{p-1} (-1)^d \sum_{r=1}^{p} \sum_{s=1}^{p} e\left(\frac{r(a-c) + s(b-d)}{p}\right) =$$

$$\frac{1}{p^2} \sum_{r=1}^{p-1} \sum_{s=1}^{p-1} \left(\sum_{\substack{a=1 \\ ab \equiv 1 \bmod p}}^{p-1} \sum_{b=1}^{p-1} e\left(\frac{ra + sb}{p}\right)\right) \left(\sum_{c=1}^{p-1} (-1)^c e\left(\frac{-rc}{p}\right)\right) \left(\sum_{d=1}^{p-1} (-1)^d e\left(\frac{-sd}{p}\right)\right) =$$

$$\frac{1}{p^2} \sum_{r=1}^{p-1} \sum_{s=1}^{p-1} S(r,s;p) \frac{1 - e\left(\frac{-r}{p}\right)}{1 + e\left(\frac{-r}{p}\right)} \frac{1 - e\left(\frac{-s}{p}\right)}{1 + e\left(\frac{-s}{p}\right)} \leqslant$$

$$\frac{1}{p^2} \sum_{r=1}^{p-1} \sum_{s=1}^{p-1} |S(r,s;p)| \frac{1}{\left|\sin \dfrac{\pi(p-2r)}{2p}\right|} \frac{1}{\left|\sin \dfrac{\pi(p-2s)}{2p}\right|} \leqslant$$

$$\sum_{r=1}^{p-1} \sum_{s=1}^{p-1} \frac{\sqrt{p}}{|p-2r||p-2s|} \leqslant p^{\frac{1}{2}} \ln^2 p.$$

这就证明了估计(2).

现在，对于任意整数 $1 \leqslant a \leqslant p-2$，注意到同余 $\overline{a} + \overline{a+1} \equiv \overline{a} + \overline{a}\,\overline{\overline{a}+1} \equiv \overline{a} + (\overline{a}+1)\overline{\overline{a}+1} - \overline{\overline{a}+1} \equiv \overline{a} + 1 - \overline{\overline{a}+1} \bmod p$，则有恒等式

$$\overline{a} + \overline{a+1} = \overline{a} + 1 + p - \overline{\overline{a}+1},$$

或者

$$(-1)^{\overline{a}+\overline{a+1}} = (-1)^{p-2\overline{\overline{a}+1}} (-1)^{\overline{a}+1+\overline{\overline{a}+1}} = -(-1)^{\overline{a}+1+\overline{\overline{a}+1}}. \tag{4}$$

结合式(2)、式(4)和模 p 简化剩余系的性质，有

$$\sum_{a=1}^{p-2} (-1)^{\overline{a}+\overline{a+1}} = -\sum_{a=1}^{p-2} (-1)^{\overline{a}+1+\overline{\overline{a}+1}} = 1 - \sum_{a=1}^{p-1} (-1)^{a+\overline{a}} = O(p^{\frac{1}{2}} \ln^2 p).$$

这就证明了引理 2.

引理 3　设 p 是一个奇素数，那么有估计 $\sum_{m=1}^{p-1} e\left(\dfrac{-m}{p}\right)\left(\sum_{a=1}^{p-1}(-1)^{a+\bar{a}}e\left(\dfrac{ma}{p}\right)\right)^2 = O(p^{\frac{3}{2}}\ln^2 p)$.

证明　由式（2）、式（3）和引理 2，有

$$\sum_{m=1}^{p-1} e\left(\frac{-m}{p}\right)\left(\sum_{a=1}^{p-1}(-1)^{a+\bar{a}}e\left(\frac{ma}{p}\right)\right)^2 = \sum_{m=0}^{p-1} e\left(\frac{-m}{p}\right)\left(\sum_{a=1}^{p-1}(-1)^{a+\bar{a}}e\left(\frac{ma}{p}\right)\right)^2 - \left(\sum_{a=1}^{p-1}(-1)^{a+\bar{a}}\right)^2 =$$

$$\sum_{a=1}^{p-1}\sum_{b=1}^{p-1}(-1)^{a+b+\bar{a}+\bar{b}}\sum_{m=0}^{p-1}e\left(\frac{m(a+b-1)}{p}\right) - \left(\sum_{a=1}^{p-1}(-1)^{a+\bar{a}}\right)^2 =$$

$$p\sum_{\substack{a=1\\a+b\equiv 1\bmod p}}^{p-1}\sum_{b=1}^{p-1}(-1)^{a+b+\bar{a}+\bar{b}} - \left(\sum_{a=1}^{p-1}(-1)^{a+\bar{a}}\right)^2 = p\sum_{a=2}^{p-1}(-1)^{p+1+\bar{a}+\overline{p+1-a}} - \left(\sum_{a=1}^{p-1}(-1)^{a+\bar{a}}\right)^2 =$$

$$p\sum_{a=1}^{p-2}(-1)^{\overline{a+1}+\overline{p-a}} - \left(\sum_{a=1}^{p-1}(-1)^{a+\bar{a}}\right)^2 = -p\sum_{a=1}^{p-2}(-1)^{\overline{a+1}+\bar{a}} - \left(\sum_{a=1}^{p-1}(-1)^{a+\bar{a}}\right)^2 = O(p^{\frac{3}{2}}\ln^2 p).$$

这就证明了引理 3.

2　定理的证明

首先证明定理 1. 事实上，由式（2）、式（3）、引理 3 和 $S(p)$ 的定义，有

$$S(p) = \frac{1}{p}\sum_{a=1 a\in N(p)}^{p-1}\sum_{b=1 b\in N(p)}^{p-1}\sum_{m=0}^{p-1}e\left(\frac{m(a+b-1)}{p}\right) =$$

$$\frac{1}{p}\sum_{a=1 a\in N(p)}^{p-1}\sum_{b=1 b\in N(p)}^{p-1}1 + \frac{1}{p}\sum_{m=1}^{p-1}e\left(\frac{-m}{p}\right)\left(\frac{1}{2}\sum_{a=1}^{p-1}(1-(-1)^{a+\bar{a}})e\left(\frac{ma}{p}\right)\right)^2 =$$

$$\frac{1}{p}|N(p)|^2 + \frac{1}{4p}\sum_{m=1}^{p-1}e\left(\frac{-m}{p}\right)\left(-1-\sum_{a=1}^{p-1}(-1)^{a+\bar{a}}e\left(\frac{ma}{p}\right)\right)^2 =$$

$$\frac{1}{p}\left(\frac{1}{2}p + O(p^{\frac{1}{2}}\ln^2 p)\right)^2 - \frac{1}{4p} + \frac{1}{2p}\sum_{a=1}^{p-1}\sum_{m=1}^{p-1}(-1)^{a+\bar{a}}e\left(\frac{m(a-1)}{p}\right) +$$

$$\frac{1}{4p}\sum_{m=1}^{p-1}e\left(\frac{-m}{p}\right)\left(\sum_{a=1}^{p-1}(-1)^{a+\bar{a}}e\left(\frac{ma}{p}\right)\right)^2 =$$

$$\frac{1}{4}p + \frac{1}{4p}\sum_{m=1}^{p-1}e\left(\frac{-m}{p}\right)\left(\sum_{a=1}^{p-1}(-1)^{a+\bar{a}}e\left(\frac{ma}{p}\right)\right)^2 + O(p^{\frac{1}{2}}\ln^2 p) =$$

$$\frac{1}{4}p + O(p^{\frac{1}{2}}\ln^2 p).$$

这就证明了定理 1.

下面证明定理 2. 由 D. H. Lehmer 数的定义、式（2）以及引理 2，有

$$H(p) = \frac{1}{4}\sum_{a=1}^{p-2}(1-(-1)^{a+\bar{a}})(1-(-1)^{a+1+\overline{a+1}}) =$$

$$\frac{1}{4}(p-2) - \frac{1}{4}\sum_{a=1}^{p-2}(-1)^{a+\bar{a}} - \frac{1}{4}\sum_{a=1}^{p-2}(-1)^{a+1+\overline{a+1}} - \frac{1}{4}\sum_{a=1}^{p-2}(-1)^{\bar{a}+\overline{a+1}} =$$

$$\frac{1}{4}p + O(p^{\frac{1}{2}}\ln^2 p).$$

定理 2 的证明完成.

参考文献：

［1］　R K Guy. Unsolved problems in Number Theory［M］. Second Edition. New York：Springer-Verlag，1994.

［2］　Zhang Wenpeng. On a problem of D. H. Lehmer and its generalization［J］. Composito Mathematica，1993(86)：307-316.

［3］　Zhang Wenpeng. A problem of D. H. Lehmer and its generalization (II)［J］. Compositio Mathematica，1994(91)：47-56.

［4］　Zhang Wenpeng. A problem of D. H. Lehmer and its mean square value formlua［J］. Japanese Journal of Mathematics，2003(29)：109-116.

［5］　Tom M Apostol. Introduction to Analytic Number Theory［M］. New York：Springer-Verlag，1976.

[6]　A V Malyshev. A generalization of Kloosterman sums and their estimates(in Russian) [J]. Vestnik Leningrad University,1960(15):59-75.

[7]　S Chowla. On Kloosterman's sum [J]. Norkse Vid Selbsk Fak Frondheim,1967(40):70-72.

[8]　T Estermann. On Kloostermann's sums [J]. Mathematica,1961(8):83-86.

On the D. H. Lehmer Number and Related Problems

ZHAO Jian-hong[1] , LI Xiao-xue[2]

(1. *Department of Teachers Education,Lijiang Teachers College,Lijiang 674199,Yunnan,China;*

2. School of Science,Xïan Aeronautical University,Xïan 710077,China)

Abstract：Let p be an odd prime. For each integer a with $1 \leqslant a \leqslant p-1$,it is clear that there exists one and only one b with $1 \leqslant b \leqslant p-1$ such that $ab \equiv 1 \bmod p$. Let $N(p)$ denotes the set of all integer solutions a of the congruence equation $ab \equiv 1 \bmod p$ for $1 \leqslant a,b \leqslant p-1$ in which a and b are of opposite parity, $S(p)$ denotes the number of all solutions $a,b \in N(p)$ such that $a+b \equiv 1 \bmod p$. Using the analytic method and the properties of Gauss sums, it studied the related problems on the D. H. Lehmer problem, and proved that there exist two integers $a,b \in N(p)$ such that $a+b \equiv 1 \bmod p$,then a sharp asymptotic formula for $S(p)$ was given.

Key words：Lehmer's problem; asymptotic formula; analytic formula; Gauss sums

【责任编辑 金淑兰】

（上接第 11 页）

A Modified Invasive Weed Algorithm to Solving Continuous Space Optimization Problems

ZHU Li-na[1] , LI Shuang[2]

(1. *Department of Basic,Zhengzhou Vocational College of Industrial Safety,Zhengzhou 455000,China;*

2. School of Mathematics and Information Science,Henan Normal University,Xinxiang 453007,Henan,China)

Abstract：The paper presents a modified metaheristic based on the invasive weed optimization (IWO) algorithm. The modified invasive weed optimization (MIWO) method aims to tackle the dilemma of easily trapping in local optimum and premature convergence when solving continuous space optimization problems. Firstly,an opposition-based learning technique is employed to generate initial solution with high quality. Meanwhile,the proposed MIWO determines the number of each seed considering both its fitness and age. A divide-and-conquer technique is introduced to enhance the exploring ability of original IWO. Moreover,MIWO adopts individual replacement mechanism to avoid premature convergence. Finally,the simulations were conducted on five benchmark functions,and the results are compared to those obtained by other metaheuristic. The experiment demonstrates that the modified algorithm has efficiently enhanced its exploring ability.

Key words：invasive weeds algorithm; reverse learning; divide and conquer thought; optimization problem; continuous space

【责任编辑 金淑兰】

第 48 卷第 12 期　　　　　　数学的实践与认识　　　　　　Vol.48, No. 12

2018 年 6 月　　　MATHEMATICS IN PRACTICE AND THEORY　　　June, 2018

椭圆曲线 $y^2 = (x+2)(x^2-2x+43)$ 的整数点

李玉龙 [1], 赵建红 [2,*], 万 飞 [1]

(1. 红河学院 教师教育学院, 云南　蒙自 661199)

(2. 丽江师范高等专科学校 教师教育学院, 云南　丽江 664199)

摘　要: 利用初等方法证明了椭圆曲线 $y^2 = (x+2)(x^2-2x+43)$ 仅有整数点 $(x, y) = (-2, 0)$.

关键词: 椭圆曲线; 整数点; 同余; Legendre 符号

椭圆曲线的整数点是数论和算术代数几何学中基本而又重要的问题, 关于椭圆曲线

$$y^2 = (x+a)(x^2-ax+b), a, b \in Z \tag{1}$$

的整数点问题, 目前结论主要集中在 $a = \pm 2$ 上, 相关结论为:

$a = -2$ 时, 方程 (1) 为

$$y^2 = (x-2)(x^2+2x+b), b \in Z \tag{2}$$

1987 年, D. Zagier[1] 提出了 $b = 31$ 时椭圆曲线 (2) 的整数点的问题; 2009 年, 祝辉林和陈建华 [2] 运用代数数论和 p-adic 分析方法找出了 $b = 31$ 时椭圆曲线 (2) 的全部整数点; 2010 年, 吴华明 [3] 运用初等方法给出了 $b = 31$ 时椭圆曲线 (2) 的全部整数点; 同年, 贺艳峰 [4] 在她的博士论文中同样运用初等数论方法给出了 $b = 31$ 时椭圆曲线 (2) 的全部整数点; 2014 年, 管训贵 [5] 运用初等方法给出了当 $b = 36s^2 - 5(s \in Z^+, 2 \nmid s)$ 是素数时椭圆曲线 (2) 的整数点.

$a = 2$ 时, 方程 (1) 为

$$y^2 = (x+2)(x^2-2x+b), b \in Z \tag{3}$$

2016 年, 过静 [6] 运用初等数论方法给出了 $b = 31$ 时椭圆曲线 (3) 的全部整数点; 2017 年, 杜先存 [7] 运用初等方法给出了当 $b = 36s^2 - 5(s \in Z^+, 2 \nmid s)$ 是素数时椭圆曲线 (3) 的整数点. 对于 $b \equiv 3 \pmod 8$ 为素数的情况目前无相关结论, 本文对 $a = 2, b = 43$ 时, 方程 (3) 的情况进行讨论, 得出了以下结论.

定理　椭圆曲线

$$y^2 = (x+2)(x^2-2x+43) \tag{4}$$

仅有整数点为 $(x, y) = (-2, 0)$.

收稿日期: 2017-08-03

资助项目: 云南省科技厅应用基础研究计划青年项目 (2017FD166)

* 通信作者

证明 显然 $(x,y) = (-2,0)$ 是椭圆曲线 (4) 的整数点, 设 (x,y) 是椭圆曲线 (4) 的异于 $(-2,0)$ 的整数点. 因为 $\gcd(x+2, x^2-2x+43) = \gcd(x+2, 51) = 1$ 或 3 或 17 或 51, 故 (4) 式可分解为以下 4 种可能的情形:

情形 I $x+2 = u^2$, $x^2-2x+43 = v^2$, $y = uv$

情形 II $x+2 = 3u^2$, $x^2-2x+43 = 3v^2$, $y = 3uv$

情形 III $x+2 = 17u^2$, $x^2-2x+43 = 17v^2$, $y = 17uv$

情形 IV $x+2 = 51u^2$, $x^2-2x+43 = 51v^2$, $y = 51uv$

其中 $\gcd(u,v) = 1$, $u, v \in Z$.

下面分别讨论这 4 种情形下椭圆曲线 (4) 的整数点的情况.

情形 I 因为 $u^2 \equiv 0,1,4(\bmod 8)$, 所以 $x = u^2 - 2 \equiv 2,6,7(\bmod 8)$. 因此 $x^2-2x+43 \equiv 3,6(\bmod 8)$, 而 $v^2 \equiv 0,1,4(\bmod 8)$, 故有 $3,6(\bmod 8) \equiv x^2-2x+43 = v^2 \equiv 0,1,4(\bmod 8)$, 显然矛盾, 故该情形 I 不成立.

情形 II 由 $x^2-2x+43 = 3v^2$, 得 $(x-1)^2+42 = 3v^2$, 将 $x+2 = 3u^2$ 代入, 得 $(3u^2-3)^2+42 = 3v^2$, 即

$$3(u^2-1)^2 + 14 = v^2 \tag{5}$$

(5) 式两边取模 7, 得

$$3(u^2-1)^2 \equiv v^2 (\bmod 7) \tag{6}$$

因为 Legendre 符号值 $(\frac{3(u^2-1)^2}{7}) = -1$, 而 Legendre 符号值 $(\frac{v^2}{7}) = 1$, 故 (6) 式不成立, 因此 (6) 式无解, 所以情形 II 不成立.

情形 III 因为 $u^2 \equiv 0,1,4(\bmod 8)$, 所以 $x = 17u^2 - 2 \equiv 2,6,7(\bmod 8)$. 因此 $x^2-2x+43 \equiv 3,6(\bmod 8)$, 而 $v^2 \equiv 0,1,4(\bmod 8)$, 因此有 $3,6(\bmod 8) \equiv x^2-2x+43 = 17v^2 \equiv 0,1,4(\bmod 8)$, 显然矛盾, 所以情形 III 不成立.

情形 IV 当 $2 \nmid u$ 时有 $u^2 \equiv 1(\bmod 4)$, 故有 $x = 51u^2 - 2 \equiv 1(\bmod 4)$, 因此 $x^2-2x+43 \equiv 2(\bmod 4)$, 而 $v^2 \equiv 0,1(\bmod 4)$, 则 $51v^2 \equiv 0,3(\bmod 4)$, 故有 $2(\bmod 4) \equiv x^2-2x+43 = 51v^2 \equiv 0,3(\bmod 4)$, 矛盾, 因此 $2 \nmid u$ 不成立, 所以 $2|u$. 令 $u = 2w, w \in Z$, 则 $x+2 = 51u^2$ 为 $x+2 = 204w^2$, 代入 $x^2-2x+43 = 51v^2$ 得 $(204w^2-3)^2+42 = 51v^2$, 配方得 $(12w^2-1)^2+672w^4 = v^2$, 即:

$$(v+12w^2-1)(v-12w^2+1) = 672w^4 \tag{7}$$

因为 $\gcd(v+12w^2-1, v-12w^2+1) = \gcd(24w^2-2, v-12w^2+1) = \gcd(2(12w^2-1), v-(12w^2-1))$. 又因为 $2|u$, 故由 $x+2 = 51u^2$ 知 $2|x$, 则由 $x^2-2x+43 = 51xv^2$ 得 $2+v$, 故 $2|[v-(12w^2-1)]$. 所以 $\gcd(2(12w^2-1), v-(12w^2-1)) = 2\gcd(12w^2-1, v-(12w^2-1)) = 2\gcd(12w^2-1, v)$. 设 $\gcd(12w^2-1, v) = d$, 则 $d|v, d|(12w^2-1)$, 故由 (7) 知 $d|672w^4$.

又 $\gcd(12w^2-1, 672w^4) = \gcd(12w^2-1, 2^3 \times 3 \times 7) = \gcd(12w^2-1, 7) = 1$ 或 7, 若 $\gcd(12w^2-1, 12k+7) = 7$, 则有 $12w^2-1 \equiv 0(\bmod 7)$, 即 $12w^2 \equiv 1(\bmod 7)$, 而 Legendre 符号值 $(\frac{12w^2}{7}) = (\frac{3}{7})(\frac{4w^2}{7}) = (\frac{3}{7}) = -1$, Legendre 符号值 $(\frac{1}{7}) = 1$, 故 $\gcd(12w^2-1, 12k-7) = 7$ 不成立, 所以 $\gcd(12w^2+1, 12k+7) = 1$. 因此 $\gcd(v+12w^2-1, v-12w^2+1) = 2$. 又 $672 = 2^3 \times 3 \times 7$,

故 (7) 式可分解为:

$$v + 12w^2 - 1 = 2ra^4, v - 12w^2 + 1 = \frac{336}{r}b^4, w = ab, \gcd(a,b) = 1, \gcd(r, \frac{168}{r}) = 1 \qquad (8)$$

其中 $r = 1, 2^3, 3, 2^3 \times 3, 2^3 \times 3 \times 7, 2^3 \times 7, 3 \times 7, 7$

由 (8) 的前两式, 得

$$12w^2 - 1 = ra^4 - \frac{168}{r}b^4 \qquad (9)$$

对 (9) 式两边取模 3, 得

$$-1 \equiv ra^4 - \frac{168}{r}b^4 (\text{mod} 3) \qquad (10)$$

对 (10) 式两边取模 4, 得

$$-1 \equiv ra^4 - \frac{168}{r}b^4 (\text{mod} 4) \qquad (11)$$

当 $r = 1, 7$ 时, (11) 式为

$$-1 \equiv a^4 (\text{mod} 3) \qquad (12)$$

因为 Legendre 符号值 $(\frac{-1}{3}) = -1$, Legendre 符号值 $(\frac{a^4}{3}) = 1$, 故 (12) 式不成立, 因此 $r = 1, 7$ 时 (8) 式不成立, 即情形 IV 不成立.

当 $r = 3, 21$ 时, (10) 式为

$$-1 \equiv b^4 (\text{mod} 3) \qquad (13)$$

因为 Legendre 符号值 $(\frac{-1}{3}) = -1$, Legendre 符号值 $(\frac{b^4}{3}) = 1$, 故 (13) 式不成立, 因此 $r = 3, 21$ 时 (8) 式不成立, 即情形 IV 不成立.

当 $r = 2^3 \times 3, 2^3 \times 7$ 时, (11) 式为

$$-1 \equiv b^4 (\text{mod} 4) \qquad (14)$$

因为 $b^2 \equiv 0, 1(\text{mod} 4)$, 故 $b^4 \equiv 0, 1(\text{mod} 4)$, 因此 (11) 式为 $-1 \equiv 0, 1(\text{mod} 4)$, 矛盾, 故 (13) 式不成立, 因此 $r = 2^3 \times 3, 2^3 \times 7$ 时 (8) 式不成立, 即情形 IV 不成立.

当 $r = 2^3 \times 3 \times 7$ 时, (9) 式为 $12w^2 - 1 = 168a^4 - b^4$, 将 (8) 式的 $w = ab$ 代入得 $12a^2b^2 - 1 = 168a^4 - b^4$, 配方得

$$(6a^2 + b^2)^2 - 204a^4 = 1 \qquad (15)$$

令 $s = 6a^2 + b^2$, 则 (15) 式为

$$s^2 - 204a^4 = 1 \qquad (16)$$

令 $t = 2a^2$, 则 (16) 式为

$$s^2 - 51t^2 = 1 \qquad (17)$$

设方程 (17) 有正整数解 (s, t), 则方程 (16) 有正整数解 $(s, t) = (s, 2a^2)$. 又因为 Pell 方程 (17) 的基本解为 $(50, 7)$, 则方程 (17) 的全部正整数解可表为:

$$s_n + t_n\sqrt{51} = (50 + 7\sqrt{51})^n, n \in Z^+$$

由此可知方程 (17) 的全部正整数解满足

$$(6a^2 + b^2) + 2a^2\sqrt{51} = (50 + 7\sqrt{51})^n, n \in Z^+ \qquad (18)$$

由 (18) 式有

$$2a^2 = \begin{cases} \sum\limits_{i=1}^{\frac{n+1}{2}} \binom{n}{2i-1} \times 50^{n-2i+1} \times 7^{2i-1} \times 51^{i-1}, & 2 \nmid n \\ \sum\limits_{i=1}^{\frac{n}{2}} \binom{n}{2i-1} \times 50^{n-2i+1} \times 7^{2i-1} \times 51^{i-1}, & 2 \mid n \end{cases}$$

即

$$2a^2 = \begin{cases} \sum\limits_{i=1}^{\frac{n+1}{2}} \binom{n}{2i-1} \times 2^{n-2i+1} \times 5^{2(n-2i+1)} \times 7^{2i-1} \times 51^{i-1}, & 2 \nmid n \\ \sum\limits_{i=1}^{\frac{n}{2}} \binom{n}{2i-1} \times 2^{n-2i+1} \times 5^{2(n-2i+1)} \times 7^{2i-1} \times 51^{i-1}, & 2 \mid n \end{cases}$$

也即

$$2a^2 = \begin{cases} 7^n \times 51^{\frac{n-1}{2}} + \sum\limits_{i=1}^{\frac{n-1}{2}} \binom{n}{2i-1} \times 2^{n-2i+1} \times 5^{2(n-2i+1)} \times 7^{2i-1} \times 51^{i-1}, & 2 \nmid n \\ \sum\limits_{i=1}^{\frac{n}{2}} \binom{n}{2i-1} \times 2^{n-2i+1} \times 5^{2(n-2i+1)} \times 7^{2i-1} \times 51^{i-1}, & 2 \mid n \end{cases} \tag{19}$$

因为 (19) 式左边为偶数, 而当 $2 \nmid n$ 时 (19) 式右边为奇数, 故 $2 \nmid n$ 时 (19) 式不成立, 因此 (19) 式要成立需 $2 \mid n$, 由文 [8] 中的定理知 (16) 要有解需 $n = 2$.

当 $n = 2$ 时, 由 (18) 式得 $(6a^2 + b^2) + 2a^2\sqrt{51} = (50 + 7\sqrt{51})^2 = 4999 + 700\sqrt{51}$, 因此有 $6a^2 + b^2 = 4999, 2a^2 = 700$, 则 $a^2 = 350$, 显然无整数解, 故 $r = 2^3 \times 3 \times 7$ 时 (8) 式不成立, 即情形 IV 不成立.

当 $r = 2^3$ 时, (9) 式为 $12w^2 - 1 = 8a^4 - 21b^4$, 将 (9) 的 $w = ab$ 代入得

$$12a^2b^2 - 1 = 8a^4 - 21b^4 \tag{20}$$

由 (20) 式知 $2 \nmid b$, 故 $b^2 \equiv 1 \pmod 8$, 因此 (9) 式两边取模 8 得

$$4a^2 - 1 \equiv 3 \pmod 8 \tag{21}$$

由 (21) 式知 $2 \nmid a$, 故 $a^2 \equiv 1, 9 \pmod{16}$. 又 $2 \nmid b$, 故 $b^2 \equiv 1, 9 \pmod{16}$, 因此 (20) 式两边取模 16 得

$$11 \equiv 3 \pmod{16} \tag{22}$$

(22) 式显然不成立, 故 (20) 式不成立. 因此 $r = 2^3$ 时 (8) 式不成立, 即情形 IV 不成立.

综上所述定理得证.

参考文献

[1] Zagier D. Lager Integral Point on Elliptic Curves [J]. Math Comp, 1987, 48: 425-436

[2] Zhu H L, Chen J H. Integral point on $y^2 = x^3 + 27x - 62$[J].J Math Study, 2009, 42(2): 117-125

[3] 吴华明. 椭圆曲线 $y^2 = x^3 + 27x - 62$ 的整数点 [J]. 数学学报中文版, 2010, 53(1): 205-208.

[4] 贺艳峰. 数论函数的均值分布及整点问题的研究 [D]. 西北大学博士论文, 2010: 20-25.

[5] 管训贵. 椭圆曲线 $y^2 = x^3 + (p-4)x - 2p$ 的整数点 [J]. 数学进展, 2014, 43(4): 521-526.

[6] 过静. 椭圆曲线 $y^2 = x^3 + 27x + 62$ 的整数点 [J]. 重庆师范大学学报 (自然科学版), 2016, 33(5): 50-53.

[7] 杜先存, 赵建红, 万飞. 椭圆曲线 $y^2 = (x+2)(x^2 - 2x + p)$ 的整数点 [J]. 西南大学学报 (自然科学版), 2017, 39(6): 69-73.

[8] Togbé A, Voutier P M, and Walsh P G. Solving a family of Thue equations with an application to the equation$x^2 - Dy^4 = 1$[J]. Acta Arith, 2005, 120(1): 39-58.

The Integral Points on the Elliptic Curve$y^2 = (x+2)(x^2 - 2x + 43)$

LI Yu-long[1], ZHAO Jian-hong[2], WAN Fei[1]

(1. College of Teacher Education, Honghe University, Mengzi 661199, China)

(2. College of Teacher Education, LiJiang Teachers College, Lijiang 664199, China)

Abstract: Using elementary number theory methods, the elliptic curve in title has no positive integer points were proved.

Keywords: elliptic curve; integer point; congruence; Legendre symbol

~~~~~~

全国偏联系数专题讲研班通知

偏联系数是集对分析的一个前沿. 自 2005 年提出以来 (赵克勤, 偏联系数【C】, 中国人工智能进展 2005, 北京, 邮电大学出版社, 2005, 883-884), 已在航空维修、地铁施工、火灾施工、火灾预防、区域创新、技术预警、教育评估、环境保护、水文水资源、管理决策、卫生统计、系统风险分析等领域得到应用. 但由于受文献传播的局限, 偏联系数的应用中存在一些程度不等的问题, 这些问题有可能会导致结论与实际严重不符. 为此定于 2018 年 8 月 25-28 日在集对分析发源地诸暨市举办 2018 年第 5 期全国偏联系数专题讲研班, 由集对分析暨偏联系数创始人赵克勤研究员作系统讲解, 有意参加者请提前报名.

联系电话：18258527568

Email:spacnm@163.com

联系人：赵老师

全国集对分析学术研讨会组委会

诸暨市联系数学研究所

第 48 卷第 13 期
2018 年 7 月

数学的实践与认识
MATHEMATICS IN PRACTICE AND THEORY

Vol.48, No. 13
July, 2018

椭圆曲线 $y^2 = (x - 6)(x^2 + 6x + 19)$ 的正整数点

万 飞 [1], 赵建红 [2,*], 李玉龙 [1]

(1. 红河学院 教师教育学院, 云南 蒙自 661199)

(2. 丽江师范高等专科学校 教师教育学院, 云南 丽江 674199)

摘 要: 利用初等方法证明了椭圆曲线 $y^2 = (x - 6)(x^2 + 6x + 19)$ 无正整数点.

关键词: 椭圆曲线; 正整数点; 同余; 递归序列

椭圆曲线的整数点是数论中一个很重要的问题, 椭圆曲线

$$y^2 = (x + a)(x^2 - ax + p) , a, p \in Z \tag{1}$$

的整数点问题, 目前结论主要集中在 $a = \pm 2, \pm 6$ 上. $a = 2$ 时的结论主要集中在文 [1-2].
$a = -2$ 时的结论主要集中在文 [3-7]. $a = 6$ 时的结论主要集中在文 [8]. $a = -6$ 时的结论主
要集中在文 [9].

对于 $a = -6, p = 19$ 时 (1) 的整数点问题, 目前还没有相关结论, 本文得出了以下结论:

定理 椭圆曲线

$$y^2 = (x - 6)(x^2 + 6x + 19) \tag{2}$$

无正整数点.

1 相关引理

引理 [10] 设 D 是一个非平方的正整数, 则方程 $x^2 - Dy^4 = 1$ 至多有 2 组正整数解
(x, y). 如果 $x^2 - Dy^4 = 1$ 恰有两组正整数解, 则当 $D = 2^{4s} \times 1785$, 其中 $s = 0, 1$ 时, $(x_1, y_1) =$
$(169, 2^{1-s})$ 且 $(x_2, y_2) = (6525617281, 2^{1-s} \times 6214)$; 当 $D \neq 2^{4s} \times 1785$ 时, $(x_1, y_1) = (u_1, \sqrt{v_1})$
且 $(x_2, y_2) = (u_2, \sqrt{v_2})$ 这里 (u_n, v_n) 是 Pell 方程 $u^2 - Dv^2 = 1$ 的正整数解. 如果方程
$x^2 - Dy^4 = 1$ 仅有 1 组正整数解 (x, y) 且正整数 n 适合 $(x, y^2) = (u_n, v_n)$, 则当 n 是偶数时,
必有 $n = 2$; 当 n 是奇数时, 必有 $n = 1$ 或 p, 这里 p 是适合 $p \equiv 3 \pmod 4$ 的素数.

2 定理证明

证明 设 $(x, y), x, y \in Z^+$ 是椭圆曲线 (2) 的正整数点. 由 $gcd(x - 6, x^2 + 6x + 19) =$
$gcd(x - 6, 91) = 1$ 或 7 或 13 或 91, 知 (2) 式可分解为以下 4 种可能的情形:

收稿日期: 2017-11-27

资助项目: 云南省科技厅应用基础研究计划青年项目 (2017FD166)

* 通信作者

情形 I $x - 6 = u^2$, $x^2 + 6x + 19 = v^2$, $y = uv$.

情形 II $x - 6 = 7u^2$, $x^2 + 6x + 19 = 7v^2$, $y = 7uv$.

情形 III $x - 6 = 13u^2$, $x^2 + 6x + 19 = 13v^2$, $y = 13uv$.

情形 IV $x - 6 = 91u^2$, $x^2 + 6x + 19 = 91v^2$, $y = 91uv$.

其中 $gcd(u,v) = 1$, $u, v \in Z$. 下面分别讨论这四种情形下方程 (2) 的整数解.

情形 I 因为 $u^2 \equiv 0, 1, 4(mod8)$, 所以 $x = u^2 + 6 \equiv 2, 6, 7(mod8)$, 因此 $x^2 + 6x + 19 \equiv x^2 + 6x + 3 \equiv 3, 6(mod8)$, 而 $v^2 \equiv 0, 1, 4(mod8)$, 故有 $3, 6(mod8) \equiv x^2 + 6x + 19 = v^2 \equiv 0, 1, 4(mod8)$, 显然矛盾, 故该情形下椭圆曲线 (2) 无整数点.

情形 II 因为 $u^2 \equiv 0, 1, 4(mod8)$, 所以 $x = 7u^2 + 6 \equiv 2, 5, 6(mod8)$, 因此 $x^2 + 6x + 19 \equiv x^2 + 6x + 3 \equiv 2, 3(mod8)$, 而 $v^2 \equiv 0, 1, 4(mod8)$, 故 $7v^2 \equiv 0, 4, 7(mod8)$, 因此有 $2, 3(mod8) \equiv x^2 + 6x + 19 = 7v^2 \equiv 0, 4, 7(mod8)$, 显然矛盾, 故该情形下椭圆曲线 (2) 无整数点.

情形 III 因为 $u^2 \equiv 0, 1, 4(mod8)$, 所以 $x = 13u^2 + 6 \equiv 2, 3, 6(mod8)$, 因此 $x^2 + 6x + 19 \equiv x^2 + 6x + 3 \equiv 3, 6(mod8)$, 而 $v^2 \equiv 0, 1, 4(mod8)$, 故 $13v^2 \equiv 0, 4, 5(mod8)$, 因此有 $3, 6(mod8) \equiv x^2 + 6x + 19 = 13v^2 \equiv 0, 4, 5(mod8)$, 显然矛盾, 故该情形下椭圆曲线 (2) 无整数点.

情形 IV 当 $2 \nmid u$ 时有 $u^2 \equiv 1(mod8)$, 则有 $x = 91u^2 + 6 \equiv 3u^2 + 6 \equiv 1(mod8)$, 故有 $x^2 + 6x + 91 \equiv 2(mod8)$, 而 $v^2 \equiv 0, 1, 4(mod8)$, $91v^2 \equiv 0, 3, 4(mod8)$, 故有 $2(mod8) \equiv x^2 + 6x + 19 = 91v^2 \equiv 0, 3, 4(mod8)$, 显然矛盾, 因此 $2 \nmid u$ 不成立, 所以 $2|u$. 令 $u = 2m, m \in Z$, 则 $x - 6 = 91u^2$ 为 $x - 6 = 364m^2$, 代入 $x^2 + 6x + 19 = 91v^2$ 得 $(36m^2 + 1)^2 + 160m^4 = v^2$, 即:

$$(v + 36m^2 + 1)(v - 36m^2 - 1) = 160m^4 \tag{3}$$

又 $gcd(v + 36m^2 + 1, v - 36m^2 - 1) = gcd(72m^2 + 2, v + 36m^2 + 1) = gcd(2(36m^2 + 1), v + 36m^2 + 1))$. 因为 $2|u$, $gcd(u, v) = 1$, 所以 $2 \nmid v$, 故 $2|(v + 36m^2 + 1)$. 故 $gcd(2(36m^2 + 1), v + 36m^2 + 1) = 2gcd(36m^2 + 1, v + 36m^2 + 1) = 2gcd(36m^2 + 1, v)$. 设 $gcd(36m^2 + 1), v) = d$, 则 $d|v$, $d|(36m^2 + 1)$, 故由 (3) 知 $d|160m^4$. 因为 $gcd(36m^2 + 1, 160) = gcd(36m^2 + 1, 2^5 \times 5) = gcd(36m^2 + 1, 5) = 1$ 或 5.

当 $gcd(36m^2 + 1, 96m^4) = 1$ 时, $gcd(36m^2 + 1, v) = 1$, 故 $gcd(v + 36m^2 + 1, v - 36m^2 - 1) = 2$. 又 $160 = 2^5 \times 5$, 故 (3) 式可分解为:

$$v + 36m^2 + 1 = 2ga^4, \quad v - 36m^2 - 1 = \frac{80}{g}b^4, \quad m = ab \tag{4}$$

其中 $gcd(a, b) = 1$, $gcd(g, \frac{40}{g}) = 1$, $g = 1, 5, 2^3, 2^3 \times 5$.

由 (4) 的前两式, 得

$$36m^2 + 1 = ga^4 - \frac{40}{g}b^4 \tag{5}$$

对 (5) 式两边取模 4, 得

$$1 \equiv ga^4 - \frac{40}{g}b^4 (mod4) \tag{6}$$

当 $g = 2^3, 2^3 \times 5$ 时, (6) 式为

$$1 \equiv -b^4 (mod4) \tag{7}$$

因为 $b^2 \equiv 0, 1 (mod 4)$, 故 $b^4 \equiv 0, 1 (mod 4)$, 又 $b^4 \equiv 3 (mod 4)$, 矛盾, 因此 $g = 2^3, 2^3 \times 5$ 时 (3) 式不成立, 即情形 IV 不成立.

当 $g = 1$ 时, (5) 式为 $36m^2 + 1 = a^4 - 40b^4$, 将 (4) 式的 $m = ab$ 代入得 $36a^2b^2 + 1 = a^4 - 40b^4$, 配方得

$$(a^2 + 18b^2)^2 - 364b^4 = 1 \tag{8}$$

令 $s = a^2 - 18b^2, s \in Z^+$, 此时 (8) 式为

$$s^2 - 364a^4 = 1 \tag{9}$$

令 $r = 2a, r \in Z^+$, 则 (9) 式为

$$s^2 - 91r^2 = 1 \tag{10}$$

由引理 1 得方程 (9) 至多有一组正整数解 (s, a), 且若方程 (9) 有正整数解 (s, a), 则方程 (10) 有正整数解 $(s, r) = (s, 2a^2)$. 因为 Pell 方程 (10) 的基本解为 $(1574, 165)$, 其全部正整数解可表为: $s_n + r_n\sqrt{91} = (1574 + 165\sqrt{91})^n, n \in Z^+$.

由此可知方程 (9) 的全部正整数解满足:

$$(a^2 - 18b^2) + a^2\sqrt{364} = (a^2 - 18b^2) + 2a^2\sqrt{91} = (1574 + 165\sqrt{91})^n, n \in Z^+ \tag{11}$$

则有 $a^2 - 18b^2 = s_n, 2a^2 = r_n, n \in Z^+$. 容易验证下式成立:

$$r_{n+2} = 3148r_{n+1} - r_n, r_0 = 0, r_1 = 165 \tag{12}$$

对递归序列 (12) 取模 2, 得周期为 2 的剩余类数列 $0, 1, 0, 1, \cdots$, 且仅当 $n \equiv 1 (mod 2)$ 时, 有 $r_n \equiv 1 (mod 2)$; 仅当 $n \equiv 0 (mod 2)$ 时, 有 $r_n \equiv 0 (mod 2)$. 因为 $2a^2 = r_n$, 所以 r_n 为偶数, 故 (11) 式要成立需 $n \equiv 0 (mod 2)$.

又由引理 1 知 (11) 式要成立需满足 $n = 2$ 或 $2 \nmid n$, 因此 $n = 2$. 由 (11) 式得 $(a^2 - 18b^2) + a^2\sqrt{364} = (1574 + 165\sqrt{91})^2 = 4954951 + 519420\sqrt{91}$, 因此有 $18a^2 - b^2 = 4954951, a^2 = 259710$, 显然无整数解, 故方程 (9) 无整数解, 因此 $g = 1$ 时 (3) 式不成立, 即情形 IV 不成立.

当 $g = 2^3$ 时, (5) 式为 $36m^2 + 1 = 8a^4 - 5b^4$, 将 (4) 式的 $m = ab$ 代入得 $36a^2b^2 + 1 = 8a^4 - 5b^4$, 即 $72a^2b^2 + 2 = 16a^4 - 10b^4$, 配方得

$$(4a^2 - 9b^2)^2 - 91b^4 = 2 \tag{13}$$

(13) 式两边取模 13, 得

$$(4a^2 - 9b^2)^2 \equiv 2 (mod 13) \tag{14}$$

因为 Legendre 符号值 $\left(\frac{2}{13}\right) = -1$, 故 (14) 式不成立, 因此方程 (13) 无整数解, 所以 $g = 2^3$ 时 (3) 式不成立, 即情形 IV 不成立.

当 $gcd(36m^2 + 1, v) = 5$ 时, 有 $gcd(v + 36m^2 + 1, v - 36m^2 - 1) = 10$. 又 $160 = 2^5 \times 5$, 故 (3) 式可分解为以下两种情形:

情形 i $v + 36m^2 + 1 = 10ha^4, v - 36m^2 - 1 = \frac{16}{h}b^4$.

情形 ii $v - 36m^2 - 1 = 10ha^4, v + 36m^2 + 1 = \frac{16}{h}b^4$.

其中 $m = ab, gcd(a, b) = 1, gcd(h, \frac{16}{h}) = 1, h = 1, 2^3$.

情形 i 因为 $gcd(v + 36m^2 + 1, v - 36m^2 - 1) = 10$, 故由 $v - 36m^2 - 1 = \frac{16}{h}b^4$ 知 $5|b$, 设 $b = 5c, c \in Z$, 则由前两式得

$$36m^2 + 1 = 5ha^4 - \frac{5000}{h}c^4 \tag{15}$$

(15) 式两边取模 4, 得

$$1 \equiv ha^4 - \frac{5000}{h}c^4 \pmod 4 \tag{16}$$

当 $h = 2^3$ 时, (16) 式为

$$1 \equiv -c^4 \pmod 4 \tag{17}$$

因为 $c^2 \equiv 0, 1 \pmod 4$, 故 $c^4 \equiv 0, 1 \pmod 4$, 又 $c^4 \equiv 3 \pmod 4$, 矛盾, 故 (17) 式不成立, 因此 $h = 2^3$ 时 (3) 式不成立, 即情形 IV 不成立.

当 $h = 1$ 时, (15) 式为 $36m^2 + 1 = 5a^4 - 5000c^4$, 将 $m = ab$ 及 $b = 5c$ 代入得 $900a^2c^2 + 1 = 5a^4 - 5000c^4$, 即

$$5a^4 - 5000c^4 - 900a^2c^2 = 1 \tag{18}$$

对 (18) 两边取模 5, 得 $0 \equiv 1 \pmod 5$, 不成立, 故 $h = 1$ 时 (3) 式不成立, 即情形 IV 不成立.

情形 ii 因为 $gcd(v + 36m^2 + 1, v - 36m^2 - 1) = 10$, 故由 $v + 36m^2 + 1 = \frac{16}{h}b^4$ 知 $5|b$, 设 $b = 5c, c \in Z$, 则由前两式得

$$36m^2 + 1 = \frac{5000}{h}c^4 - 5ha^4 \tag{19}$$

(19) 式两边取模 4, 得

$$1 \equiv \frac{5000}{h}c^4 - ha^4 \pmod 4 \tag{20}$$

当 $h = 1$ 时, (20) 式为

$$1 \equiv -a^4 \pmod 4 \tag{21}$$

因为 $a^2 \equiv 0, 1 \pmod 4$, 故 $a^4 \equiv 0, 1 \pmod 4$, 又 $a^4 \equiv 3 \pmod 4$, 矛盾, 故 (21) 式不成立, 因此 $h = 1$ 时 (3) 式不成立, 即情形 IV 不成立.

当 $h = 2^3$ 时, (15) 式为 $36m^2 + 1 = 625c^4 - 40a^4$, 将 $m = ab$ 及 $b = 5c$ 代入得 $900a^2c^2 + 1 = 625c^4 - 40a^4$, 即

$$625c^4 - 40a^4 - 900a^2c^2 = 1 \tag{22}$$

对 (22) 两边取模 5, 得 $0 \equiv 1 \pmod 5$, 不成立, 故 $h = 2^3$ 时 (3) 式不成立, 即情形 IV 不成立.

综上所述, 定理得证.

参考文献

[1] 过静. 椭圆曲线 $y^2 = x^3 + 27x + 62$ 的公解 [J]. 重庆师范大学学报 (自然科学版), 2016, 33(5): 50-53.

[2] 杜先存, 赵建红, 万飞. 椭圆曲线 $y^2 = (x + 2)(x^2 - 2x + p)$ 的整数点 [J]. 西南大学学报 (自然科学版), 2017, 39(6): 69-73.

[3] Zagier D. Lager Integral Point on Elliptic Curves [J]. Math Comp, 1987, 48: 425-436.

[4] Zhu H L, Chen J H. Integral point on $y^2 = x^3 + 27x - 62$ [J]. Math Study, 2009, 42(2): 117-125.

[5] 吴华明. 椭圆曲线 $y^2 = x^3 + 27x - 62$ 的整数点 [J]. 数学学报中文版, 2010, 53(1): 205-208.

[6] 贺艳峰. 数论函数的均值分布及整点问题的研究 [D]. 西北大学博士论文, 2010: 20-25.

[7] 管训贵. 椭圆曲线 $y^2 = x^3 + (p - 4)x - 2p$ 的整数点 [J]. 数学进展, 2014, 43(4): 521-526.

[8] 过静. 椭圆曲线 $y^2 = x^3 - 21x + 90$ 的正整数解 [J]. 数学的实践与认识, 2017, 47(8): 288-291.

[9] 李亚卓, 崔保军. 关于椭圆曲线 $y^2 = x^3 - 21x - 90$ 的正整数解 [J]. 延安大学学报: 自然科学版, 2015, 34(3): 14-15.

[10] Togbé A, Voutier P M, and Walsh P G. Solving a family of Thue equations with an application to the equation. Acta Arith, 2005, 120(1): 39-58.

The Integral Points on the Elliptic Curve
$y^2 = (x - 6)(x^2 + 6x + 19)$

WAN Fei[1], ZHAO Jian-hong[2], LI Yu-long[1]

(1. College of Teacher Education, Honghe University, Mengzi 661199, China)

(2. College of Teacher Education, LiJiang Teachers College, Lijiang 674199, China)

Abstract: Using elementary number theory methods, the elliptic curve in title has no positive integer points were proved.

Keywords: elliptic curve; integer point; congruence; recursive sequence.

第 47 卷第 20 期
2017 年 10 月

数学的实践与认识
MATHEMATICS IN PRACTICE AND THEORY

Vol.47, No. 20
Oct., 2017

关于 Pell 方程组 $x^2 - 3y^2 = 1$ 与 $y^2 - Dz^2 = 1$ 的解

过 静 [1]，赵建红 [2]，杜先存 [3,*]

(1. 江西科技师范大学 数学与计算机科学学院, 江西 南昌 330013)
(2. 云南丽江师范高等专科学校 数学与计算机科学系, 云南 丽江 674199)
(3. 红河学院 教师教育学院, 云南 蒙自 661199)

摘 要: 设 $D = 2p_1 \cdots p_s (1 \leq s \leq 4)$, $p_1, \cdots, p_s$ 是互异的奇素数. 证明了：Pell 方程组 $x^2 - 3y^2 = 1, y^2 - Dz^2 = 1$ 除开 $D = 2 \times 7, 2 \times 3 \times 5 \times 7 \times 13$ 外, 仅有平凡解 $(x, y, z) = (\pm 2, \pm 1, 0)$.

关键词 Pell 方程组; 基本解; 整数解; 奇素数

Pell 方程组

$$x^2 - D_1 y^2 = 1 \text{ 与 } y^2 - D_2 z^2 = 1 \tag{1}$$

的整数解一直受到人们的关注. 对于方程组 (1) 的解, Ljunggren W[1] 得出了 $(D_1, D_2) = (2, 3)$ 时方程组 (1) 仅有正整数解 $(x, y, z) = (3, 2, 1)$; Pan Jiayu, Zhang Yuping, Zou Rong[2] 讨论了 $D_1 = 8$ 时方程组 (1) 的解的情况; 乐茂华 [3] 给出了方程组 (1) 有正整数解的一个充要条件. 本文主要讨论 $D_1 = 3$, D_2 为偶数时方程组 (1) 的解的情况. 文中用 $p(a)$(a 为正整数, p 为素数) 表示正整数 a 中含素数 p 的最高次数.

1 关键性引理

引理 1[4] 设 $D \in Z^+$ 且不是一个完全平方数, (a, b) 为方程 $x^2 - Dy^2 = 1$ 的最小解, 则 $x^2 - Dy^2 = 1$ 的任一组解可以表示为：$x + y\sqrt{D} = \pm(a + b\sqrt{D})^n, n \in Z$.

引理 2[5] 设 a, b 是 $x^2 - Dy^2 = 1$ 的基本解, 则有下面递归序列成立：

$$\begin{cases} x_{n+2} = 2ax_{n+1} - x_n, y_{n+2} = 2ay_{n+1} - y_n; \\ x_0 = 1, x_1 = a; y_0 = 0, y_1 = b \end{cases}$$

引理 3[6] 设 $d \in Z^+$ 且不是一个完全平方数, (x_1, y_1) 为方程 $x^2 - dy^2 = 1$ 的最小解, 则 $x^2 - dy^2 = 1$ 的全部正整数解 (x, y) 可以由下式给出：

$$\begin{cases} x_n = \frac{1}{2}[(x_1 + \sqrt{d}y_1)^n + (x_1 - \sqrt{d}y_1)^n], \\ y_n = \frac{1}{2\sqrt{d}}[(x_1 + \sqrt{d}y_1)^n - (x_1 - \sqrt{d}y_1)^n] \end{cases} \quad n \in N.$$

引理 4 设 Pell 方程 $x^2 - 3y^2 = 1$ 的全部整数解为 (x_n, y_n), $n \in Z$, 则对任意 $n \in Z$, 有 $y_n^2 = y_{n-1}y_{n+1} + 1$.

收稿日期: 2017-02-18
资助项目: 江西省教育厅科学技术研究项目 (GJJ160728); 云南省应用基础研究计划项目 (Y0120160010); 红河学院校级教改项目 (JJJG151010)
* 通信作者

证明 因为 Pell 方程 $x^2 - 3y^2 = 1$ 的最小解为 $(x_1, y_1) = (2, 1)$，则 $x_1 + y_1\sqrt{3} = 2 + \sqrt{3}$，$x_1 - y_1\sqrt{3} = 2 - \sqrt{3}$. 则由引理 2，得

$$y_n^2 = [\frac{(2+\sqrt{3})^n - (2-\sqrt{3})^n}{2\sqrt{3}}]^2$$

$$= \frac{(2+\sqrt{3})^{2n} + (2-\sqrt{3})^{2n} - 2}{12}, \quad y_{n+1}y_{n-1} = \frac{(2+\sqrt{3})^{n+1} - (2-\sqrt{3})^{n+1}}{2\sqrt{3}} \cdot$$

$$\frac{(2+\sqrt{3})^{n-1} - (2-\sqrt{3})^{n-1}}{2\sqrt{3}} = \frac{(2+\sqrt{3})^{2n} + (2-\sqrt{3})^{2n} - 14}{12}$$

因此

$$y_{n+1}y_{n-1} + 1 == \frac{(2+\sqrt{3})^{2n} + (2-\sqrt{3})^{2n} - 14}{12} + 1 = \frac{(2+\sqrt{3})^{2n} + (2-\sqrt{3})^{2n} - 2}{12} = y_n^2$$

命题得证.

引理 5 设 Pell 方程 $x^2 - 3y^2 = 1$ 的全部整数解为 (x_n, y_n), $n \in Z$, 则对任意 $n \in Z$, 有 $y_{2n} = 2x_n y_n$.

证明 因为 Pell 方程 $x^2 - 3y^2 = 1$ 的最小解为 $(x_1, y_1) = (3, 1)$, 则 $x_1 + y_1\sqrt{3} = 2 + \sqrt{3}$, $x_1 - y_1\sqrt{3} = 2 - \sqrt{3}$. 则由引理 2, 得 $y_{2n} = \frac{(2+\sqrt{3})^{2n} - (2-\sqrt{3})^{2n}}{2\sqrt{3}}$, $x_n = \frac{(2+\sqrt{3})^n + (2-\sqrt{3})^n}{2}$, $y_n = \frac{(2+\sqrt{3})^n - (2-\sqrt{3})^n}{2\sqrt{3}}$, 则

$$2x_n y_n = \frac{(2+\sqrt{3})^n + (2-\sqrt{3})^n}{2} \cdot \frac{(2+\sqrt{3})^n - (2-\sqrt{3})^n}{2\sqrt{3}} = \frac{(2+\sqrt{3})^{2n} - (2-\sqrt{3})^{2n}}{2\sqrt{3}} = y_{2n}$$

命题得证.

引理 6[7] 设 Pell 方程 $x^2 - 3y^2 = 1$ 的全部整数解为 (x_n, y_n), $n \in Z^+$, 则有对任意的 $m, k \in Z^+$, 若 $\gcd(m, k) = d$, 则有

(I) $\gcd(y_k, y_m) = y_d$;

(II) 当 $2 \big| \frac{mk}{d^2}$ 时, $\gcd(x_k, x_m) = 1$; 当 $2 \nmid \frac{mn}{d^2}$ 时, $\gcd(x_k, x_m) = x_d$;

(III) 当 $2 \nmid \frac{m}{d}$ 时, $\gcd(x_k, y_m) = 1$; 当 $2 \big| \frac{m}{d}$ 时, $\gcd(x_k, y_m) = x_d$.

引理 7[4] 设 p 是一个奇素数, 则丢番图方程 $x^4 - py^2 = 1$ 除开 $p = 5, x = 3, y = 4$ 和 $p = 29, x = 99, y = 1820$ 外, 无其他的正整数解.

引理 8[8] 当 $a > 1$ 且 a 是平方数时, 方程 $ax^4 - by^2 = 1$ 至多有一组正整数解.

引理 9[9] 若 D 是一个非平方的正整数, 则方程 $x^2 - Dy^4 = 1$ 至多有两组正整数解, 而且方程恰有两组正整数解的充要条件是 $D = 1785$ 或 $D = 28560$ 或 $2x_0$ 和 y_0 都是平方数, 这里 (x_0, y_0) 是方程 $x^2 - Dy^2 = 1$ 的基本解.

引理 10[10] 设 D 是一个非平方的正整数, 则方程 $x^2 - Dy^4 = 1$ 至多有 2 组正整数解 (x, y). 如果恰有两组正整数解, 则当 $D = 2^{4s} \times 1785$, 其中 $s \in \{0, 1\}$ 时, $(x_1, y_1) = (169, 2^{1-s})$ 且 $(x_2, y_2) = (6525617281, 2^{1-s} \times 6214)$; 当 $D \neq 2^{4s} \times 1785$ 时, $(x_1, y_1) = (u_1, \sqrt{v_1})$ 且 $(x_2, y_2) = (u_2, \sqrt{v_2})$, 这里 (u_n, v_n) 是 Pell 方程 $U^2 - DV^2 = 1$ 的正整数解.

引理 11 设 Pell 方程 $x^2 - 3y^2 = 1$ 的全部整数解为 (x_n, y_n), $n \in Z$, 则对任意 $n \in Z$, x_n, y_n 具有如下性质:

(I) x_n 为平方数当且仅当 $n = 0$;

(II) $\frac{x_n}{2}$ 为平方数当且仅当 $n = 1$ 或 $n = -1$;

(III) y_n 为平方数当且仅当 $n = 2$ 或 $n = 1$ 或 $n = 0$.

证明 设 $(x_n, y_n)(n \in Z)$ 是 Pell 方程 $x^2 - 3y^2 = 1$ 的整数解.

(I) 若 $x_n = a^2$, 代入原方程得 $a^4 - 3y^2 = 1$. 由引理 7 知, 方程 $a^4 - 3y^2 = 1$ 仅有平凡解 $(a, y) = (\pm 1, 0)$, 从而 $x_n = a^2 = 1$, 因此 $n = 0$. 反之, 显然.

(II) 若 $\frac{x_n}{2} = a^2$, 代入原方程得 $4a^4 - 3y^2 = 1$. 由引理 8 得, $4a^4 - 3y^2 = 1$ 至多有一组正整数解, 又 $(1,1)$ 为 $4a^4 - 3y^2 = 1$ 的正整数解, 故 $4a^4 - 3y^2 = 1$ 仅有整数解 $(a, y) = (\pm 1, \pm 1)$, 此时 $x_n = 2a^2 = 2$, 从而 $n = 1$ 或 $n = -1$. 反之, 显然.

(III) 若 $y_n = b^2$, 代入原方程得 $x^2 - 3b^4 = 1$. 因为 $x^2 - 3b^4 = 1$ 的基本解为 $(x_1, b_1) = (2, 1)$, 则有 $2x_1 = 4$. 由引理 9 知, $x^2 - 3b^4 = 1$ 有 2 组正整数解. 故由引理 10 知, $x^2 - 3b^4 = 1$ 有组正整数解 $(x, b) = (2, 1), (7, 2)$ 及平凡解 $(x, b) = (1, 0)$, 从而 $y_n = b^2 = 4, 1, 0$, 因此 $n = 2$ 或 $n = 1$ 或 $n = 0$. 反之, 显然.

2 定理及证明

定理 若 $p_1, \cdots, p_s(1 \le s \le 4)$ 是互异的奇素数, 则当 $D = 2p_1 \cdots p_s, 1 \le s \le 4$ 时, Pell 方程组

$$x^2 - 3y^2 = 1 \quad \text{与} \quad y^2 - Dz^2 = 1 \tag{2}$$

的整数解的情况为:

(i) $D = 2 \times 3 \times 5 \times 7 \times 13$ 时方程组 (2) 有整数解 $(x, y, z) = (\pm 362, \pm 209, \pm 4)$ 和平凡解 $(x, y, z) = (\pm 2, \pm 1, 0)$;

(ii) $D = 2 \times 7$ 时方程组 (2) 有整数解 $(x, y, z) = (\pm 26, \pm 15, \pm 4)$ 和平凡解 $(x, y, z) = (\pm 2, \pm 1, 0)$;

(iii) $D \ne 2 \times 7, 2 \times 3 \times 5 \times 7 \times 13$ 时方程组 (2) 仅有平凡解 $(x, y, z) = (\pm 2, \pm 1, 0)$.

证明 因为 Pell 方程 $x^2 - 3y^2 = 1$ 的最小解为 $(x_1, y_1) = (2, 1)$, 则由引理 1 知, Pell 方程 $x^2 - 3y^2 = 1$ 的全部整数解为 $x_n + y_n\sqrt{3} = (2 + \sqrt{3})^n, n \in Z$.

由引理 2 得以下递归序列成立:

$$x_{n+2} = 4x_{n+1} - x_n, x_0 = 1, x_1 = 2 \tag{3}$$

$$y_{n+2} = 4y_{n+1} - y_n, y_0 = 0, y_1 = 1 \tag{4}$$

对递归序列 (3) 取模 2, 得周期为 2 的剩余类序列 $0, 1, 0, 1, \cdots$, 则当 $n \equiv 1 (\bmod 2)$ 时有 $x_n \equiv 0 (\bmod 2)$, 当 $n \equiv 0 (\bmod 2)$ 时有 $x_n \equiv 1 (\bmod 2)$, 故有 $x_{2n} \equiv 1 (\bmod 2), x_{2n+1} \equiv 0 (\bmod 2)$.

对递归序列 (3) 取模 4, 得周期为 4 的剩余类序列 $2, 3, 2, 1, \ 2, 3, 2, 1, \ldots$, 则有对任意的 $n \in Z$, 恒有 $x_n \not\equiv 0 (\bmod 4)$.

对递归序列 (4) 取模 4, 得周期为 2 的剩余类序列 $1, 0, 1, 0, \ldots$, 则当 $n \equiv 1 (\bmod 2)$ 时有 $y_n \equiv 1 (\bmod 2)$, 当 $n \equiv 0 (\bmod 2)$ 时有 $y_n \equiv 0 (\bmod 2)$, 故有 $y_{2n} \equiv 0 (\bmod 2), y_{2n+1} \equiv 1 (\bmod 2)$.

设 $(x, y, z) = (x_n, y_n, z), n \in Z$ 为方程 (2) 的整数解, 由引理 4 得

$$Dz_n^2 = y_n^2 - 1 = y_{n-1}y_{n+1} \tag{5}$$

显然当 $n = -1$ 或 $n = 1$ 时有 $Dz_n^2 = y_{n-1}y_{n+1} = 0$, 此时可以得到方程 (2) 有平凡解 $(x, y, z) = (\pm 5, \pm 2, 0)$.

设 $(x, y, z) = (x_{n+1}, y_{n+1}, z)$, $n \in N^+$ 为方程 (2) 的正整数解, 由引理 4 得,

$$Dz^2 = y_{n+1}^2 - 1 = y_n y_{n+2}. \tag{6}$$

由 $y_{2n+1} \equiv 1(\bmod 2)$ 得, n 为正奇数时, $y_n \equiv y_{n+2} \equiv 1(\bmod 2)$, 故 $2(y_n y_{n+2}) = 0$, 而 $2(D) = 1$, 故 $2(Dz^2)$ 为奇数, 矛盾, 所以 n 只能为正偶数, 设 $n = 2m, m \in Z^+$, 则式 (6) 成为 $Dz^2 = y_{2m+1}^2 - 1 = y_{2m} y_{2m+2} = 4x_m x_{m+1} y_m y_{m+1}$, 即

$$Dz^2 = 4x_m x_{m+1} y_m y_{m+1} \tag{7}$$

2.1 m 为正偶数时, 设 $m = 2^t p$ ($t \in Z^+$, p 为正奇数), 则式 (7) 成为

$$Dz^2 = 4x_{2^t p} x_{2^t p+1} y_{2^t p} y_{2^t p+1} \tag{8}$$

反复运用引理 5, 得式 (8) 可化为

$$Dz^2 = 2^{2+t} x_{2^t p+1} x_{2^t p} x_{2^{t-1} p} \cdots x_{2p} x_p y_{2^t p+1} y_p \tag{9}$$

由 $x_{2n+1} \equiv 0(\bmod 2)$, 知式 (9) 可化为

$$Dz^2 = 2^{4+t} x_{2^t p} x_{2^{t-1} p} \cdots x_{2p} y_{2^t p+1} y_p \cdot \frac{x_p}{2} \cdot \frac{x_{2^t p+1}}{2} \tag{10}$$

由引理 6 的 (I) 知 $y_{2^t p+1}, y_p$ 互素; 由引理 6 的 (II) 知 $\frac{x_{2^t p+1}}{2}, x_{2^t p}, x_{2^{t-1} p}, \cdots, x_{2p}, \frac{x_p}{2}$ 两两互素; 由引理 6 的 (III) 知 $\frac{x_{2^t p+1}}{2}, x_{2^t p}, x_{2^{t-1} p}, \cdots, x_{2p}, \frac{x_p}{2}$ 分别与 $y_{2^t p+1}, y_p$ 互素. 因此有 $\frac{x_{2^t p+1}}{2}, x_{2^t p}, x_{2^{t-1} p}, \cdots, x_{2p}, \frac{x_p}{2}, y_{2^t p+1}, y_p$ 两两互素.

因为 $x_{2n} \equiv 1(\bmod 2)$, $y_{2n+1} \equiv 1(\bmod 2)$, 故 $x_{2^t p}, x_{2^{t-1} p}, \cdots, x_{2p}, y_{2^t p+1}, y_p$ 均为奇数; 又因为 $x_{2n+1} \equiv 0(\bmod 2)$, 而 $x_n \not\equiv 0(\bmod 4)$, 故 $\frac{x_{2^t p+1}}{2}$, $\frac{x_p}{2}$ 均为奇数. 又 $2(D) = 1$, 故 $2(Dz^2)$ 为奇数, 而 $D(2^{4+t} x_{2^t p} x_{2^{t-1} p} \cdots x_{2p} y_{2^t p+1} y_p \cdot \frac{x_p}{2} \cdot \frac{x_{2^t p+1}}{2}) = 4 + t$, 所以 t 只能为正奇数.

由引理 11 的 (II) 知 $\frac{x_p}{2}$ 为平方数仅当 $p = 1$ 或 $p = -1$; 由引理 11 的 (III) 知 y_p 为平方数仅当 $p = 1$; 由引理 11 知对于任意的正奇数 p, $\frac{x_{2^t p+1}}{2}, x_{2^t p}, x_{2^{t-1} p}, \cdots, x_{2p}, y_{2^t p+1}$ 均不为平方数, 故 $p \neq 1$ 时 $\frac{x_{2^t p+1}}{2}, x_{2^t p}, x_{2^{t-1} p}, \cdots, x_{2p}, \frac{x_p}{2}, y_{2^t p+1}, y_p$ 均不为平方数.

又由式 (3) 知仅当 $p = 1$ 或 $p = -1$ 时 $\frac{x_p}{2} = 1$, 而对于任意的正奇数 p, $\frac{x_{2^t p+1}}{2}, x_{2^t p}, x_{2^{t-1} p}, \cdots, x_{2p}$ 均恒不为 1; 由式 (4) 知仅当 $p = 1$ 时 $y_p = 1$, 而对于任意的正奇数 p, $y_{2^t p+1}$ 均不为 1. 故当 $p > 1$ 为正奇数时, $\frac{x_{2^t p+1}}{2}, x_{2^t p}, x_{2^{t-1} p}, \cdots, x_{2p}, \frac{x_p}{2}, y_{2^t p+1}, y_p$ 为 $t+4$ 个不为 1 的奇数, 故 $\frac{x_{2^t p+1}}{2}, x_{2^t p}, x_{2^{t-1} p}, \cdots, x_{2p}, \frac{x_p}{2}, y_{2^t p+1}, y_p$ 至少为 D 提供 $t+4$ 个奇素因子, 又 t 为正奇数, 则 $t+4 \geq 5$, 此时式 (10) 右边至少为 D 提供 5 个互异的素因子, 矛盾.

当 $p = 1$, $t \neq 1$ 时 $\frac{x_p}{2}$ 与 y_p 为平方数, 且 $\frac{x_p}{2} = \frac{x_1}{2} = 1$, $y_p = y_1 = 1$, 此时式 (10) 为

$$Dz^2 = 2^{4+t} x_{2^t p} x_{2^{t-1} p} \cdots x_{2p} y_{2^t p+1} \cdot \frac{x_{2^t p+1}}{2} \tag{11}$$

此时 $x_{2^t p}, x_{2^{t-1} p}, \cdots, x_{2p}, y_{2^t p+1}, \frac{x_{2^t p+1}}{2}$ 至少为 D 提供 $t+2$ 个素因子. 又因为 $t \neq 1$, t 为正奇数, 故 $t \geq 3$, 因此 $t+2 \geq 5$, 此时 $x_{2^t p}, x_{2^{t-1} p}, \cdots, x_{2p}, y_{2^t p+1}, \frac{x_{2^t p+1}}{2}$ 至少为 D 提供 5 个互异的素因子, 矛盾.

当 $p = 1$, $t = 1$ 时, (10) 为 $Dz^2 = 2^5 x_2 y_1 y_3 \cdot \frac{x_1}{2} \cdot \frac{x_3}{2} = 2^5 \times 7 \times 1 \times 15 \times 1 \times 13 = 2^5 \times 3 \times 5 \times 7 \times 13$, 则 $D = 2 \times 3 \times 5 \times 7 \times 13$, $z = 4$, 故方程组 (2) 的正整数解为 $(x, y, z) = (362, 209, 4)$, 从而得 $D = 2 \times 3 \times 5 \times 7 \times 13$ 时 (2) 的全部整数解为 $(x, y, z) = (\pm 362, \pm 209, \pm 4)$, $(\pm 2, \pm 1, 0)$.

2.2 当 m 为正奇数时, 设 $m = 2^t p - 1$ ($t \in Z^+$, p 为正奇数), 则式 (7) 成为

$$Dz^2 = 4x_{2^t p} x_{2^t p-1} y_{2^t p} y_{2^t p-1} \tag{12}$$

仿 2.1 证明可得, 仅当 $p = 1$, $t = 1$ 时式 (12) 有非平凡解, 此时式 (12) 为 $Dz^2 = 4x_2x_1y_2y_1 = 4 \times 7 \times 2 \times 4 \times 1 = 2^5 \times 7$, 则 $D = 2 \times 7, z = 4$, 故方程组 (2) 的正整数解为 $(x, y, z) = (26, 15, 4)$, 从而得 $D = 2 \times 7$ 时 (2) 的全部整数解为 $(x, y, z) = (\pm 26, \pm 15, \pm 4)$, $(\pm 2, \pm 1, 0)$.

综上, 定理得证.

参考文献

[1] Ljunggren W. Litt om Simultane Pellske Ligninger [J]. Norsk Mat Tidsskr, 1941, 23: 132-138.

[2] Pan Jiayu, Zhang Yuping, Zou Rong. The Pell Equations $x^2 - 8y^2 = 1$ 和 $y^2 - Dz^2 = 1$[J]. Chinese Quarterly Journal of Mathematics, 1999, 14(1): 73-77.

[3] 乐茂华. 关于联立 Pell 方程方程组 $x^2 - 4D_1y^2 = 1$ 和 $y^2 - D_2z^2 = 1$[J]. 佛山科学技术学院学报: 自然科学版, 2004, 22(2): 1-3+9.

[4] 柯召, 孙琦. 谈谈不定方程 [M]. 哈尔滨: 哈尔滨工业大学出版社, 2011: 15, 64.

[5] 赵天. 关于不定方程 $x^3 \pm 2^{3n} = 3Dy^2$[D]. 重庆师范大学, 2008: 9.

[6] 单, 余红兵. 不定方程 [M]. 合肥: 中国科学技术大学出版社, 1991: 90.

[7] 陈永高. Pell 方程组 $x^2 - 2y^2 = 1$ 与 $y^2 - Dz^2 = 4$ 的公解 [J]. 北京大学学报, 自然科学版: 1994, 30(3): 298-302.

[8] 乐茂华. 一类二元四次 Diophantine 方程 [J]. 云南师范大学学报: 自然科学版, 2010, 30(1): 12-17.

[9] Walsh G A. note on a theorem of Ljunggren and the diophantine equations $x^2 - kxy^2 + y^4 = 1$or4[J]. Arch Math, 1999, 73(2): 504-513.

[10] Togbé A, Voutier P M, and Walsh P G. Solving a family of Thue equations with an applic- ation to the equation $x^2 - Dy^4 = 1$. Acta Arith, 2005, 120(1): 39-58.

On the Pell Equations $x^2 - 3y^2 = 1$ and $y^2 - Dz^2 = 1$

GUO Jing[1], ZHAO Jian-hong[2], DU Xian-cun[3]

(1. Jiangxi Science and Technology Normal University, Nanchang 330013, China)
(2. Lijiang Teachers College, Lijiang 674199, China)
(3. College of Teacher Education, Honghe University, Mengzi 661199, China)

Abstract: Let$D = 2p_1 \cdots p_s(1 \leq s \leq 4)$, $p_1, \cdots, p_s$are diverse odd primes. In this paper, the following conclusion are proved: the Pell equations$x^2 - 3y^2 = 1$and$y^2 - Dz^2 = 1$has only trivial solution $(x, y, z) = (\pm 2, \pm 1, 0)$with the exceptions that $D = 2 \times 7, 2 \times 3 \times 5 \times 7 \times 13$.

Keywords: the system of Pell equations; fundamental solution; integer solution; odd prime

注释：

［1］One kind hybrid character sums and their upper bound estimates 原文是 Zhao，JH（Zhao Jianhong）与 Wang，X（wang，xiao）发表在 JOURNAL OF INEQUALITIES AND AP-PLICATIONS 上并被 SCI 数据库检索的文章，文献号是 160，DOI：10. 1186/s13660 − 018 − 1753 − 4，入藏号：WOS：000437367800001，Web of Science 类别：Mathematics，Applied；Mathematics，研究方向：Mathematics，IDS 号：GL7HG，ISSN：1029 − 242X.

［2］Some symmetric identities involving Fubini polynomials and Euler numbers 原文是 Zhao，JH（Zhao Jianhong）与 Chen，ZY（Chen Zhuoyu）发表在 SYMMETRY-BASEL 上并被 SCI 数据库检索的文章，Volume：10，期：8，文献号是 303，DOI：10. 3390/sym10080303，出版年：AUG 2018，入藏号：WOS：00044248660005，Web of Science 类别：Multidisciplinary Sciences，研究方向：Science & Technology-Other Topics，IDS 号：GR3JU，ISSN：2073 − 8994.

［3］Elliptic Curve Integral Points on $y^2 = x^3 + 19x + 46$ 原文是 Zhao，JH（Zhao Jian-hong）发表在 Materials Science and Engineering 上的并被 EI CompendexWeb 数据库检索的文章，Volume：382（2018），期：5，文献号是 052038，DOI：10. 1088/1757 − 899X/382/5/052038，出版年：July 2018，入藏号：WOS：20183105622724，ISSN：17578981.

［4］On the Diophantine equations $x^2 − 27y^2 = 1$ and $y^2 − 2^pz^2 = 25$ 原文是 Zhao，JH（Zhao Jianhong）发表在 Earth and Environmental Science 上的并被 EI CompendexWeb 数据库检索的文章，Volume：153（2018），期：5，文献号是 042001，DOI：10. 1088/1755 − 1315/153/4/042001，出版年：June 2018，入藏号：WOS：20183105636221，ISSN：17551307.

［5］On the Diophantine equations $x^2 − 7y^2 = 1$ and $y^2 − Dz^2 = 9$ 原文是 Zhao，JH（Zhao Jianhong）发表在 Materials Science and Engineering 上的并被 EI CompendexWeb 数据库检索的文章，Volume：382（2018），期：5，文献号是 052038，DOI：10. 1088/1757 − 899X/382/5/052038，出版年：July 2018，入藏号：WOS：20183105622724，ISSN：17578981.

［6］Elliptic Curve Integral Points on $y^2 = x^3 + 3x − 14$ 原文是 Zhao，JH（Zhao Jianhong）发表在 Earth and Environmental Science 上的并被 EI CompendexWeb 数据库检索的文章，Volume：128（2018），期：5，文献号是 012108，DOI：10. 1088/1757 − 1315/128/1/012108，出版年：March 2018，入藏号：WOS：20181705041214，ISSN：17551307.

［7］On the Diophantine equations $x^2 − 47y^2 = 1$ and $y^2 − Pz^2 = 49$ 原文是 Zhao，JH（Zhao Jianhong）与 Yang，Lixing 发表在 Materials Science and Engineering 上的并被 EI CompendexWeb 数据库检索的文章，Volume：382（2018），期：5，文献号是 052037，DOI：10. 1088/1757 − 899X/382/5/052037，出版年：July 2018，入藏号：WOS：20183105622723，ISSN：17578981.